LARGE-SCALE SOLAR POWER SYSTEMS

This book is a comprehensive discussion and economic analysis of large-scale solar power systems, specifically referencing critical issues related to design construction and financing. The book provides practical design, installation, and financing guidelines for large-scale commercial and industrial solar power projects. Engineering design and construction methodologies as well as economic analysis provide a step-by-step walk-through of all aspects of solar power systems. Design methodologies outline the specific requirements of solar and electrical design and construction documentation in meticulous detail, which can readily be applied to ground mount, roof mount, building integrated (BIPV), and carport-type solar power projects. In view of the importance of solar power systems as a viable present and future energy resource, the book includes a dedicated chapter on smart grid transmission and large-scale energy storage systems.

Dr. Peter Gevorkian is President of Vector Delta Design Group, Inc., an international electrical engineering and solar power consultancy, and CEO of Solar Analytic Solutions, Inc., an engineering R & D organization. He has designed more than 50 MW of photovoltaic systems for health care, aviation, and regional utility companies and holds numerous patents. Dr. Gevorkian has also developed advanced econometric analytical software for large-scale solar power design and financing. Dr. Gevorkian is an active member of the Canadian and California Professional Engineers and California Society of Energy Engineers.

Over his 40-year career, Dr. Gevorkian has received numerous honors and awards including the AIA 2007 Engineering Merit Award for Renewable Energy Systems and Exceptional Contribution to the Advancement of Solar Power Co-generation in Building Design, the AIA 2007 Design Honor Award for Outstanding Engineering Design for the Metropolitan Water District's Museum of Water & Life, the AIA 2008 Honor Award for Outstanding Design Achievement in Solar Power Engineering, and the AIA 2009 Honor Award for Excellence in Solar Power Design.

Dr. Gevorkian is the author of numerous technical papers and technical books on renewable energy systems design. His books include *Sustainable Energy Systems in Architectural Design*, *Sustainable Energy System Engineering*, *Solar Power Systems in Building Design*, *Alternative Energy Systems in Building Design*, *Large Scale Solar Photovoltaic System Design*, and *Large Scale Solar Power Construction & Economics*.

Sustainability Science and Engineering Series

Series Editor
Professor I. S. Jawahir
James F. Hardymon Chair in Manufacturing Systems
University of Kentucky, College of Engineering

Peter Gevorkian, *Large-Scale Solar Power Systems: Construction and Economics*

Large-Scale Solar Power Systems

Construction and Economics

Peter Gevorkian

CAMBRIDGE
UNIVERSITY PRESS

CAMBRIDGE UNIVERSITY PRESS
Cambridge, New York, Melbourne, Madrid, Cape Town,
Singapore, São Paulo, Delhi, Mexico City

Cambridge University Press
32 Avenue of the Americas, New York, NY 10013-2473, USA

www.cambridge.org
Information on this title: www.cambridge.org/9781107027688

First published 2012

Printed in the United States of America

A catalog record for this publication is available from the British Library.

Library of Congress Cataloging in Publication data
Gevorkian, Peter.
  Large-scale solar power systems : construction and economics / Peter Gevorkian.
    p. cm.
  Includes bibliographical references and index.
  ISBN 978-1-107-02768-8
  1. Solar power plants. I. Title.
  TK1545.G48 2012
  621.31′244–dc23   2012015677

ISBN 978-1-107-02768-8 Hardback

Contents

Preface

In the past decade, deployment of solar power systems has been misunderstood and hampered mainly by lack of appreciation of the technology, as well as by skeptics who without knowledge of solar power system technology and its economics have flooded the media with opinions that are inaccurate and misrepresentative regarding significant benefits of the technology in terms of its significant impact on global warming and its inevitable future economic significance.

In previously published books, I have discussed implications of global dependency on fossil fuel use and the consequences of global warming, which due to the topics' significance, are once more been discussed in this book.

About This Book

During years of practice as a research and design engineer, I have come to realize that some of the greatest obstacles hindering accelerated development of the US solar power industry can be attributed to several factors. First and foremost, even though in the past decade the solar power industry has been advancing at an accelerated rate, higher educational institutions and more specifically engineering studies have not kept up with the solar power system technologies. It is noteworthy to recognize that solar power technology, unlike classical engineering, is not compartmentalized as a singular discipline such as structural, mechanical, or electrical disciplines; rather, it is a symbiosis of multiple disciplines that necessitate fundamental understanding of the principles of several engineering fundamentals that range from environmental studies, soils and civil engineering, electrical and electronics engineering, several fundamental understanding of physical principles of material sciences, optics, and engineering economics.

At present there are no higher learning institutions in the United States that offer a complete four- or five-year curriculum in Solar Power and Alternative Energy Systems Engineering specifically tailored to encompass all sciences and disciplines referenced earlier.

Furthermore, it should be noted that large-scale solar power technology deployment entails significant sums of investment; in order to appreciate and justify the validity of solar power system financial transactions, banking and financial lending institutions also need a fundamental understanding of solar power system technologies which unlike conventional investment associated with capital equipment loans or financing may offer a uniquely positive perspective regarding return on investment which otherwise could be ignored and dismissed as high risk financial transactions.

Therefore in the author's opinion the best way to promote the use of solar power as a sustainable energy design is through proper education of key professionals, such as architects, engineers, banking and investment personnel, and program managers whose opinions always affect project financing and development.

Furthermore, in two earlier books, *"Sustainable Energy Systems in Architectural Design"* and *"Sustainable Energy Systems Engineering,"* I attempted to introduce architects, engineers, and scientists to a number of prevailing renewable energy technologies and their practical use, in the hopes that a measure of familiarity and understanding would encourage their deployment.

This book has been specifically written to serve as a pragmatic resource for solar photovoltaic power system financing. When writing the manuscript, I attempted to minimize unnecessary mathematics and related theoretical photovoltaic physics, by only covering real-life, straightforward design methodologies are commonly practiced in the industry.

As scientists, engineers, and architects, as well as finance professionals and economists, we have throughout the last few centuries been responsible for the elevation of human living standards and contributed to the advancement of technology while ignoring the devastating side effects to the global ecology. In the process of creating betterment and comforts of life, we have tapped into the most precious nonrenewable energy resources, miraculously created over the life span of our planet, and have been misusing them in a wasteful manner to satisfy our most rudimentary energy needs.

It should also be noted that, before it is too late, as responsible citizens of our global village, it is high time that we assume individual and collective responsibility for resolving today's environmental issues and ensuring that future life on earth will continue to exist as nature intended.

Acknowledgments

I would like to thank my colleagues and individuals who have encouraged me to write this book. I am especially grateful to all agencies and organizations that provided photographs and allowed use of some textual material. Special thanks to Mary Olson Kanian, Eric Knight, Krispin Leydon, Carla Gharibian, Arlen Gharibian, Jeffrey Trucksess, and Annette Malekandrasians for preliminary review of various chapters of the book.

I would also like to acknowledge Kurt Kelly, chief technology officer of Firefly Technology, for providing helpful documents on foam type energy storage battery technologies. Finally, special thanks to my dear friend Dr. Jagdish Doshi for his support and encouragement and recommendation to write a special chapter on large-scale energy storage technologies.

1 Global Warming and Climate Change

Ever since the Industrial Revolution, human activities have constantly changed the natural composition of Earth's atmosphere. Concentrations of trace atmospheric gases, nowadays termed "greenhouse gases," are increasing at an alarming rate. There is conclusive evidence that the consumption of fossil fuels, the conversion of forests to agricultural land, and the emission of industrial chemicals are the principal contributing factors to air pollution.

According to the National Academy of Sciences, the Earth's surface temperature has risen by about 1 degree Fahrenheit in the past century, with accelerated warming occurring in the past three decades. According to statistical reviews of the atmospheric and climatic records there is substantial evidence that global warming over the past 50 years is directly attributable to human activities.

Under normal atmospheric conditions, energy from the Sun controls the Earth's weather and climate patterns. Heating of the Earth's surface resulting from the Sun radiates energy back into space. Atmospheric greenhouse gases, including carbon dioxide (CO_2), methane (CH_4), nitrous oxide (N_2O), troposphere ozone (O_3), and water vapor (H_2O), trap some of this outgoing energy, retaining it in the form of heat, somewhat like a glass dome. This process is referred to as the GREENHOUSE EFFECT.

Without the Greenhouse Effect, the Earth's surface temperature would be roughly 30 degrees Celsius (54 degrees Fahrenheit) cooler than it is today – too cold to support life. Reducing greenhouse gas emissions is dependent on a reduction in the amount of fossil fuel–fired energy that we produce and consume.

Fossil fuels include coal, petroleum, and natural gas, all of which are used to fuel electric power generation and transportation. Substantial increases in the use of nonrenewable fuels have been a principal factor in the rapid increase in global greenhouse gas emissions. The use of renewable fuels can be extended to power industrial, commercial, residential, and transportation applications to reduce air pollution substantially.

Examples of zero-emission renewable fuels include solar, wind, geothermal, and renewably powered fuel cells. These fuel types, in conjunction with advances in energy-efficient equipment design and sophisticated energy management techniques, can reduce the risk of climate change and the resulting harmful effects on ecosystems. It should be kept in mind that natural greenhouse gases are a necessary part of sustaining life on Earth. It is the anthropogenic, or human-caused, increase in greenhouse gases that is of concern to the international scientific community and governments around the world.

Since the beginning of the modern Industrial Revolution, atmospheric concentrations of carbon dioxide have increased nearly 30%, methane concentrations have more than doubled, and nitrous oxide concentrations have risen by about 15%. These increases in greenhouse gas emissions have enhanced the heat-trapping capability of Earth's atmosphere.

Fossil fuels that are burned to operate electric power plants, run cars and trucks, and heat homes and businesses are responsible for about 98% of U.S. carbon dioxide emissions, 24% of U.S. methane emissions, and 18% of U.S. nitrous oxide emissions. Increased deforestation, landfills, large agricultural production, industrial production, and mining also contribute a significant share of emissions. In 2000, the United States produced about 25% of total global greenhouse gas emissions, the largest contribution by any country in the world.

Estimating future emissions depends on demographic, economic, technological, policy, and institutional developments. Several emissions scenarios based on differing projections of these underlying factors have been developed. It is estimated that by the year 2100, in the absence of emission control policies, carbon dioxide concentrations will be about 30–150% higher than they are today.

Increasing concentrations of greenhouse gases are expected to accelerate global climate change. Scientists expect that the average global surface temperature could rise an additional 1–4.5 degrees Fahrenheit within the next 50 years and 2.2–10 degrees Fahrenheit over the next century, with significant regional variation. Records show that the 10 warmest years of the 20th century all occurred in the last 15 years of that century. The expected impact of this warming trend includes the following:

Water Resources

Warming-induced decreases in mountain snowpack storage will increase winter stream flows (and flooding) and decrease summer flows. This, along with an increased evaporation and aspiration rate, is likely to cause a decrease in water deliveries.

Agriculture

The agricultural industry will be adversely affected by lower water supplies and increased weather variability, including extreme heat and drought.

Forestry

An increase in summer heat and dryness is likely to result in forest fires, an increase in the insect population, and disease.

Electrical Energy

Increased summer heat is likely to cause an increase in the demand for electricity, namely, increased reliance on air-conditioning. Reduced snowpack is likely to decrease the availability of hydroelectric supplies.

Regional Air Quality and Human Health

Higher temperatures may worsen existing air quality problems, particularly if there is a greater reliance on fossil fuel–generated electricity. Increased temperatures would also increase health risks for segments of the population.

Rising Ocean Levels

Thermal expansion of the ocean and glacial melting are likely to cause a 0.5 to 1.5 m (2 to 4 ft) rise in sea level by 2100.

Natural Habitat

Rising ocean levels and reduced summer river flow are likely to reduce coastal and wetland habitats. These changes could also adversely affect spawning fish populations. A general increase in temperatures and an accompanying increase in summer dryness could also adversely affect wild land plant and animal species.

Scientists calculate that without considering feedback mechanisms, a doubling of carbon dioxide would lead to a global temperature increase of 1.2 degrees Celsius (2.2 degrees Fahrenheit). However, the net effect of positive and negative feedback patterns appears to be substantially more warming than would be caused by the change in greenhouse gases alone.

Pollution Abatement Consideration

According to a 1999 study report by the U.S. Department of Energy (DOE), 1 kilowatt of energy produced by a coal-fired electrical power generating plant requires about 5 lb of coal. Likewise, the generation of 1.5 kWh of electrical energy per year requires about 7,400 lb of coal, which in turn produces 10,000 lb of carbon dioxide (CO_2).

Roughly speaking, the calculated projection of the power demand for the project totals about 2,500 to 3,000 kWh. This will require between 12 and 15 million pounds of coal, thereby producing about 16 to 200 million pounds of carbon dioxide. Solar power, if implemented as previously discussed, will substantially minimize the air pollution index. The Environmental Protection Agency (EPA) will soon be instituting an air pollution indexing system that will be factored into all future construction permits. All major industrial projects will be required to meet and adhere to these air pollution standards and offset excess energy consumption by means of solar or renewable energy resources.

Energy Escalation Cost Projection

According to an Energy Information Administration data source published in 1999, California consumes just as much energy as Brazil or the United Kingdom. The entirety of global crude oil reserves are estimated to last about 30 to 80 years, and more than 50% of the nation's energy is imported from abroad. It is thus inevitable that energy costs will surpass historical cost escalation averaging projections. The growth of fossil fuel consumption is illustrated in Figure 1.1. It is estimated that within

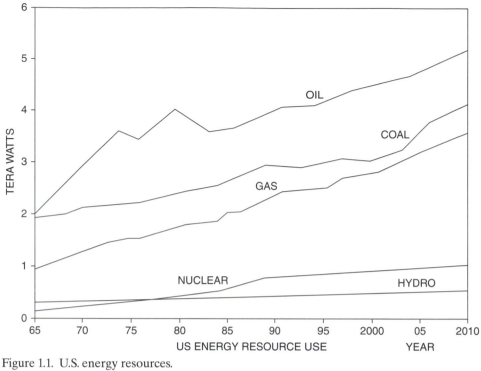

Figure 1.1. U.S. energy resources.
Source: DOE

the next decade producers will increase the cost of nonrenewable energy sources by approximately 4–5%.

When compounded with a general inflation rate of 3%, over the next decade average energy costs could be expected to rise at a rate of about 10–12% per year. This cost increase does not take into account other inflation factors, such as regional conflicts, embargoes, and natural catastrophes.

Solar power cogeneration systems, if designed and monitored properly, require minimal maintenance and are more reliable than any manmade power generation devices. The systems have an actual life span of 35 to 50 years and are guaranteed by the manufacturers for a period of 25 years. It is my opinion that in a near-perfect geographic setting, the integration of these systems into mainstream architectural design will not only enhance design aesthetics but also generate considerable savings and mitigate adverse effects on the ecosystem and global warming.

Social and Environmental Concerns

Nowadays, we do not think twice about leaving lights on or turning off the television or computers, which oftentimes run for hours. Most people believe that energy is infinite, but in fact, that is not the case. The global consumption of fossil fuels, which supply us with most of our energy, is steadily rising. In 1999, it was discovered that of 97 quads of energy used (a quad is $3 \times 1,011$ kWh), 80 quads were from coal, oil, and natural gas. The imminent depletion of fossil fuel sources within a few generations requires us to be prepared with novel and alternative sources of energy. In reality, as early as 2020, we could witness serious energy deficiencies (see Figure 1.2 for

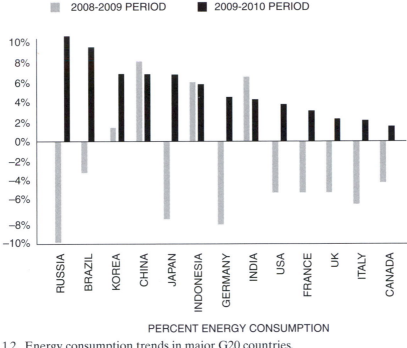

Figure 1.2. Energy consumption trends in major G20 countries.
Source: Enerdata

energy consumption trends in major G20 countries). Therefore, interest in renewable sources of energy, including wind, solar, and hydropower, has become a hot topic.

Although they may seem novel, renewable fuels are by no means a recent phenomenon. In fact, the Industrial Revolution was launched with renewable fuels. Until energy conferences in 1973 and 1974, the United States and the rest of the world had been using energy without serious concern for years. This was when energy conservation issues were brought to the attention of the industrialized world. Ever since, we have been forced to realize that the global supply of fossil fuels would one day run out and that we would have to find alternate sources of energy.

An extensive report published by the DOE in 1999 disclosed that by the year 2020, there will be a 60% increase in carbon dioxide emissions, which will create a serious strain on the environment and further aggravate the dilemma with greenhouse gasses. Figure 1.3 shows carbon dioxide generation by various industries and Figure 1.4 shows the growth of carbon dioxide in the atmosphere over a period of two centuries.

A reduction in energy consumption may seem a simple solution. However, this would not be feasible. There is a correlation between high electricity consumption (4,000 kWh per capita) and a high Human Development Index (HDI), which measures quality of life. In other words, there is a direct correlation between quality of life and amount of energy used. This is one of the reasons that the standard of living in industrialized countries is better than it is in third world countries, where there is very little access to electricity. In 1999, the United States possessed 5% of the world's population and produced 30% of the gross world product. The U.S. also consumed 25% of the world's energy and emitted 25% of its carbon dioxide.

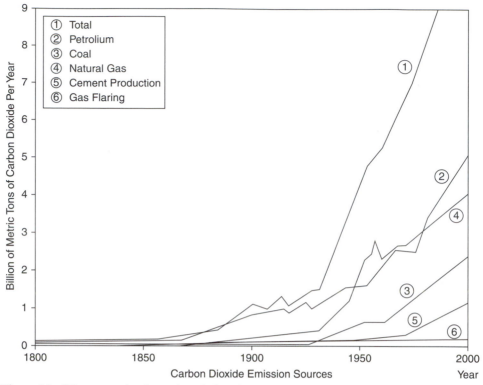

Figure 1.3. CO_2 generation by various industries.

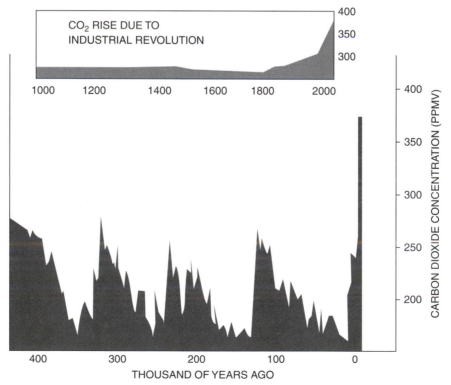

Figure 1.4. CO_2 generation over thousands of years.

It is not hard to imagine what countries like China and India, with their burgeoning populations and economic growth, can do to the state of global ecology.

The most significant feature of solar energy is that it is clean energy that does not harm the environment. Solar power use does not emit any of the extremely harmful greenhouse gases that contribute to global warming. A small amount of pollution is generated when solar panels are produced, but it is minuscule in comparison to the pollution associated with fossil fuels. The Sun is also a free source of energy. As technology advances, solar energy will become ever more economically feasible because the price of photovoltaic modules will decrease. One of the primary concerns with solar power is that it is not energy on demand and that it only works during the day and during very sunny weather. The only way to overcome this problem is to build storage facilities that save up some of the energy in batteries. However, this adds more to the cost of solar energy.

A Few Facts about Coal-Based Electric Power Generation

Presently, the most abundant fossil fuel resource available in the United States is coal. Coal-based electric power generation represents about 50% of energy used and is the largest source of environmental pollution. Coal burned in boilers generates an abundance of CO_2, SO_2 (sulfur dioxide), NO_2 (nitrogen dioxide), arsenic, cadmium, lead, mercury, soot particles, and tons of coal ash, which pollute the atmosphere and water. At present, 40% of the world's CO_2 emissions are from coal burning power plants.

The coal industry in the United States has recently developed the advertising slogan "Opportunity Returns" as part of an attempt to convey an unsubstantiated message to the public that a new clean coal gasification technology, assumed to be superclean, is on the horizon and will provide safe energy for the next 250 years. Whatever the outcome of the promised technology, at present coal-fired electric power generation plant construction is on the rise and 120 power generation plants are currently under construction.

So-called clean coal Integrated Gasification Combined Cycle (IGCC) technology converts coal into synthetic gas, which is supposed to be as clean as natural gas and 10% more efficient when used to generate electricity. The technology is expected to increase plant power efficiency by 10%, produce 50% less solid waste, and reduce water pollution by 40%. Despite all of the coal power energy production improvements, the technology will remain a major source of pollution.

Coal Power Generation Industry Facts

- By the year 2030, it is estimated that coal-based electrical power generation will represent a very large portion of the world's power and provide 1,350 thousand megawatts of electric energy, which in turn will inject 572 billion tons of CO_2 into the atmosphere. This is equal to the amount of pollution generated over the past 250 years.
- In the United States, it takes 20 pounds of coal to generate sufficient energy requirements per person. In total, this represents approximately 1 billion tons of coal. The percentage of coal-based energy production in the United States will be 50%, as compared to 40% in China and 10% in India.

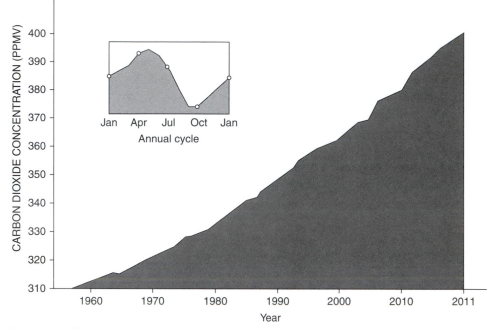

Figure 1.5. CO_2 generation over 60 years.

- By the year 2030, global energy demand is projected to double.
- Worldwide coal resources are estimated to be at 1 trillion tons. The United States holds 25% of these total resources, or about 270 billion tons. China has 75 billion tons of coal, which is expected to provide 75 years of coal-based electric energy.
- Cheap electric power generated from coal will require large national expenditures to mitigate environmental pollution and related public health problems, which will translate into medical bills for treating asthma, emphysema, heart attacks, and cancer. Pollution will also lead to global ecological demise and genetic changes in plant and animal life.
- According to a study conducted by Princeton University, the effects of U.S. coal-fired electric power generation plants on public health would add $130 billion per megawatt hour of energy. At a cost of $0.12/ kWh per megawatt of power, this is equal to $120.00
- The Kyoto Protocol, which has been ratified by 162 nations, calls for cutting greenhouse gases by 5.2% by 2012. China and developing nations like India, which are exempt, are estimated to generate twice the expected amount of atmospheric pollution. The United States and Australia have not ratified the Kyoto Protocol. See Figure 1.5 for CO_2 generation over 60 years.

In conclusion, it should be noted that as responsible citizens considering investment in solar power or renewable energy systems, we should also consider the negative effects of indirect cost burdens on the global economy, which would be required to mitigate the deterioration of the global ecosystem and the consequences of environmental pollution on human health, welfare, and lifestyle for present and future generations.

To be a skeptic without expert knowledge about a subject is always easier than to be knowledgeable, since acquiring knowledge requires one to expend considerable time studying, exploring, and analyzing a subject, which requires an inquisitive and alert mind and perseverance in discovering the truth. It is only the few who can measure up to such standards, and they are the ones whom we call scholars. However, skepticism based on a solid foundation is the essence of every new human intellectual discovery.

2 Solar Power and Sustainable Energy Technologies and Their Impact on Global Economy

Introduction

In the past couple of decades, solar power as well as sustainable energy system deployment have been misunderstood and largely dismissed as viable sources of sustainable energy. More specifically, the valuation of solar power system technology as a viable source of clean energy has in some instances been hampered, mainly by a lack of appreciation of the technology's impact on the national and global economy and its significant long-term benefits to human welfare and global ecology.

When considering the trade-off of inexpensive power generated by coal and other fossil fuels versus degradation of the ecosystem, which is essential to the sustenance of human and animal life, we must also measure the costs of sustainable energy and its significance in relation to the survival of life itself.

Unfortunately, competing business interests and market forces have recently flooded the news and television media with negative opinions and notions in regard to the viability of sustainable energy systems. These reports are frequently inaccurate and wholly misrepresentative of technologies that hold great promise for significant long-term benefits to the global ecology and economy. As a result, alternative energy systems, and more specifically *solar power technologies,* have been dismissed and labeled as expensive and inefficient solutions unable to meet the ever-increasing demand for global energy. Yet coal-fired electrical energy advertisements have taken center stage and have been branded as "clean energy" that could solve today's and tomorrow's world energy shortcomings.

In a book entitled *Solar Power in Building Design*, recently published by McGraw-Hill, I wrote a forward discussing the implications of global dependency on fossil fuels and their contribution to global warming. In it, I attempted to illustrate a number of concerns that urgently require further discussion. Without a profound appreciation of the dire ecological consequences of continued fossil fuel energy use, and lacking the validation of sustainable energy generation technologies as *essential* in their impact on the human lifestyle, it is highly unlikely that we will be able to achieve an equitable balance to maintain a reasonably viable lifestyle for future generations.

Energy Production and Ecology

Since the dawn of agriculture and civilization, human beings have hastened deforestation and its subsequent impact on climatic and ecological conditions. Deforestation and the increasing use of fossil fuel energy diminish the natural recycling of carbon dioxide gases. This accelerates and increases the inversion layer that traps the reflected energy of the Sun. The augmented inversion layer has resulted in elevated atmospheric temperature, giving rise to global warming. This in turn has caused the melting of polar ice, substantial changes to climatic conditions, and depletion of the ozone layer.

Within a couple of centuries, the unchecked effects of global warming will not only change the makeup of the global landmass, but also affect human lifestyle on the planet.

Continued melting of the polar ice caps will raise sea levels, which will gradually reach some habitable areas of global shorelines. It will also result in unpredictable climatic changes, such as unusual precipitation, floods, hurricanes, and tornados.

In view of the rapid expansion of the world's economies, particularly those of developing countries with large populations such as China and India, the demand for fossil fuel and construction materials will become severe. Within the next few decades, if continued at the present projected pace, the excessive demand for fossil fuel energy resources, such as crude oil, natural gas, and coal, will undoubtedly result in the irreversible collapse and demise of the ecology of our planet. Even today, China's enormous demand for energy and construction materials has resulted in considerable cost escalations for crude oil, construction steel, and lumber, the production and movement of which require the expenditure of fossil fuel energy.

Developing countries are the most efficient consumers of energy, since every scrap of material, paper, plastic, metal cans, rubber, and even common trash is recycled and reused. However, when the 2.3 billion combined inhabitants of China and India attain a higher margin of families with middle-class incomes, new demand for electricity, manufacturing, and millions of automobiles will undoubtedly change the balance of ecological and social stability to a level beyond imagination.

The United States is the wealthiest country in the world. With 5% of the world's population, the country presently consumes 25% of global aggregate energy. As a result of its economic power, the United States enjoys one of the highest standards of living and excellence in medical care, resulting in optimal levels of human longevity. The relative affluence of the country on the whole has enabled its population to squander and waste its low-cost energy for decades.

Most of the consumption of fossil fuel energy is a result of inefficient and wasteful transportation and electric power generation technologies. Because of the lack of comprehensive energy control policies and the lobbying efforts of special interest groups, research and development funding to accelerate sustainable and renewable energy technologies has been neglected.

In order to curb the waste of fossil fuel energy, it is imperative that our nation as a whole, from politicians and educators to the general public, be made aware of the dire consequences of our nation's energy policies and make every effort to promote the use of all available renewable energy technologies so that we can reduce the demand for nonrenewable energy and safeguard the environment for future generations.

The deterioration of our planet's ecosystem and atmosphere cannot be ignored or considered a matter that is not of immediate concern. According to scientists, our planet's ozone layer has been depleted by about 40% over the past century, and greenhouse gases have altered meteorological conditions. Unfortunately, the collective social consciousness of our society's educated masses has not concerned itself with the disaster awaiting future generations and continues to ignore the seriousness of the situation.

The Significance of Education and Awareness about Sustainable Energy Systems

During years of practice as a research and design engineer, I have come to realize that some of the greatest obstacles hindering accelerated development of the U.S. solar power industry can be attributed to several key factors. First and foremost, even though the solar power industry has been advancing at an accelerated rate in the past decade, higher educational institutions and more specifically engineering studies have not kept up with solar power system technologies. It is noteworthy to recognize that solar power technology, unlike classical engineering, is not compartmentalized as a singular discipline as the structural, mechanical, and electrical disciplines are. Rather, it is a symbiosis between multiple disciplines that necessitates a fundamental understanding of several engineering principles, including environmental studies, soils, and civil, electrical, and electronics engineering, as well as a fundamental understanding of the physical principles of material sciences, optics, and engineering economics.

At present, there are no higher learning institutions in the United States that offer a complete four- or five-year curriculum in solar power and alternative energy systems engineering as part of a formal degree that offers an education program specifically tailored to encompass all sciences and disciplines referenced earlier.

Furthermore, it should be noted that large-scale solar power technology deployments entail significant sums of investment in order to appreciate and justify the validity of solar power system financial transactions. Banking and financial lending institutions would also benefit from having a fundamental understanding of solar power system technologies, which, unlike conventional investment associated with capital equipment loans or financing, may offer a uniquely positive perspective regarding return on investment, which otherwise could be ignored and dismissed as a high-risk financial transaction.

The key to promoting the use of solar power as a sustainable and viable energy source lies in the proper education of key professionals, such as architects, engineers, banking and investment personnel, and program managers, all of whose opinions always affect project financing and development.

As scientists, engineers, and architects as well as finance professionals, educators, and political leaders, we have throughout the last few centuries been responsible for the elevation of human living standards and contributed to the advancement of technology. We have succeeded in putting a man on the Moon while ignoring the devastating deterioration of global ecology. In the process of advancing living standards and the subsequent comforts of life, we have tapped into the most precious nonrenewable energy resources, miraculously created over the life span of our planet, and have been misusing them in a wasteful manner to satisfy our most rudimentary energy needs.

It is high time that we, as responsible citizens of our global village, assume individual and collective responsibility to resolve today's environmental issues before it is too late, ensuring that future life on Earth will continue to exist as nature intended.

Solar Power Energy Production and Its Impact on the U.S. Economy

The benefits of any energy source must be viewed not only in terms of economics, but also in terms of its short- and long-term effects on ecology and human lifestyle.

For instance, ever since the development of nuclear technology, all industrialized nations of the world have expended large portions of their wealth and resources on the development of various types of nuclear power technologies without evaluating the consequences of plutonium waste disposal and other such extreme environmental and ecological hazards, which became evident in the Three Mile Island incident, Chernobyl, Japan, and other regions. Now, as a result of public concern, Germany's nuclear power plants are scheduled to close down by 2022.

Similarly, the consequences associated with the construction of hydroelectric dams and their negative impact on local ecology and the economy have been studied and exemplified in the construction of the Aswan Dam in Egypt, which destroyed local ecology and promoted the perpetuation of the West Nile virus.

Even though the consequences of coal use and hydrocarbon-based energy production have been known since the Industrial Revolution, the abundance of coal resources and the enormous economic engine created by the industry have perpetuated the myth that coal is clean. In fact, statistics show that there is a direct nexus between a coal-fired economy and negative impact on human health. Air pollution is not the only by-product of coal mining. Coal extraction and processing involve the considerable use of water, which becomes contaminated with numerous carcinogens such as chromium, arsenic, and rare earth elements whose leachates find their way into underground water reservoirs. Moreover, recent scientific studies have found that underground mining causes damage to the geology of mining regions, which has resulted in tremors and mud geysers. This has been evidenced in Australia.

Concerns that merit attention include issues associated with energy demand in the context of human population growth, the availability or lack of energy resources and their effect on human life, and the impact of energy resources on the economies of both developed and underdeveloped nations. Additional issues of concern relate to long-term national security, global energy distribution, and numerous other factors that directly relate to the availability of global sustainable energy resources.

There is no question that the health and welfare of a modern industrialized world are intrinsically related to the availability of electrical energy. Likewise, it is obvious that cheap and plentiful energy ameliorates living standards as well as the health and welfare of nations. What also needs to be questioned are the limits and capacities of global energy resources in meeting the needs of a rapidly expanding human population

The responses to all of the preceding questions are complex and the solutions are as yet unknown. What is of paramount concern is that access to energy resources is of vital importance to all nations. Therefore, demand and supply economics dictate that economic prosperity and national security, as well as national welfare, are intrinsically and directly related to the availability of energy resources.

A brief overview of available energy resources currently used includes coal, hydroelectric power, crude oil, natural gas, and nuclear power. As discussed previously, the use of coal and fossil fuels of one sort or another has, over the past couple of centuries, resulted in significant atmospheric and oceanic water pollution and has caused nearly irreversible ecological damage. As a result, the use of fossil-based electrical power generation is gradually being diminished and substituted by use of natural gas. Ultimately, there are no fossil fuel reserves on the planet, including natural gas and even coal reserves, which could sustain the accelerated demand for energy by the expanding world population. Therefore, even if we consider the plentitude of today's coal resources (which generate 50% of our nation's electrical energy), they will eventually, within several generations, be fully exhausted, and their use as a feedstock in generating electrical power will diminish substantially.

Hydroelectric power generation resources, which represent approximately 11% of U.S. electrical power generation resources, have, as a result of geological formation of the land, reached their maximum capacity. This leaves us with nuclear power generation, the potential of which, as discussed, is on the road to decline.

When considering present and future alternatives for electrical energy generation, we are left with but a few difficult choices, which include the reduction and curtailment of energy use, population growth control, and the *advancement of alternative clean energy production technologies*. The first two alternatives are nearly impossible to control since it is not practical to create a global parity in energy use and consumption between the "have" and "have not" nations of the world; nor is it logical to assume that the world's population will somehow cease to grow and level in the future. Therefore, the only obvious alternative is to develop newer and more ecologically friendly energy generation technologies that can sustain life on Earth up until such time as we can develop more advanced technologies, such as fusion reactors that would enable us to harvest electrical energy from the fusion reaction process.

Other alternative energy generation technologies, such as wind, solar, geothermal, oceanic power, and biofuels, which at present represent less than 3% of electrical power generation in the United States, will ultimately offer valuable alternatives as a sustainable means of generating significant amounts of sustainable energy for future generations, especially in view of diminishing fossil fuel resources.

The Significance of Feedstock in Generating Sustainable Energy

At this juncture, it is important to examine the inevitability of the challenges facing humanity within the context of energy production and its impact on the survival of the human race. As discussed, all technologies associated with electrical power generation, whether based on fossil fuel, atomic energy, or biofuel, require a constant supply of feedstock, which converts condensed forms of energy like coal or hydrocarbon into heat. This in turn is used to generate steam that drives electromechanical electrical power generators.

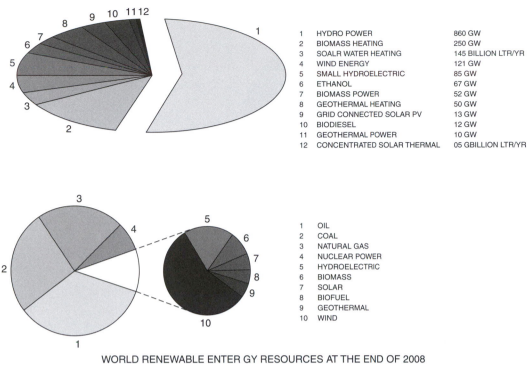

1	HYDRO POWER	860 GW
2	BIOMASS HEATING	250 GW
3	SOALR WATER HEATING	145 BILLION LTR/YR
4	WIND ENERGY	121 GW
5	SMALL HYDROELECTRIC	85 GW
6	ETHANOL	67 GW
7	BIOMASS POWER	52 GW
8	GEOTHERMAL HEATING	50 GW
9	GRID CONNECTED SOLAR PV	13 GW
10	BIODIESEL	12 GW
11	GEOTHERMAL POWER	10 GW
12	CONCENTRATED SOLAR THERMAL	05 GBILLION LTR/YR

1	OIL
2	COAL
3	NATURAL GAS
4	NUCLEAR POWER
5	HYDROELECTRIC
6	BIOMASS
7	SOLAR
8	BIOFUEL
9	GEOTHERMAL
10	WIND

WORLD RENEWABLE ENTER GY RESOURCES AT THE END OF 2008

Figure 2.1. World energy resources at the end of year 2008.
Source: DOE

On the other hand, hydroelectric, wind, geothermal, oceanic current, and tidal power, as well as solar photovoltaic and solar-thermal energies, use natural dynamic forces such as wind energy, the flow of water currents, and tidal power to create electricity through rotational electromechanical devices that do not require fuel or feedstock (Figures 2.1 and 2.2). Sustainable energy technologies such as photovoltaic solar cells or solar-thermal systems are the only technologies that, instead of feedstock, make use of the plentiful and abundant energy of the Sun in the form of photons that are absorbed and converted into electrical energy.

In other words, solar power system–based technologies can be considered the most natural form of energy harvesting process, offering unlimited power generation, so as long as the Sun shines on the surface of the globe. As such, solar power is the only technology that can, regardless of human population growth, offer unlimited energy production.

Even though solar power generation may be considered to be an evolving technology, within just a couple of decades, a steady advancement in the efficiency of energy production and in the exceptional longevity and life expectancy of photovoltaic solar cells has consolidated the foundations of what has in a short time become a multibillion-dollar global industry.

The Significance of Solar and Sustainable Energy Technologies to the Human Lifestyle

In an era of modern technology, the maintenance of lifestyle, health, and welfare, as well as human values at present and in the near-future, will be more and more

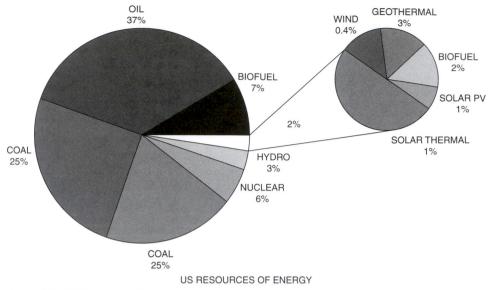

Figure 2.2. U.S. sources of energy.
Source: DOE

dependent upon the gross national product and per capita disposable income of the countries of the world, all of which will be driven by the availability of an abundance of energy resources.

The shortage of national energy resources will inevitably result in a deficit in national wealth and national security as well as in individuals' disposable income. Nations without sufficient energy resources will inevitably experience unprecedented hardship in the form of lower living standards and economic depression, which will erode the makeup of their present standards of living.

This downgrade in economic standards will ultimately lead to lowered national security and global readjustment to account for subsequent economic disparity. This in turn will result in global economic and political tensions, all of which we are experiencing today.

In the last several decades, because of the lack of cohesive national energy and educational policies, the United States has gradually lost its edge as a global industrial and economic power. This situation could be considered a direct result of the laxity in national technology development and the loss of manufacturing capacity.

Solar power technology, like thousands of American innovations, has been nourished and nurtured by the Chinese, Japanese, and Germans, whose economies have been trending upward at a constant rate. In a short couple of decades, China has become the second largest economy in the world, and in another decade, its economy will surpass that of the United States.

Chinese solar and wind energy technologies, which were in their infancy a decade ago, have taken center stage and have flooded the U.S. and global markets. To circumvent the "Made in America" or "Buy American Act," Chinese solar industries have expanded their manufacturing facilities within the country and soon will be overtaking the fledgling American solar power industry, which, in its inception, had never been given significant recognition or encouragement by any quarter of the private or public sector.

Even today, American ingenuity dominates the highest plateau for the invention and manufacture of the most efficient solar power technology in the world. Therefore, it is high time to empower the *American* sustainable energy industries in the United States and consider solar and alternative energy technologies as a significant and vital source for future energy production.

Sustainable Electrical Energy and Cost Parity with Grid Power

In the past couple of decades, the worldwide growth and deployment of solar power in residential, commercial, and industrial applications have been tied to the economics and cost of electrical energy production. Industrial countries such as Japan and Germany, with their limited natural energy resources, high energy tariffs, and national concern for global environmental pollution (resulting from fossil fuel and nuclear energy use), have concluded that alternative solar power energy production is cost competitive with conventional power generation. Both of those countries, whose electrical tariffs are five to six times higher than those of the United States, have found justification for the promotion and development of commercial solar power technologies.

In the early 1990s, Japan initiated the large-scale production of solar photovoltaic systems and was soon followed by Germany. Over a single decade of substantial national investments in research and product development by both countries, there resulted the establishment of viable solar power photovoltaic technologies and associated solar power conversion technologies such as inverters, and solar power batteries. Fledgling solar power industries that started in Japan and Germany have now become multibillion-dollar industries. Today, both countries, as well as Spain, have taken the lead in the manufacture and deployment of solar power technologies.

In view of the industrial economic potential, China during the last ten years, with considerable governmental subsidies, developed an extensive global solar power industrial enterprise, which has resulted in its domination of the world PV production market. Cheap labor and government subsidies of solar power technology in China have recently driven the cost of solar power PV production down to a level that has resulted in a 30–40% reduction in the cost of solar power deployment and energy production.

In addition to the cost reduction of conventional photovoltaic technologies, novel advancements in high efficiency concentrated solar power (CSP) technologies in the United States and overseas have further reduced the cost of solar electrical energy production to a level that, in large-scale multimegawatt installations, is yielding a unit cost of electrical energy per kilowatt that is approaching that of conventional electrical power generated by coal-fired or fossil-fired plants. In view of the consistent cost escalation of grid electricity and in anticipation of the considerable investment required to upgrade the U.S. grid network ("smart grid" system), it is anticipated that in the near-future, the cost of solar power energy production may even prove to be more economical than the cost of grid power (see Figure 2.3).

Even today, the seemingly relatively higher cost of solar power installation, if properly understood and analyzed in detail and in view of the longevity and reliability of the technology, could readily provide ample economic justification as a viable long-term investment vehicle.

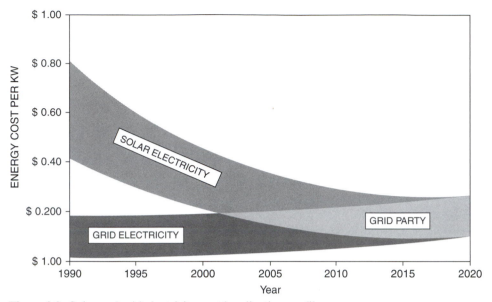

Figure 2.3. Solar and grid electricity cost levelization profile.
Source: DOE

In conclusion, it should be noted that investments in solar or any viable sustainable energy technologies, such as wind and biofuel, have greater merit in the context of their value proposition, particularly when considering the larger picture of global population expansion, inevitable greater energy demand, and the dire consequences of dismissing the impact of the continued use of fossil fuels.

From a national security and technology development perspective, as a nation we must also consider whether we, in the immediate future, would be content to assume a *lesser* leadership role in evolving sustainable energy technologies and *lower* economic development expectations in the highly competitive global economies of the future.

3   Overview of Solar Power System Technology

Introduction

In the past two decades there has been considerable advancement in solar photovoltaic power generation system deployment. Nevertheless, for numerous reasons financing institutions (because of lack of in-depth appreciation of the technology) have consistently refused to take advantage of great investment opportunities that are inherent in medium to large-scale solar power cogeneration systems.

This chapter is intended to clarify some of the ambiguities of solar power photovoltaic system technologies by familiarizing the reader with the fundamental aspects of solar power technology in lay terms and highlighting all aspects of cost components that typify various types of solar power systems.

The financial analysis methodology reflected in this paper is based on pragmatic, real life design, construction, and unique financing scenarios, which are rarely discussed or analyzed by financial banking or lending institutions and investors.

Unlike well-established technologies such as construction, manufacturing, transportation, and energy sector financing, solar power photovoltaic technologies are relatively new. Regardless, this somewhat unfamiliar sector of the industry has been advancing at great pace despite the lack of a financial performance legacy and lack of a specific analytical methodology. Lending institutions have not had the opportunity to study intricacies of the solar power technology in order to develop and establish specific lending policies that are unique and unlike those of any other sectors of the economy.

Solar Photovoltaic Power Generation Overview

Unlike any power generation system technologies, solar photovoltaic (PV) power systems are essentially solar power energy harvesting systems that absorb packets of solar energy called photons. These photons are used to energize electrons in a semiconductor device, which creates flow of electrons or electrical current. The flow of current in a conductive medium such as copper or aluminum wires, when it encounters resistance, develops voltage or pressure. This phenomenon is the fundamental basis of electrical power and can be analogized to water flowing through a hose. Depending on the orifice of the hose, constriction of the flow, referred to as pressure, when multiplied with the current is defined as power. If we designate

Figure 3.1. Mined quartz crystal.

I = current and R = resistance and P = power, the power production formula can be represented as $P = R \times I$. In electrical engineering or physics, the term "I" stands for electrical current (amperes), "R" stands for resistance of a medium (ohms), and "W" stands for power (watts).

At this stage it is important to pause and take note of the unique characteristics of a solar PV system as an electrical energy generation system.

All conventional electrical power that turns the world economies and sustains our modern way of life, with the exception of hydroelectric power, which provides 12% of the electrical power in the United States, requires a fuel or a feedstock. Close to 20% of the U.S. electrical energy is generated by nuclear power, more than 50% by coal fire power plants, about 15% by natural gas, and the remainder by geothermal, wind, and other energy resources. Figure 3.1 is a photograph of mined quartz crystal.

Since the energy feedstocks of solar PV power systems are the solar photons that impact the PV modules in the form of irradiance, solar power generation can be considered a perpetual energy generation technology, which can generate electrical power as long as the system is operational.

One of the key characteristics of the solar PV technology, which also differs from the rest of the electrical power plants, including wind and geothermal plants, is that the majority of high-efficiency PV solar power technologies are fabricated from silicon or glass based materials and are sealed and encapsulated hermetically. This allows for the manufacturers to guarantee longevity of the product for 25 years. In any sector of the manufacturing industry, no mechanical devices are known to be guaranteed for more than a period of 5 to 10 years. Figure 3.2 shows photograph of a purified silicon ingot.

Another unique feature of PV technology is that if maintained very minimally, solar power installations may last several decades beyond their 25-year warrantee

Figure 3.2. A purified silicon ingot used in photovoltaic solar power technology.

period. Such longevity of a power generation lifecycle is unknown in any other industry, since electrical turbines used in all types of electrical power generation plants are constructed from electro-mechanical rotating equipment that require frequent or periodic maintenance and replacement. Likewise nuclear plants, upon the expiration of certain operational periods, require costly and extensive rework and refurbishment.

Therefore, when loaning capital for construction of electrical energy plants or equipment purchase, conventional accounting procedures that are applied take into consideration return on investment. This is formulated in fundamental equations that include the cost of capital equipment, equipment life cycles, maintenance and operation costs, equipment salvage values at the end of the plant's operational life cycle, and energy production economic returns.

Figure 3.3 depicts a solar photovoltaic power generation diagram.

Three often overlooked significant factors that are unique to solar power system financing are:

- Anticipated system extended life expectancy, which could be 40–50 years
- Absolutely minimal overhead and maintenance requirement
- Dynamic salvage value of the system at the end of the contractual life cycle, which is unlike any other capital investment common to the industry as a whole.

To develop a viable financial model for solar photovoltaic systems, the accounting structure and formulation, in addition to the conventional methodology of cost accounting, must take into consideration numerous factors that constitute the anatomy of each particular solar power project. These include the type of solar power project (for example, a roof mount, ground mount, carport, or simply just a building integrated solar power system), the solar power platform specifics, types of technology used, design methodology and logistics, system reliability, installation

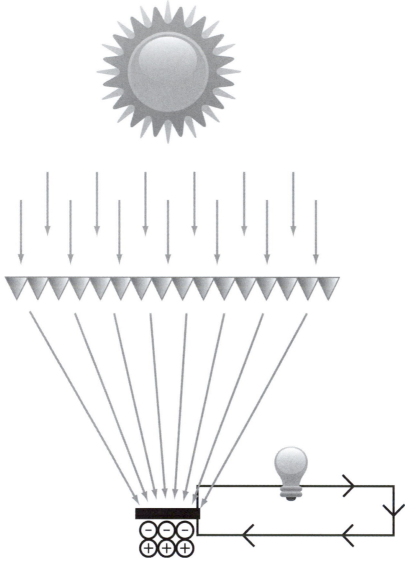

Figure 3.3. Solar photovoltaic power generation.

and integration procedures, system operation and maintenance, electrical energy tariff and associated cost escalation, types of financing, methodology of accounting, and much more. All of these factors may be unknown to financial lending or banking institutions. Figure 3.4 depicts resources of energy use.

Solar Photovoltaic Power System Knowledge Base

In order to develop a pragmatic financial model, solar power technology lending institutions must have minimal levels of appreciation of the following:

1. Fundamentals of solar power technology
2. Basic understandings of solar physics and effects of environmental variations on solar PV system output performance

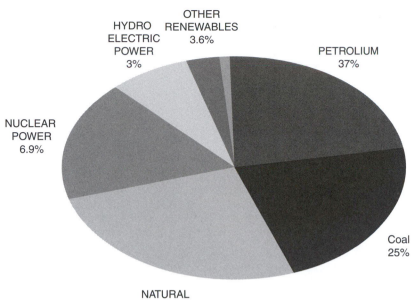

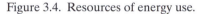

Figure 3.4. Resources of energy use.

3. Basic understanding of the types of solar power technologies such as mono-silicon, poly-silicon, amorphous silicon, film technology and implication of cost differential, and longevity and viability of each type of installation
4. Familiarity with basic types of solar power system installation on various types of platforms such as rooftop, ground mount, carport, and building integrated systems, and the characteristic differences among them
5. Implications and the importance of a thorough solar power system feasibility study
6. Familiarity with the logistics of solar power system installation of each
a. Type of solar power system installation
7. Implication of solar power system integration with the electrical system and grid connectivity
8. Some understanding of types of electrical tariffs, net metering, time of use, and their implication for electrical energy cost
9. Some knowledge of the electrical power use and implications of low and peak power use and the associated penalties
10. Fundamental issues associated with global warming and implications of fossil fuel use
11. Fair assessment of the availability of global fossil fuel resources and their effect on energy cost escalation
12. Basic awareness of energy cost escalation associated with electrical transmission line renovation and implications of the smart grid system for future solar power energy generation; also, present and near future percentage projected cost increase of electrical power
13. Minimal knowledge of implications of good engineering design, project management, system test and integration methodologies
14. Fundamental components of solar power system cost structures and implications of labor and material components for the overall solar power system cost

Conventional Solar Cells   UNI-SOLAR® Laminates

Figure 3.5. Types of solar power technologies.

15. Implications of solar power energy insolation and energy production vis-a-vis electrical power use in various industries
16. Familiarity with solar power output performance and economic contributions throughout PV system life cycle
17. Understanding of solar power output performance and energy production profiles for each type of technology
18. Knowledge of life cycle cost of solar power system, such as maintenance, repair, and energy peak power optimization
19. Familiarity with solar power system rebates, state and federal tax incentive programs
20. Fundamental knowledge of engineering econometric analysis and the economics of capital investment
21. Fundamental understanding of differences between conventional capital equipment and solar power system depreciation; this issue is of paramount importance and must be clearly understood by owners and lenders to appreciate the superior value of long-term investment advantages of the solar PV power systems

Figure 3.5 depicts several types of solar photovoltaic technologies. Figure 3.6 is a photograph of a solar power carport. Figure 3.7 is a photograph of a roof mount solar power system.

22. Familiarity with types of investments and the different advantages and disadvantages among cash purchase or self-financing, lease option to buy and various methodologies and structures of Power Purchase Agreements (PPAs), municipal lease structure, and much more
23. Familiarity with various types of economic analytical profiling and accounting tools specific to solar power technology assessment and investment such as NREL PV Watts, SAM, and other commercial programs

Figure 3.6. Solar power carport system.
Source: Vector Delta Design Group, Inc.

Figure 3.7. A roof mount solar power system.
Source: Vector Delta Design Group, Inc.

24. Familiarity with DOE reports regarding technology advancement and economic profile of global solar power technologies
25. Impact of solar power system use on global warming, atmospheric and environmental pollution, and its long-term implications for the lifestyle of future generations
26. As mentioned, all of the cited issues have significant implications for the formulation of an economic model for solar power system deployment and use;

Figure 3.8. Robotic PV module assembly.
Source: Martifer

considering the extent of knowledge required to formulate viable econometrics, lending institutions, bankers, and clients must equally have fundamental understanding as to how each of the parameters adds to or subtracts from the economic value of the capital investment

Figure 3.8 is a photograph of a robotic PV module assembly.

A Few Words about Solar Power System Energy Production

As discussed previously, solar photovoltaic power technologies are essentially electrical energy power generation devices that absorb the Sun's emanated energy in the form of photons and convert them into electrical power. Simply stated, solar power cells (as long as they are exposed to solar irradiance) will produce electrical energy.

In view of Earth's daily rotation around its axis every 24 hours (altitudes), solar energy impacts the surface of the Earth for several daylight hours. Since daylight hours vary from month to month (even hour to hour) solar power production of PV systems takes place during daylight. Depending on a geographic location, the average daily solar daylight hours (the sum of total yearly daylight from January to December divided by 365 days) varies from one spot to the next. The average daily solar exposure is referred to as daily insolation. For instance, daily insolation in Los Angeles, California, is about 5.42 hours, whereas in Palm Springs, California, it is about 6 hours.

Maximum harvesting of solar energy by solar cells also takes place when solar rays or photons impact the surface of the PV modules. Since the Earth's axis of rotation with reference to the Sun is 23.5 degrees, and the surface of planet Earth curves up and down from the equator to the North and South Poles (latitude), then in order to maximize solar energy capture, solar PV panels must be tilted at an angle that allows solar rays to impact them at right angles. Figure 3.9 depicts Earth's solar declination angle variations.

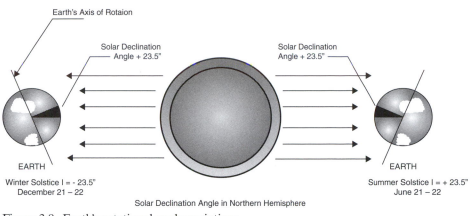

Figure 3.9. Earth's rotational angle variations.

Another important factor to take into consideration is that while the Earth rotates around the Sun in an oval path, its rotational axis sways back and forth by 23.5 degrees. This swinging pattern of the Earth's axis effectively results in the Sun's angle's rising and falling from month to month. For example, in mid-June the Sun appears to be directly above in the sky (its zenith point), whereas in December its angle appears to be lower.

As we know, the Sun rises in the morning from the east and sets in the west. Therefore, to harvest the maximum amount of energy, a PV module must be mounted on a mechanism that that during daylight hours tilts from east in the morning and swings toward maximum west at sunset. In addition, in view of the rise and fall of the solar angle, the platform must also change its latitude tilt angle a few degrees in summer for a maximum amount of solar power and require an automated tracking system angle in the wintertime. On the basis of the preceding principles of solar physics, this implies that solar power installation and power production may be subject to the type of support platforms that PV modules can be installed on. As a result, a solar PV module could be deployed on three types of platforms, namely:

- Fixed angle platforms, where PV modules are mounted on a stationary structure tilted in one angular position
- A single axis motorized tracking platform, which automatically tilts a group of PV modules east at dawn to west at dusk
- A dual axis tracking motorized platform, which automatically tracks the solar angle on a continuous basis year round

In general, single axis solar power trackers produce an average of 25% more annual energy than fixed axis stationary solar power systems. Likewise, dual axis solar power tracker type systems produce 25% more annual energy than single axis trackers and approximately 50% more energy than fixed axis solar power systems.

Figure 3.10 depicts daily solar travel path, Figure 3.11 depicts solar zenith angle, Figure 3.12 depicts annual solar height, and Figure 3.13 shows monthly solar percent irradiance. Figure 3.14 is a graphic curve that shows solar energy efficiency loss due to PV module misalignment.

A specific issue to be taken into consideration is that fixed angle and single axis solar power systems require considerable platforms space, whereas dual axis

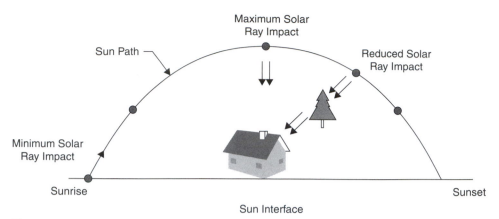

Figure 3.10. Daily solar travel path.

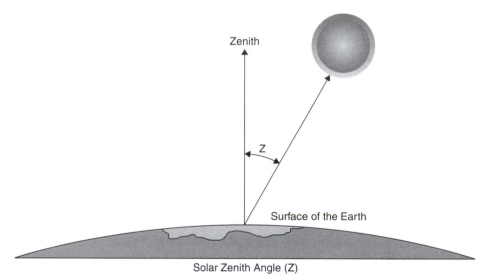

Figure 3.11. Solar zenith angle.

trackers, because of their hydraulic support structures, occupy extremely minimal space. For example, a 1-megawatt fixed angle or single axis solar power system may require 5–6 acres of land, whereas a dual axis tracker system platform producing 50-kilowatt power per tracker may occupy no more than a circle of 20 foot prints of 3–4 feet in diameter footing. Figure 3.15 is a photograph of a fixed angle solar power system. Figure 3.16 is a photograph of a single axis tracking solar power farm. Figure 3.17 is a photograph of a dual axis solar photovoltaic concentrator system.

Solar photovoltaic system power production

As referenced previously, solar photovoltaic power modules produce electrical energy during sunlight hours. The bell shaped curve shown in Figure 3.14 depicts energy output production of a typical solar power system from sunrise to sunset. It should be noted that the shaded area of the bell curve represents the maximum or peak power output energy production period that coincides with the intensity

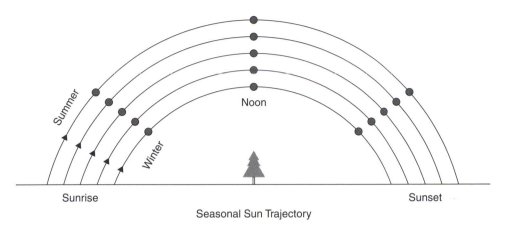

Figure 3.12. Annual solar height path.

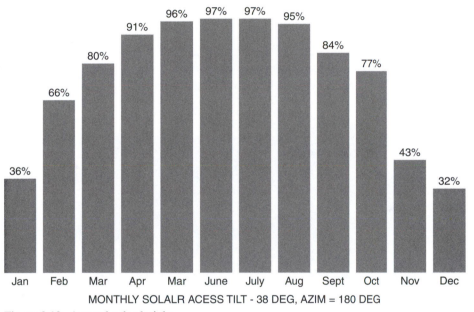

Figure 3.13. Annual solar height.

of solar irradiance. For instance, the peak power production time in California during summertime occurs between the hours of 11 am and 5 pm. Such peak power production occurs in exactly the same hours that coincide with maximum grid power use by residential, commercial, and industrial consumers.

In grid connected solar power systems, the amount of power produced or harvested from the Sun results in net reduction of power that would otherwise be used from the grid. The effect of overall power consumption by solar power is generally referred to as Peak Power Shaving. Figure 3.18 depicts a daily solar power energy production curve. Figure 3.19 shows solar power energy as a peak power shaving energy cogeneration system.

The example shown in Figure 3.19 makes assumptions based on a commercial or industrial entity that operates 10 hours a day, has a steady 2 megawatt hours of daily

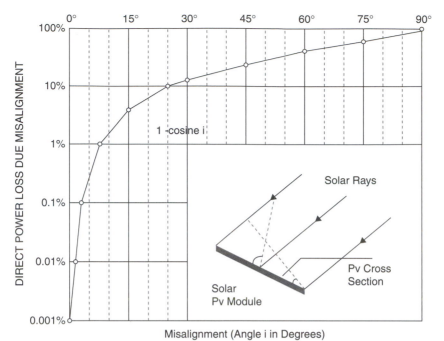

Figure 3.14. Solar power energy efficiency loss due to misalignment of photovoltaic modules.

Figure 3.15. A fixed angle solar power system.

electrical energy demand, and has a 1-megawatt (1000 kW) solar power cogeneration system. In the figure (from left to right shown) is the following tabulation:

    Column 1 – Months of the year from January to December
    Column 2 – Insolation (average solar irradiance) hours in each month
    Column 3 – Days in each month
    Column 4 – Electrical energy demand of user

Figure 3.16. A single axis tracking solar power system

Figure 3.17. A dual axis tracking solar power system. *Source:* Photo courtesy of AMONIX

Column 5 – Monthly grid electrical energy use without solar power cogeneration
Column 6 – Monthly solar power energy generated (represented as a negative value)
Column 7 – Net monthly energy consumption

The green curve shows the net effect of grid peak electrical power reduction in the critical hot months of summertime when air-conditioning systems operate, consuming close to 50–60% of commercial energy.

Overview of photovoltaic system solar power output performance characteristics

In view of special performance characteristics of solar PV cells, solar power systems could be considered as virtual distributed power generation centers that could be

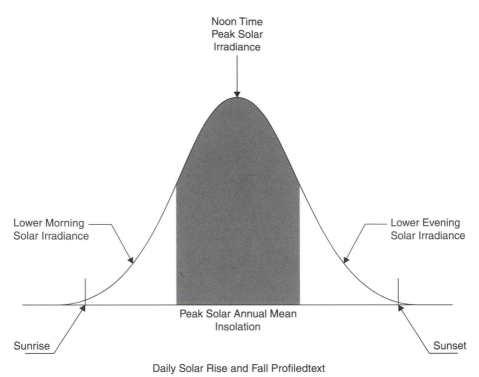

Figure 3.18 Daily solar power energy production curve

located on any property. These centers could be owned and operated as local power generators or cogenerators of electrical energy.

To understand specific operation performance characteristics of solar cells better, users, owners, and finance and lending organizations must have a fundamental understanding of specific operation and performance characteristics of solar cells. Unlike any other technology, they are subject to numerous environmental parameters that affect power output production over a project life cycle. Some of the parameters affecting power output performance of solar PV power systems are the following:

- Temperature – Solar PV cells are manufactured from semiconductor materials such as silicon (Si) that conduct electrical current (flow of electrons) in a crystalline medium that has ohmic or electrical resistive characteristic. When cold, the crystalline medium becomes less resistive and when hot it becomes highly resistive to the flow of electrical current. In other words, in sunny yet cold temperatures, PV modules produce more power than in hot sunny weather.
- Ambient humidity – When operating in a humid location, a solar power module can dissipate heat by evaporating adjacent water particles. This results in the removal of heat from the panel, therefore somewhat adding to the operational efficiency.
- Wind – In windy conditions the panels are cooled, so they may slightly improve their efficiency.
- The reflection of sunlight from adjacent hills located in various locations (except in the front) and lakes, the local solar platform (called the albedo effect), as well as solar flares, may cause PV modules to produce more energy.

SOLAR POWER CONTRIBUTION CALCULATION EXAMPLE						kW/hr
ASSUME AN ENTITY WHICH HAS A 2 MW.HR ENERGY USE			TOTAL DAILY KW			20,000
AND 10 HRS OF DAILY OPERATION			SOLAR HOURLY			1000

MONTH	SOLAR INSOLATION HRS ZIP-92392	DAYS PER MONTH	AC POWER Kw/hr	GRID ENERGY USE	SOLAR ENEGRY	NET POWER USE
JAN	3.56	31	20000	620,000	110,358	509,642
FEB	4.3	29	20000	580,000	124,698	455,302
MAR	5.81	31	20000	620,000	180,107	439,893
APR	7.02	30	20000	600,000	210,597	389,403
MAY	7.91	31	20000	620,000	245,207	374,793
JUN	8.34	30	20000	600,000	250,196	349,804
JUL	8.16	31	20000	620,000	252,956	367,044
AUG	7.64	31	20000	620,000	236,837	383,163
SEP	6.57	30	20000	600,000	197,097	402,903
OCT	5.31	31	20000	620,000	164,608	455,392
NOV	4.15	30	20000	600,000	124,498	475,502
DEC	3.35	31	20,000	620,000	103,849	516,151
	TOTALS			7,320,000	2,201,009	5,118,991
PERCENT CONTRIBUTION		30%				

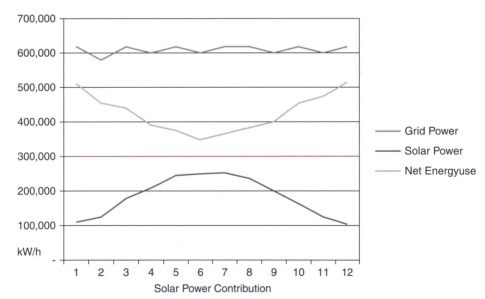

Figure 3.19. Solar power energy as a peak power shaving energy cogeneration system.

- Air mass – Air mass is the volume of air that occupies the space between solar power PV modules and the stratosphere. The larger the volume of air or the distance, the more particulates, gases, and water vapor there are, resulting in more solar ray refraction.

It should be noted that when solar panels are tested in a laboratory environment, called Standard Test Conditions (STC), they are subjected to a barometric pressure

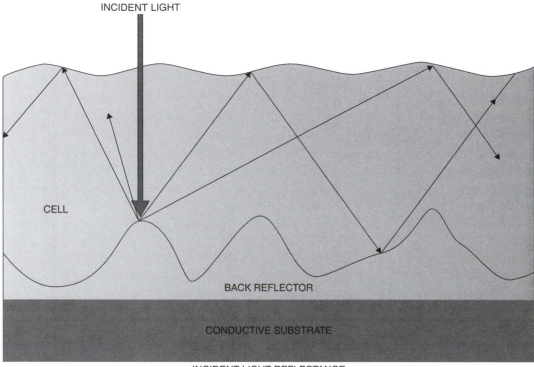

Figure 3.20. Incident light and reflectance.

equal to what exists at sea level and a temperature of 25 centigrade. As such any deviation of temperature or pressure from STC will have an effect on power output performance of a PV module.

It should also be noted that 1 square meter of solar energy, impacting at a sea level measured at 25 centigrade, produces approximately 1000 watts, which forms a terrestrial solar power irradiance reference. A square meter of surface at the stratosphere results in 1314 watts of measured energy.

Therefore, the amount of solar power energy impacting on the surface of the Earth is subject to the distance that photons must travel from the stratosphere to the PV module. Figure 3.20 is a graphic depiction of incident light and reflectance

- Barometric pressure – As a result of the phenomena discussed, barometric pressure variation (which represents the presence or absence of air mass) can cause obstruction of solar irradiance and reduce or increase power production. Therefore, climatic variation that results in high or low atmospheric pressure does affect power power output production of a solar power system.
- Clouds – Evidently any obstruction of sunlight by clouds, however invisible (such as haze), causes refraction of solar rays and results in solar power output production
- Shading – Any object such as a building, tree, or hill during a certain time of the day or a season could cause obstruction of the Sun's rays. In some instances tall buildings located at a distance of a mile may unnoticeably cast a shadow

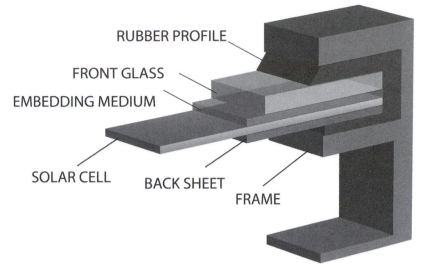

RUBBER PROFILE

FRONT GLASS

EMBEDDING MEDIUM

SOLAR CELL   BACK SHEET

FRAME

Figure 3.21. A solar photovoltaic panel cross section.

Figure 3.22. Solar PV module system installation.

on a solar power system. Therefore, shading analysis becomes an important consideration when installing solar power systems.

- Solar power modules are essentially constructed from silicon semiconductor devices that are sandwiched and laminated between two layers of strong hail impact–resistant glass or medium. The laminated solar cells are in turn assembled in aluminum frames and hermitically sealed with a lamination material called Tedlar. Tedlar prevents any moisture penetration between the laminates and exposure of the semiconductor cells to radical oxygen. Such a fabrication methodology, as well as the nature of silicon material, results in exceptional longevity of the solar cells and PV modules. However, because of the crystalline nature of the device, a slow deterioration of the laminate material does result in slight performance deterioration of the PV modules. In general, manufacturers guarantee solar power output annual deterioration not to exceed 1–1.5%. Solar PV modules generally have a guarantee of 25 years replacement warranty by the manufacturers. It should be noted that solar power PV modules constructed from amorphous or film technologies have a larger percent annual degradation. Figure 3.21 depicts a solar photovoltaic panel cross section. Figure 3.22 is a photograph of a PV module railing support structure.

4   Solar Power System Economics

Electrical Energy Use and Tariffs

When using electrical energy, whether for residential, commercial or industrial clients, we enter into a contractual agreement to pay a certain price for consuming a unit of electrical energy, called 1 kilowatt hour. For example, 1 kilowatt hour of electrical energy can be considered the energy used by a commercial toaster or a coffee brewer of 1 kilowatt for 1 hour or the energy used to light 10 100-watt lamp bulbs for 1 hour.

Unit cost of energy is contingent upon numerous factors such as:

- Type of fuel, such as coal, natural gas, or uranium
- Cost of preparation and processing of fuel stock, such as mining or purification processes
- Availability of the fuel stock resources
- Transportation costs of fuel stock to power generation centers
- Operation costs of power generation station
- Wear and tear and life expectancy, or life cycle, of the power generation equipment such as turbines
- Cost of electrical power transmission from the generation plant to a point of use
- Cost of operation and maintenance of power transmission lines
- Cost of power voltage transformation
- Loss of energy transformation from point to point

  - Amount of power use by various clients, such as bulk or cyclic power
  - Time of power use such as daytime (peak power), nighttime (off-peak power), or in between the high and low peaks (midpeak power)
  - Type of contract agreement for power use, referred to as time of use
  - Penalties associated for exceeding upper contractual use limits of power, referred to as peak power rates
  - Type of electrical power produced from alternative or fossil fuels
  - Power generation and distribution subsidies by governments
  - Redundancy of power sources
  - Power bulk purchase prices between power generation and power distribution entities

Anatomy of Solar Power System Costing

Preparing an accurate cost structure for any solar power system, whether small or large, requires several key components: a thorough knowledge of solar power system technology, considerable experience in engineering design, fundamental knowledge of econometric and financial analysis, proper knowledge of local and federal rules and regulations, and, most important, understanding of the client's present and future energy cogeneration needs.

Responsibilities of Solar Power Design Consultant

Turnkey solar power system design and integration (described later) are intended to provide the reader an overview of a comprehensive engineering design and construction scope of services that must be provided to achieve seamless solar power system installation.

In view of the multidisciplinary nature of photovoltaic technologies, integrated system design and construction necessitates diverse multidisciplinary engineering proficiency in the electrical, solar, structural, environmental, system integration, and construction management disciplines.

The following are the design and construction scope of engineering and construction services that must be provided to realize a successful turn-key solar power system project. All of the services outlined are totally independent yet complementary, and in most instances indispensible for successful project execution. The engineering services outlined in the following correspond to various phases of typical solar power design and construction:

- Phase I – Solar Power System Feasibility Study
- Phase II – Detailed Solar Power System Design
- Phase III – Solar Power System Integration Supervision
- Phase IV – Solar Power System Project Management
- Phase V – Solar Power Project Supervision and Commissioning
- Phase VI – Customer Technical Training
- Phase VII – Solar Power System Maintenance

Phase I – Solar Power System Feasibility Study

The solar power feasibility study scope of engineering and technical services provided must include the following:

A. The engineering design of a solar power system must always commence with a comprehensive project site survey study. A thorough feasibility study enables engineering consultants to assess solar power system requirements and platform energy production potential. The site evaluation must include the following:

- Solar platform topology study for rooftop, carport, and ground mount solar power systems
- Platform shading analysis
- Study and evaluation of the site electrical system and grid connectivity configuration

- Site electrical power demand loading analysis
- Optional electrical energy audit

B The preliminary solar platform topology mapping design and infrastructure study must encompass the following:

- Solar power platform topology mapping analysis
- Alternative technology evaluation that would be best suited for specific types of solar platforms (i.e., roof mount, building integrated, carport, or ground mount)
- Analysis of PV module support structures for various types of platforms
- Preliminary study of underground conduit runs and solar interplatform interconnectivity
- Econometric analysis and modeling of the solar power project, which include solar projected construction cost estimates, solar power system output generation, financial contribution for the life cycle of the project, and analysis of financial options

C The feasibility study report provided must include a detailed outline of the site survey discoveries, solar power system energy output production analysis, solar platform topographic solar power system overlays, discussions regarding specific types of solar power technologies that are best suited for the project, financial analysis, and recommendations. Customer provided documents required for conducting a feasibility study may include the following:

- Electrical engineering documents consisting of built single line diagrams, switchgear riser diagrams, panel schedules, site plan, architectural plans, locations and equipment layout diagrams of all electrical and power distribution rooms and enclosures
- Electrical bills for the last 12 months
- Roof plans (preferably showing roof mount mechanical equipment and vents)
- Site underground conduit plans
- Energy audit documentation and reports (if any)

The previously referenced solar power feasibility study should also involve a number of site investigations and meetings with the client's technical and maintenance personnel.

Phase II – Detailed Solar Power Design

Final design scope of work provided must be composed of the following engineering services:

- Review and evaluation of all pertinent existing site plans, electrical service demand load, and electrical design documentation
- Evaluation and analysis of the client's electrical power demand distribution system as well as existing grid interface equipment; these include metering, power distribution, and transformer stations
- Evaluation and study of all possible site solar power platforms such as roof mount, carport, building integrated photovoltaic BIPV, or ground mount systems
- Analysis of power production potential of each solar power platform within the project site

- Evaluation and recommendation of best use of technologies for various solar platforms such as ground mount fixed angle and single axis and dual axis tracking, carports, and roof mount solar photovoltaic power cogeneration systems
- Economic analysis of cost-effective technology alternatives best suited for the project
- Customer assistance for completion and filing of rebate application forms (if applicable)
- Final design scope of engineering services including platform shading analysis (if applicable) and preparation of complete set of electrical design and construction documentation
- Generation of project specific solar power system design specification
- Assistance with bid equipment and material evaluation (if applicable)

In the event of a roof mount and ground mount solar power system installation requiring civil, structural engineering intervention, we provide detailed engineering code compliant plans and documentation. These may be required to meet structural, seismic, and wind shear requirements.

Phase III – Solar Power System Integration Supervision

Solar power system integration (construction) supervision must involve an integrated coordination effort among the engineering design team, the construction team, project management, and the client's technical personnel. The following are the specific scope of work and effort that should be included in solar power system construction:

- Study of engineering plans and documents
- Material takeoff
- Preparation of shop drawing
- Material and equipment procurement
- Workforce mobilization and site specific construction crew training
- Construction and project coordination procedures
- Site logistics, which include material storage, site office space, site assembly location, material handling, material transportation, equipment rental, and environmental compliance procedures
- Establishment of Occupational Safety and Health Administration (OSHA) procedures
- Project site preparation, which may include grading, solar power support structure foundation works, roof structural reinforcement, and roofing material replacement

  - Establishment of site maintenance and security procedures
  - Establishment of construction supervision procedures
  - Establishment of integration and test procedures
  - Establishment of test and commissioning test procedures

Phase IV – Solar Power System Project Management

One of the most important prerequisites for a successful large-scale solar power system provided by a solar power engineering consultant is the establishment of an integrated project management protocol. This enables us to coordinated large-scale solar power projects. The following is a summary of project management tasks and the scope of work that must be followed throughout the life cycle of the project's design and construction:

- Project managers are required to have total familiarity with the solar power system feasibility study report and the engineering design and construction documents (completed in Phases I and II of the project)
- Project managers assume contract negotiation responsibility and project cost control
- Project managers assume responsibility to oversee manpower and material expenditures
- Project managers are responsible for the preparation of proposal documentation
- Project managers are responsible for site integration coordination and periodic report preparation
- Project managers organize weekly and monthly project review and coordination meetings with engineering management, the construction supervisor, and clients
- Project managers assume a point of contact for document submittal and reception to and from the client

Phase V – Solar Power Project Supervision and Commissioning

Solar power system supervision is an independent engineering task that is intended to provide project supervision. This effort, which may be considered as an optional service, must also be considered independent solar power construction oversight during the construction period. In essence, this task is equivalent to an objective quality assurance. Construction oversight must be carried out by a member of the engineering design team who has complete familiarity with system specification and design documents. Solar power construction supervision should be considered as an important effort and an essential responsibility of the design consultant, since it prevents construction errors, minimizes possible field retrofits, optimizes construction progress, and eliminates unnecessary construction anomalies that may be overlooked.

The scope of services provided for project construction supervision must include the following:

- Scheduled weekly site inspection
- Preparation of a weekly site inspection report; weekly site reports include narrative of the construction progress, photographs, and issues of concern; site meeting coordination highlights; project weekly construction project updates; and action measures
- Weekly site coordination meeting
- System integration test oversight
- Final test and commissioning oversight

In some instances solar power project supervision can be considered as a specialty engineering service, at which time, the consulting engineer assumes the role of the *client's technical representative.* In such a scenario, the consulting engineer must assume full responsibility for project oversight. In such a role scope of engineering services must include the following:

- Project management
- Principal construction coordination agent
- Verification of all technical documents
- Direct participation in integration test
- Direct participation in solar power system final acceptance and commissioning tests
- Customer training

Phase VI – Customer Technical Training

In general, large-scale solar power system installations involve a considerable number of technologies. As a result, the maintenance and operation of each piece of equipment and solar power system component require technical knowledge and expertise. Customer training is particularly important for the client's technical as well as maintenance personnel, who may have the ultimate maintenance responsibility for the overall solar project.

To ensure proper customer training, the engineering consultant must prepare comprehensive training material and a syllabus that is specifically tailored to address all fundamental knowledge required to train customers and technical personnel, who may have varying backgrounds and experiences. Documents and customer training courses provided must include the following:

- Review of as-built drawing set (meaning field corrected plans and drawings)
- Review of all solar power system equipment technical operation and maintenance manuals
- Review of all system equipment and material specification
- Review of test and measurement equipment
- Review of equipment test procedures and test methodology
- Review of inverter system start-up procedures
- Review of system life safety procedures, electrical shock hazard prevention
- Introduction to the system's general maintenance
- Introduction to solar power system troubleshooting guide
- Review of data acquisition system hardware and software equipment operational manuals
- Review of field maintenance and test report procedure

Solar Power System Training Curriculum (Optional)

As mentioned, familiarity with solar power system technologies is of outmost importance to a client's technical and operation personnel. In view of the fact that solar power technologies involve an orchestration of multiple disciplines, an optional educational curriculum offered by a design consultant must include a comprehensive spectrum of courses that are specifically designed to inform and educate technical operations and maintenance personnel. The following is an outline of a typical solar power educational curriculum that encompasses comprehensive coverage of all aspects of solar power technologies. The course syllabus outlined in the following reflects fundamentals of each category or type of solar power system technology that must be specifically tailored for each project:

1. Introduction to solar power system curriculum

   * Discussions on environmental pollution, global warming, and the impact of energy use on ecology

2. Principal physics of photovoltaic technologies

   * Photovoltaic effect
   * Photovoltaic physics
   * Photovoltaic technologies

3. Solar power manufacturing overview

   * Mono-crystalline technology
   * Poly-crystalline technology
   * Film technology
   * Dye sensitized solar nanotechnologies
   * Multijunction technologies
   * Concentrator technologies

4. Solar power systems system application

   * Solar power system configuration classification
   * Stand-alone, grid connected systems, hybrid configuration
   * Materials and component
   * PV modules
   * Collector boxes
   * Inverters
   * PV support system
   * Battery backup systems

5. Solar power systems application

   * Grid connected single residential applications
   * Grid connected commercial systems
   * Grid connected industrial systems
   * Stand-alone irrigation system
   * Solar farms
   * NEC code compliance
   * Electrical service and solar power integration procedure and coordination
   * System grounding considerations

6. Overview of solar power system engineering design and system integration

   * Platform analysis and power generation potential
   * Solar insolation physics
   * Shading analysis
   * Service switchgear and grid connectivity system requirements

- Service provider coordination
- Rebate system evaluation
- Feasibility study report

7. Review of specific case studies

- Residential
- Commercial
- Industrial
- Industrial application
- Solar power farms
- Agricultural solar power irrigation

8. Solar power system feasibility study

- Energy audit
- Service provider coordination and tariff studies
- Rebate and local tax exemption review
- Rebate application process
- Electrical power switchgear and metering system load burden analysis
- Geoclimatic system analysis
- LEED design
- Pollution footprint

9. Solar power costing and economics

- Energy cost escalation
- Equipment and labor costing

10. Solar power design methodology

- Roof-mount platform capacity assessment
- Ground mount platform capacity assessment
- Pathfinder shading analysis for existing and new projects
- Detail design solar power capacity evaluation procedure
- Commercial roof-mount support stanchions and hardware
- Roof-mount single ply film technology solar power

11. Solar power design methodology continued

- Megaconcentrators
- Sun tracker systems
- Case study of platinum rated building design

12. Student project design workshop

- Open forum discussion, question and answer

13. Student project design workshop continued

- Project design review and criticism
- Project design review and critic

14. Certificate of course completion award

A One-Day Executive Training Seminar (Optional)

Solar Photovoltaic Systems Technologies and Application

This seminar is specifically tailored for executives who wish to acquire familiarity with general aspects of global warming, the impact of energy use on ecology, solar power system technologies, and the implication of grid connectivity, net metering, and related econometrics.

Basic physics of photovoltaic technologies

- Photovoltaic effect
- Photovoltaic physics
- Photovoltaic technologies

Solar power manufacturing

- Mono-crystalline technology
- Poly-crystalline technology
- Film technology
- Dye sensitized solar nanotechnology
- Multijunction technology
- Concentrator technologies

Solar power system configuration

- Hybrid configuration
- Materials and components
- PV modules
- Collector boxes
- Inverters
- PV support platforms
- Battery backup systems

Solar power systems application

- Grid connected single residential application
- Grid connected commercial systems
- Grid connected industrial systems
- Stand-alone irrigation system

Overview of solar power system engineering and system integration

- Platform analysis and power generation potential
- Service switchgear and grid connectivity system requirements
- Service provider coordination
- Rebate system evaluation
- Feasibility study report

Review of specific case studies

- Residential
- Commercial

- Industrial
- Solar power farms
- Agricultural solar power irrigation

  Solar power system feasibility study

- Energy audit
- Service provider coordination and tariff studies
- Rebate and local tax exemption review
- Rebate application process
- Electrical power switchgear and metering system load burden analysis
- Geoclimatic system analysis
- Leadership in Energy Efficiency Design (LEED™) design
- Pollution footprint

  Solar power costing and economics

- Energy cost escalation
- Equipment and labor costing
- Solar power econometric analysis
- Direct purchase and Power Purchase Agreement (PPA)
- California Solar Initiative and Federal Tax exemption

Phase VII – Solar Power System Maintenance

Large-scale solar power systems, because of complexity and associated life safety issues, mandate that the technical personnel responsible for system maintenance be well trained and experienced. They must show knowledge of the integration and diagnostics troubleshooting of various PV solar power system components and equipment.

In order to achieve competence in solar power systems, the owner's technical personnel must at the outset of contract award be actively involved in the project and be assigned responsibility to take part in review of technical document submittal. Maintenance personnel must also take part in the customer training program. Furthermore, maintenance personnel must also participate in site inspection, system integration and test, and the final test and commissioning of the project.

In view of the lack of familiarity of most technical personnel with high-voltage DC electrical systems (600–1000 VDC), maintenance personnel must be specifically educated about life safety procedures and be informed about specific hazards that are associated with solar power systems.

Solar Power System Cost Analysis

The following is a summary description of analytical methodology that must be used by solar power engineering design consultants for determining integration cost calculation and financial econometrics for large-scale grid connected solar power systems.

The exercise described previously is intended to provide the private owners, as well as financial lending organizations, an accurate analysis of solar power system construction funding and lease purchase agreements or Power Purchase Agreement.

The main objective of the analysis is to formulate a project costing model that encompasses numerous cost components and variable parameters associated with labor, material, engineering design solar power performance characteristics, energy cost escalation algorithms, and other aspects. These are required to perform detailed financial calculations. Needless to say, accurate cost analysis requires considerable solar power system engineering design, system integration, construction management, and engineering econometric expertise.

It should be underlined that the significant feature of the Solar Power Econometric Analysis (SPEA) costing analytical methodology used must take into consideration the accounting methodology that is uniquely applicable to solar power system capital equipment depreciation and return on investment. In most instances, the econometric analytical methodology used by consultants must include specific computational algorithms that conform to real life solar power system integration costing, photovoltaic system characteristic performance parameters under various climatic conditions, solar platform venues, PV system intrinsic degradation, and much more. The solar power system analysis described in the following can be accomplished either by a special proprietary spreadsheet or by a commercially available software program.

The following are specific costing factors required to develop a pragmatic solar power system costing profile.

Solar Power System Construction and Integration Costing

A – Engineering Cost Component Analysis

- Engineering site survey and feasibility study
- Engineering feasibility report for various solar power system alternative technologies and support platforms
- Shading analysis
- Detailed engineering design documentation
- Solar system integration, test, and commissioning specification
- Assistance to owners of client in procuring rebates
- Project coordination
- Periodic construction supervision
- Field test and acceptance oversight
- Project coordination
- Solar power platform design
- Negative environmental impact report
- Power transmission and grid integration
- Civil and structural design
- Construction and performance bonding expenses
- Construction financing

B – Solar Power System Output Performance Analysis

This analytical exercise is required to evaluate financial contributions of solar power output performance that are influenced by PV module specification and various output power depreciation factors imposed by California Solar Initiative or National Renewable Energy Laboratories (NREL) PV Watts V.2 computation.

Parameters influencing solar power performance taken into account must include the following:

- Solar platform topology power output potential analysis, which includes principal hardware components such as PV module performance specification
- Type of solar module support structure system such as fixed angle flat PV configuration, single or dual axis solar power tracking system
- Topical solar irradiance and per diem power potential
- Solar power system life cycle power output vis-a-vis the expected system degradation
- Typical monthly system solar power output under varying insolation conditions

C – Solar Power System Financial Analysis

The following analyses are intended to provide detailed cost evaluation of a solar power system for the entire contractual life cycle period. The exercise must include step by step solar power energy cost values for each year. Parameters that influence the cost such as system operation performance, dynamic power degradation, present or contractual unit energy cost/kWh, project grid electrical energy cost escalation, rebate profile for Performance Based Incentive (PBI), initial cost investment, salvage value, and many additional factors provide the year-to-year solar power income profile throughout the life cycle of the contract.

The computations must include financial analysis of the Power Purchase Agreement (PPA). The analytical methodology used must outline critical (kWh) unit energy purchase cost to the user, percent annual cost unit energy escalation, and system life cycle maintenance component. Costing parameters considered in the econometric analysis must include the following computations:

- Yearly average AC power output and solar power energy value computation
- Dynamic extrapolation of projected unit energy cost escalation for the entire life cycle of the contract
- Dynamic depreciation of solar power output for system operational life cycle
- Life cycle power output potential
- Progressive rebate accumulation for the PBI period based on the available unit energy fund availability
- Cumulative electrical energy cost income from the end of the rebate period up until the end of the system contractual life cycle
- Integrated accumulative income from the contractual annual cost escalation factor
- Maintenance income from annual maintenance fees

D – Comparative Analysis of Grid Power Energy Expenditure versus Solar Power System Energy Output Cost

This component of the costing analysis must include computations for cumulative energy cost for both grid provided power and PPA solar power contributions during the system cycle that is presented in tabular and linear chart plots.

E – Power Purchase Agreement (PPA) Accounting

The final step of computation, the software uses all of the results to perform a financial analysis for PPA type contracts that includes the following:

- Totalized energy income from life operation
- Present value of the projected system salvage amount
- Federal tax incentive income
- State tax incentive income
- Salvage value at the end of the system life cycle
- Maintenance cost income
- Recurring annual maintenance cost
- Recurring annual data acquisition cost
- Net income value over the life cycle of operation

Example of a Solar Power System Econometrics

The following is an example of a 500-kW solar power system economic analysis. Cost formulations used in this example are based on actual costing parameters that include detail system design, manpower and material installation and integration expenses, federal and state incentives, as well as current California Public Utilities Commission (CPUC) rebates available through various electrical energy providers.

The energy cost escalation profile used in the calculation is based on the recent electrical tariff annual escalation of 10%. Additional parameters used in formulation of return on investment (ROI) are based on specific energy production characteristics associated with solar power equipment cost amortization and salvage value computation that are unique to the technology.

The econometric methodology used in this example could be applied to any type or class of solar power system, as long as appropriate cost factors are adjusted for each category of installation (i.e., roof mount, ground mount, carport)

Solar Power System Cost Analysis

Solar power system cost analysis is one of the most important components of solar power feasibility, since it often establishes cost benefit analyses and criteria for financial cost justifications of the project. Without accurate financial analysis, solar power projects will undoubtedly be subjected to serious negative consequences.

In this chapter we will discuss computational performance characteristics of various types of financial analytical methodologies and software that are being used to compute econometrics of solar power systems.

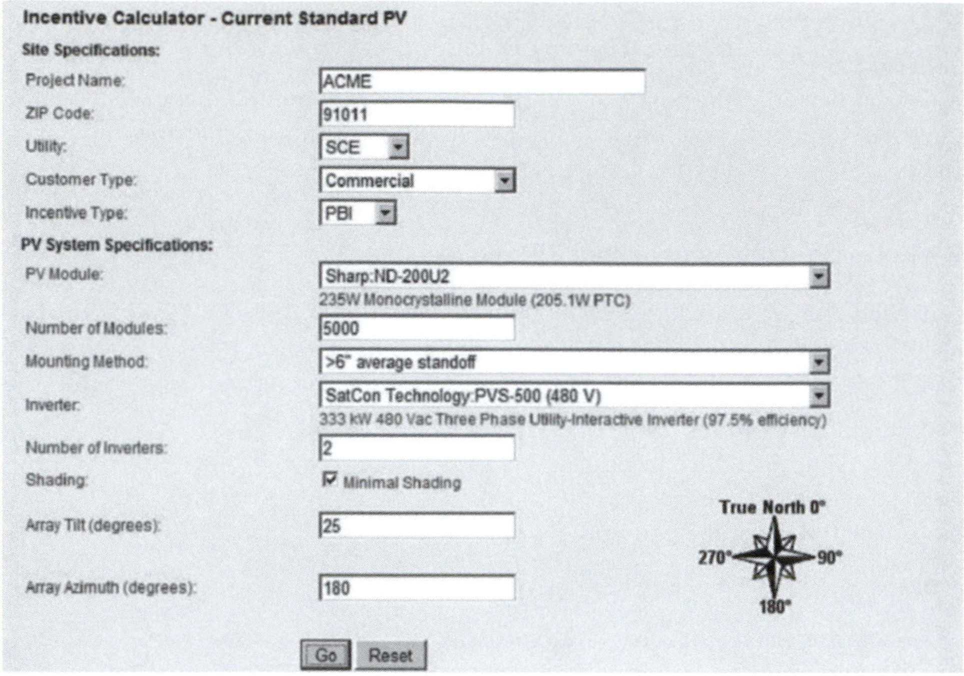

Figure 4.1. CSI solar power calculator data entry page.

California Solar Initiative Calculator

In addition to PV Watts computations, the state of California requires the use of a Web-based calculator referred to as California Solar Initiative (CSI) calculator. It essentially uses the PV Watts V.2 software engine for calculating the solar power system size. Figure 4.1 and Figure 4.2 are examples of a hypothetical project located in La Canada, California.

The CSI calculator also performs project rebate cost estimates for Expected Performance Based Initiatives (EPBIs) as well as Performance Based Initiatives (PBIs). The following are additional computation results provided by the CSI calculator:

- Annual energy production in kW/h
- Summer month power output from May to October in kW/h
- CEC AC power output in kW/h
- Capacity factor in percent
- Prevailing capacity factor in percent
- Design factor in percent
- Eligibility annual power output in kW/h
- Incentive rate per kWh in dollars
- Total CSI incentive contribution in dollars

Solar Power Costing Analytical Software

Depending on the type of financing referenced previously in this chapter, solar power cost analysis may involve a number of methodologies, which range from public

Incentive Calculator - Current Standard PV

Save as a PDF

	Proposed
Site Specifications:	
Project Name	ACME
ZIP Code	91011
City	La Canada
Utility	SCE
Customer Type	Commercial
Incentive Type	PBI
PV System Specifications:	
PV Module	Sharp:ND-200U2 200.0W STC, 173.0W PTC
Number of Modules	5000
Mounting Method	>6" average standoff
DC Rating (kW STC)	1000.0000
DC Rating (kW PTC)	865.0000
Inverter	SatCon Technology:PVS-500 (480 V)
Number of Inverters	2
Inverter Efficiency (%)	96.00 %
Shading	Minimal Shading
Array Tilt (degrees)	25
Array Azimuth (degrees)	180 **True North 0°**
Results	
Annual kWh	1,596,510
Summer Months	May-October
Summer kWh	907,099
CEC-AC Rating	830.400 kW
Capacity Factor[1]	21.947%
Prevailing Capacity Factor[2]	20.000%
Design Factor[3]	109.735%
Eligible Annual kWh[4]	1,596,510
Incentive Rate	$0.22/kWh
Incentive[5]	$1,756,161

Figure 4.2. CSI solar power calculator data output page.

domain solar power cost estimating engines such as the National Renewable Energy Laboratory's Solar Advisor Model (SAM) to a number of other commercial software packages. Solar power estimating software packages provide cost approximation based on local electrical utility tariffs, type of solar power system, weather fixed angle, and single or dual axis tracking systems. Such software estimating calculators do not take into consideration the specific project costs. They also do not consider inflation, energy cost escalation, special labor rates, transportation, and much more. In effect solar power cost calculators do not provide the accuracy required to conduct a detailed cost estimate. However, most institutional, public, and governmental agencies (as well as public utility commissions) require that all solar power calculations be based on acceptable econometric models. In fact, the SAM is considered one of the best estimating software programs available to the public free of charge. It can be accessed at https://www.nrel.gov/analysis/sam/. The following is a description of SAM as it appears in the National Renewable Energy Laboratories.

Solar Advisor Model

The Solar Advisor Model combines a detailed performance model with several types of financing (from residential to utility-scale) for most solar technologies. The solar technologies currently represented in SAM include concentrating solar power (CSP) parabolic trough, dish-Stirling, and power tower systems, as well as flat plate and concentrating photovoltaic technologies. SAM incorporates the best available models to allow analysis of the impact of changes to the physical system on the overall economics (including the levelized cost of energy). SAM development continues to add financing models and performance models to meet the needs of a growing community of users.

This comprehensive solar technology systems analysis model supports the implementation of the program's Solar America Initiative (SAI) as well as general planning for the Solar Energy Technologies Program (SETP). The use of the SAM software (together with technology and cost benchmarking, market penetration analysis, and other relevant considerations) supports the development of program priorities and direction, and the subsequent investment needed to support solar R&D activities. Most importantly, it promotes the use of a consistent methodology for analysis across all solar technologies, including financing and cost assumptions.

SAM allows users to investigate the impact of variations in physical, cost, and financial parameters to understand their impact on key figures of merit better. Figures of merit related to the cost and performance of these systems include, but are not limited to

- System output
- Peak and annual system efficiency
- Levelized cost of electricity
- System capital and operating and maintenance (O&M) costs
- Hourly system production

SAM uses a systems-driven approach (SDA) to establish the connection between market requirements and R&D efforts and how specific R&D improvements contribute to the overall system cost and performance. This SDA allows managers to allocate resources more efficiently.

System Cost Data

SAM software includes a set of sample files that contain cost data prepared to illustrate its use. The cost data are meant to be realistic, but not to represent actual costs in the marketplace. Actual costs vary, depending on the market, technology, and geographic location of a project. Because of price volatility in solar markets, the cost data in the sample files are likely to be out of date.

Photovoltaic Cost

PV cost input data in SAM are divided into two broad categories: capital and O&M costs. Capital costs are further categorized into direct and indirect costs. Direct costs are costs associated with the purchase of equipment: PV modules, inverter(s), balance-of-system (BOS), and installation costs. BOS costs are equipment costs that cannot be assigned to either the PV module or the inverter and may include such costs as mounting racks, junction boxes, and wiring. Installation costs are the labor costs associated with installing the equipment.

Indirect costs may include all other costs that are built into the price of a system such as profit, overhead (including marketing), design, permitting, shipping, and so on.

O&M costs are costs associated with a system after it is installed and are categorized into fixed and variable O&M costs. Fixed O&M costs are costs that vary with the size of the system and may include the cost of inverter replacements and periodic maintenance checks. Variable O&M costs vary with the output of the system and may be considered to be zero or very small for most PV systems.

SAM uses the total installed cost, which is the sum of direct and indirect costs, to calculate the levelized cost of energy. Since the way costs are assigned to each category does not affect the total installed cost, the user can choose to distribute profit, overhead, shipping, and other costs among the component categories. These may consist of modules, inverters, system balance components, and system installation (or it may be considered as a single value and categorized as miscellaneous).

Note that costs in the PV sample files used in the program are based on 2005 costs that are derived from the DOE Multi-Year Program Plan. Therefore, the total installed costs are intended to represent equipment purchase and labor costs plus a margin sufficient to sustain a profitable business with a reasonable return on investment.

Additional Sources of PV Cost

The California Energy Commission's Emerging Renewables Program Web site provides information about systems installed in California and includes a link to a spreadsheet of total installed system costs for systems installed throughout the states. SolarBuzz provides current and historical price data for the U.S. and around the world based on market studies. The site is a valuable resource that provides detailed statistical information about PV module prices, inverters, and the solar electricity price index.

Guide to Using the Solar Advisor Model

The Solar Power Advisor consists of the following 13 data entry and display domains:

A – System Summary – This provides a tabulated overview of the computation, which displays the information shown in the list:

Solar power system capacity in kW
Total direct cost in dollars
Total installed cost in dollars
Total installed cost per kilowatt
Solar power system life cycle in years
Projected inflation rate in percent
Applicable discount rate in percent

B – Climate – Provides information regarding the sites based on postal zip codes, which provides the following weather information and temperature data:

City
State
Time zone in Greenwich mean time (GMT)
Elevation in meters
Latitude
Longitude
Direct normal radiance in kW/m^2
Diffused horizontal radiance in kW/m^2
Dry-bulb temperature in degrees centigrade (°C)
Wind speed in meters per second (m/s)

C – Utility rate – This displays the following fields: Projected utility rate inflation in %

Up to date cost of electrical power per $/kW
Projected utility inflation rate in percent

D – Financing – This domain allows the user to make a number of project specific entries for insurance and taxes, federal depreciation options, state depreciation options, project life cycle, projected inflation rates, and real discount rates. The following are various types of options available in solar power system financing:

- Energy Savings Performance Contract (ESPC) – ESPC is the partnership between the owner and a federal government agency.
- Energy Service Company (ESCO) – ESCO is a financing system that arranges necessary financing for funding the solar PV plant and guarantees the estimated energy cost savings to the owner. This analysis determines the minimum tariff at which electric power can be sold from the solar PV plant to the owner's facility.
- Utility Energy Savings Contract (UESC) – In this arrangement, the federal agency enters into partnership with their franchised or serving utilities in order to implement energy improvements at their facilities. The utility arranges financing to cover the capital costs of the project and is repaid by the owner over the contract term and in turn provides cost savings to the owner.
- Enhanced Use Lease Contract (EUL) – EUL program refers to legislative authority that allows owners to lease underutilized land and improvements to

a developer or lessee for a term of up to 75 years. In exchange for the EUL, the developer would be required to provide the owner with *fair consideration,* such as cash, or *in-kind* considerations, as determined by the owner.

- Direct Funding – In this option, the owner provides 100% funding for the solar power project. No debt financing is assumed.

General

Analysis period (project life cycle) in years
Projected inflation rate in percent
Real discount rate

Note that discounting is a financial mechanism in which a debtor or the borrower obtains the right to delay payments to a creditor for a defined period in exchange for a charge or fee. In other words, the party that currently owes money purchases the right to delay the payment until some future date. The discount, or charge, is the difference between the original amount owed in the present and the amount that has to be paid in the future to meet the debt obligation.

Tax and Insurance

Federal tax percent of the project cost in percent
State tax as percent of the project cost in percent
Property tax
Sales tax in percent
Solar power system life cycle in years in percent
Insurance cost as percentage of the installed cost

Federal Depreciation

In this domain the user is allowed to select various choices regarding types of rebate plans, as well as years of equipment depreciation. Choice includes the following:

- No depreciation
- Modified accelerated cost recovery system (MACRS) midquarter convention
- MACRS half-year convention
- Straight line depreciation
- Custom depreciation, which is percent of the installed system cost

In view of the importance of accelerated depreciation, it is important for the Solar Advisor Model fully to appreciate depreciation choices that may be best suited for a particular project. The following discussion is intended to familiarize the reader with several depreciation methodologies allowed by the U.S. tax system. It should be noted that selection of type of depreciation must always be undertaken by an expert tax accountant.

Modified Accelerated Cost Recovery System (MACRS)

MACRS is a methodology used for recovering capitalized costs of depreciable tangible property other than natural resources. Under this system, the capitalized cost, also referred to as the *basis,* is recovered over the life cycle of a tangible property or asset by annual deductions for depreciation. The life spans of various types of property are broadly specified in the U.S. Internal Revenue Code. Various classes of asset life cycles are tabulated and published by the Internal Revenue Service (IRS). The deduction for depreciation is computed by one of two methods: declining balance or straight line depreciation. http://en.wikipedia.org/wiki/Modified_Accelerated_Cost_Recovery_System – cite_note-0

Depreciable Lives by Class

As mentioned, MACRS specifies that a taxpayer must compute tax deductions for the depreciation of tangible property or assets using specified life span and methods. Assets are divided into classes by type of asset or business in which the asset is used.

For each class, the life cycles are specified as the general depreciation system (GDS) or alternative depreciation system (ADS). Taxpayers have a choice to use either the ADS or the GDS methodology. Life cycles of assets may vary from 5 years to 20 years.

Federal Accelerated Depreciation for Solar Power Systems

Commercial and industrial systems (upon qualification) can take advantage of special solar power system depreciations (26 USC Sec. 168 – MACRS), which allow asset depreciation and amortization over a period of 5 years. Such accelerated depreciation, depending upon combined federal and state tax credits, enables investment recovery of up to 50%. It should be noted that the asset value is the total installed less the amount of the rebates received.

In general cash rebates, tax credits, and accelerated depreciation schedules are designed to facilitate short-term returns on investments. This encourages businesses to generate their own solar power. In some instances such programs can recover up to 40–70% of the total system cost in a very short period.

Note, too, that the 30 percent federal Investment Tax Credit (ITC) is calculated before any state or utility rebates. The claim is calculated after deducting state rebates, such as the net cost paid by installers, who typically collect the rebate on your behalf.

As of December 31, 2009, the IRS had not instituted clear guidance on the issue. Table 4.1 is the IRS MACRS depreciation table showing various asset classifications, asset descriptions, and associated Alternative Depreciation System (ADS) and regular depreciation (GDS) life cycles.

Depreciation Methods

Only the declining balance method and straight line method of computing depreciation are allowed under MACRS. All solar power systems installed during the current

Table 4.1. *IRS asset depreciation table*

IRS asset classes *M*	Asset description 0	ADS class life 0	GDS class life 0
00.11	Office furniture, fixtures, and equipment	**10**	**7**
00.12	Information systems: computers/ peripherals	**6**	**S**
00.22	Automobiles, taxis	**2**	**S**
00.241	Light general-purpose trucks	**4**	**5**
00.25	Railroad cars and locomotives	**15**	**7**
00.40	Industrial steam and electric distribution	**22**	**3**
01.11	Cotton gin assets	**10**	**7**
01.21	Cattle, breeding or dairy	**7**	**5**
13 00	Offshore drilling assets	**7.5**	**5**
13.30	Petroleum refining assets	**16**	10
15.00	Construction assets	**6**	**5**
20.10	Manufacture of grain and grain mill products	**17**	10
20.20	Manufacture of yarn, thread, and woven fabric	11	**7**
24.10	Cutting of timber	**B**	**5**
32.20	Manufacture of cement	**20**	**15**
20.1	Manufacture of motor vehicles	12	**7**
48.10	Telephone distribution plant	**24**	**15**
48.2	Radio and television broadcasting equipment	**6**	**5**
49.12	Electric utility nuclear production plant	**20**	**15**
49.13	Electric utility steam production plant	**28**	**20**
49.23	Natural gas production plant	**14**	7
50.00	Municipal wastewater treatment plant	24	15
57.0	Distributive trades and services	**9**	5
80.00	Theme and amusement park assets	12.5	7

year are considered "placed in service" in the middle of the tax year, referred to as the *Half Year Convention*. The method and life cycle used in depreciating an asset are an accounting method, and any deviation or change requires IRS approval.

Alternative Depreciation Systems

Alternative depreciation systems also available from the IRS pertain to certain assets that must be depreciated under an alternative depreciation system (ADS) using the ADS life cycle, which uses straight line depreciation methodology. Such depreciation only applies to assets used outside the United States and is not applied to solar power systems.

California Exemption from Property Taxes

When installing a solar power system in California, it should be noted that the state excludes all solar power system installation costs from being added to the property value of the property, thus preventing an increase in tax valuation.

Such an exemption means that a solar power system increases the value of property since the added equity is completely tax-free.

Direct Capital Costs (DCCs)

Direct capital cost (DCC) entry sheets allow the user to enter specific installed unit costs ($/kW DC) for the DC component of the solar project, the inverter, storage and transportation, balance of the hardware, and integration costs. Additional fields allow entries for solar power system engineering as percent, nonfixed or fixed cost, and miscellaneous items and sales tax. It should be noted that prior to completing the DCC, the user must complete the PVWatts calculations as discussed earlier. Prior to completing the calculations, the designer must complete the solar power system preliminary design and have an estimated value of material and labor costs. Figure 4.3 is an outline of a typical construction manpower and material cost estimate. Upon completion of PVWatts computation, the amount of total Kw/DC is automatically entered in the system power output capacity field. The remaining required entries include the following:

A – Direct PV System Cost

- Number of modules
- DC kW/unit
- Total system power in kW/DC
- Combined cost of PV and support structure per $/ kW/DC

B – Inverter System Cost

Upon completion of the preceding entries, SAM calculates the direct material cost. The row following the previous entry calculates the cost of the inverter and has the following entries:

- Inverter count
- AC power output rating in kW/AC
- Inverter cost per $/ kW/AC

C – Battery System Cost

The entries compute total cost of the inverter system. The third line identical to the inverter computes battery cost (when applicable).

- Storage battery capacity in kW/h
- Storage cost kW/AC
- cost per $/kW/hr

D – Balance of Other Costs

- Balance of the material costs
- Fixed installation cost
- Contingency
- Engineering, project management, construction supervision either as percentage of the direct cost, variable, or fixed engineering service cost

- Addition of the preceding cost result in total installed cost, which is then divided by the total kW/DC, which yields cost per installed watts $/kW/DC

Operation and Maintenance Costs

In addition to the DCC, SAM allows users to enter operation and maintenance costs for the duration of the life cycle of the project, as well as anticipated escalation rates due to inflation. Options include:

- Fixed annual cost   % Escalation rate
- Fixed cost by capacity   % Escalation rate
- Variable cost by generation % Escalation rate

Figure 4.3 shows a typical solar power construction cost estimate ledger.

E – Indirect Capital Costs

In general, solar power systems have minimal maintenance requirements. However, to prevent marginal degradation in output performance from dust accumulation, solar arrays require a biyearly rinsing with a regular water hose. Since solar power arrays are completely modular, system expansion, module replacement, and troubleshooting are simple and require no special maintenance skills. All electronic DC-to-AC inverters are modular and can be replaced with minimum downtime.

In some instances a computerized system-monitoring console can provide a real-time performance status of the entire solar power cogeneration system. Installation cost of a software-based supervisory program that features data monitoring and maintenance reporting systems must also be taken into account.

Annual System Performance

This entry allows the user an arbitrary entry for system degradation and system availability or reliability.

The last entry page includes a number of entries specific to types of federal and state investment based incentives such as:

- Performance based initiative (PBI)
- Capacity based incentive (CPI)
- Investment based incentive (IBI)

Special Costing Considerations

As mentioned previously, standard Web-based software calculators are designed to provide rough estimates of solar power production and costing that are adequate for feasibility study and rebate application. However, they lack accuracy and bear no relation to specific requirements of projects.

The following is a description of a comprehensive guide for estimating various categories of solar power systems. In order to have accurate estimating results, the solar power design engineer or the designer must have significant field installation

Typical construction cost estimate

INITIAL COSTS & CREDITS

		Hours	$150.00 per hour Rate		Total	%
	Engineering Rate					
	Site Investigation	40 $	150.00 $		6,000.00	46.88%
	Preliminary Design Coordination	24 $	150.00 $		3,600.00	28.13%
	Report Preparation	8 $	150.00 $		1,200.00	9.38%
	Travel & Accommodations	1 $	2,000.00 $		2,000.00	15.63%
	Other					0.00%
	Sub Total				12,800.00	100.00%

DEVELOPMENT

	Permits & Rebate Applications	8 $	150.00 $		1,200.00	5.41%
	Project Management	120 $	150.00 $		18,000.00	81.08%
	Travel Expenses	1 $	2,000.00 $		2,000.00	9.01%
	Other	1 $	1,000.00 $		1,000.00	4.50%
	Sub Total				22,200.00	100.00%

ENGINEERING

	PV Systems Design	90 $	150.00 $		13,500.00	10%
	Architectural Design	90 $	150.00 $		13,500.00	10%
	Structural Design	90 $	150.00 $		13,500.00	10%
	Electrical Design	420 $	150.00 $		63,000.00	48%
	Tenders & Contracting	48 $	150.00 $		7,200.00	0%
	Construction Supervision	94 $	150.00 $		14,100.00	11%
	Training Manuals	48 $	150.00 $		7,200.00	5%
	Sub Total			$	132,000.00	100%

RENEWABLE ENERGY EQUIPMENT

	PV Modules (per kWh-DC)	255 $	3,900.00 $		994,500.00	92%
	Transportation	1 $	5,000.00 $		5,000.00	0%
	Other					0%
	Tax (Equipment Only)	8.25%			82,046.25	0%
	Sub Total				1,081,546.25	100%

INSTALLATION EQUIPMENT

	PV Module Support Structure (per kWh)	255 $	500.00		127,500.00	18%
	Inverter (per kWh)	320 $	488.00		156,160.00	22%
	Electrical Materials (per kW)	320 $	250.00		80,000.00	11%
	System Installation Labor (per kWh)	320 $	1,000.00		320,000.00	45%
	Transportation	1 $	3,000.00		3,000.00	0%
	Other	0				0%
	Tax (Equipment Only)	8.25%			30,001.95	4%
	Sub Total				716,661.95	100%

Figure 4.3. Typical solar power construction cost estimate.

and solar power design experience to account for numerous cost items. These may include but are not limited to the following:

- Site survey and feasibility study
- Engineering design
- Material cost
- Civil and structural design
- Rebate contributions
- Negative environmental impact report
- Field installation labor cost
- Field test and acceptance oversight

- Project management
- Material transportation and storage
- Insurance cost (i.e., errors and omissions, construction insurance, liability insurance)
- Construction loan, construction bond, and long-term financing
- Customer personnel training
- Warranty and maintenance costs
- System life cycle profit calculation such as present value and depreciation cost

A software program designed for computing econometrics of a solar power system must provide subroutines and computational engines that will enable users to insert the numerous variables listed earlier. The software must also compute power performance characteristics and energy production costs and must include algorithms for dynamic energy cost escalation for the duration of the system life cycle. As highlighted previously, the system-costing methodology must account for capital equipment depreciation and return on investment.

Additional parameters that influence the cost are system operation performance, dynamic power degradation, present or contractual unit energy cost/kWh, project grid electrical energy cost escalation, rebate profile for PBI, initial cost investment, salvage value, and many additional factors. These provide year-to-year solar power income profiles throughout the life cycle of the contract.

Additional consideration must be given to Power Purchase Agreement (PPA) type financial type option. In particular, some of the factors that may affect a PPA may include the following:

- Yearly average AC power output and solar power energy value computation
- Dynamic extrapolation of projected unit energy cost escalation for the entire life cycle of the contract
- Dynamic depreciation of solar power output for system operational life cycle
- Life cycle power output potential
- Progressive rebate accumulation for the PBI period based on the available unit energy cost availability
- Cumulative electrical energy cost income from the end of the rebate period up until the end of the system contractual life cycle
- Integrated accumulative income from the contractual annual cost escalation factor
- Maintenance income from annual maintenance fees
- Comparative analysis of grid power energy expenditure versus solar power system energy output cost

Costing computation must also include cumulative life cycle energy cost analysis for grid connected expenses such as:

- Totalized energy income from life operation
- Present value of the projected system salvage amount
- Federal tax incentive income
- State tax incentive income
- Salvage value at the end of the system life cycle
- Maintenance cost income

- Recurring annual maintenance cost
- Recurring annual data acquisition cost
- Net income value over the life cycle of operation

Additional Cost Factors

In addition to manpower and material costs, there are several other cost components. These include life cycle cost variations of utility, operation, and maintenance expenses that are of significant importance for leased or PPA financed projects.

Electrical Energy Cost Increase

In the past several decades the cost of electrical energy production and its consistent increase have been an issue that has dominated global economics and geopolitical politics. It has affected our public policies, has been a significant factor in the gross national product (GNP) equation, has created numerous international conflicts, and has made more headlines in newsprint and on television than any other subject. Electrical energy production not only affects the vitality of international economics but is one of the principal factors that determine standards of living, health, and general well-being of the countries that produce it in abundance.

Every facet of our economy is in one way or another connected to the cost of electrical energy production. Since a large portion of global electrical energy production is based on fossil fuel fired electrical turbines, the price of energy production is therefore determined by the cost of coal, crude oil, or natural gas commodities. Figure 4.5 depicts comparative cost inflation of a utility rate at various interest rates during the life cycle of a project, and Figure 4.4 is a graphic presentation of a $0.13/kW/h utility rate at an 8% inflation rate over the life cycle of a project.

System Maintenance and Operational Costs

As mentioned earlier, solar power systems have a near-zero maintenance requirement. However, to prevent marginal degradation in output performance from dust accumulation, solar arrays require a biyearly rinsing with a regular water hose. Since solar power arrays are completely modular, system expansion, module replacement, and troubleshooting are simple and require no special maintenance skills. All electronic DC-to-AC inverters are modular and can be replaced with minimum downtime.

Further, in Chapter 6 we will discuss a computerized data acquisition and monitoring system that could provide the real-time performance status of an entire solar power cogeneration system, as well as features that allow instantaneous indication of system malfunction.

Federal Tax Credits for Commercial Use

The Energy Policy Act of 2006, which has been extended to 2016, makes provisions for a 30% Investment Tax Credit (26 USC Sec. 48). It should be noted that the tax credit is not a deduction; rather, it is a direct, dollar-for-dollar reduction from taxes owed.

The national Solar Energy Industries Association (SEIA) has a document entitled "The SEIA Guide to Federal Tax Incentives," which was compiled with the help of SEIA members and SEIA's tax attorneys and contains a great deal of specific information and advice regarding the interpretation of the new federal law. Further information can be accessed at the Web link www.SEIA.com.

Incentives for Commercial Solar Projects

Federal Grant

To promote national green energy production, U.S. Treasury grants are available for commercial solar installations. Instead of taking the 30% tax credit, businesses can instead opt to receive a cash grant equal to 30% of installed costs of a solar PV system. This grant option was made possible by the federal stimulus package that was passed in February 2009.

Thirty Percent Federal Investment Tax Credit (ITC)

Under U.S. Code Title 26 (Section 48(a)(3)), the federal government extends a corporate tax credit to businesses that invest in renewable power. The types of eligible solar technologies include solar water heat systems, solar space heat, solar thermal electric, solar thermal process heat, and photovoltaic (PV) systems. The credit or the grant is fixed at 30%. Note that the credit for businesses is not constrained by a dollar-value cap. So regardless of whether the installation of the solar power system costs $100,000 or $1 million, the businesses are permitted to take a 30% credit. In October 2008 Congress voted to extend the ITC for 8 years, through 2016.

Using the federal renewable energy credit with other incentive programs, commercial entities planning to take advantage of the federal credit (in conjunction with other incentive programs) should be aware of a few important considerations. In general, most incentives represent income on which federal income taxes are paid. As a result, most incentives do not decrease the basis on which the federal ITC is calculated. For example, a business that receives rebate money from the state government will be required to pay federal income tax on the amount, which does not affect the cost basis used to determine the 30% investment tax credit. State rebates (or buy downs), grants, and other taxable incentives fall into this category. There are also rare categories of incentives that are not taxable. An example is a nontaxable rebate from utilities. Another is a nontaxable grant. When taking these types of incentives, companies are required to reduce the system's cost basis prior to calculating the ITC amount.

For instance, if a business receives $100,000 in nontaxable utility rebates, when determining the ITC amount they must subtract the cost of the solar energy system. This would then determine the credit on this adjusted cost basis. The key phrase commonly used is *subsidized energy financing* (SEF), which broadly applies to nontaxable energy incentives. The IRS defines subsidized energy financing as "financing provided under a federal, state, or local program, a principal purpose of which is to provide subsidized financing for projects designed to conserve or produce energy."

COMPOUND INTEREST CALCULATION FOR ELECTRICAL ENERGY INFLATION

ENERGY ESACLLATION RATE	12%	11%	10%	9%	8%	7%	6%	5%	4%
PRESENT ENERGY COST /kWh	$ 0.13	$ 0.13	$ 0.13	$ 0.13	$ 0.13	$ 0.13	$ 0.13	$ 0.13	$ 0.13
YEARS	COST/ kWh	COST/ kWh	COST/ kWh	COST/ kWh	COST/ kWh	COST/ kWh	COST/ kWh	COST/ kWh	COST/ kWh
1	$ 0.15	$ 0.14	$ 0.14	$ 0.14	$ 0.14	$ 0.14	$ 0.14	$ 0.14	$ 0.14
2	$ 0.16	$ 0.16	$ 0.16	$ 0.15	$ 0.15	$ 0.15	$ 0.15	$ 0.14	$ 0.14
3	$ 0.18	$ 0.18	$ 0.17	$ 0.17	$ 0.16	$ 0.16	$ 0.15	$ 0.15	$ 0.15
4	$ 0.20	$ 0.20	$ 0.19	$ 0.18	$ 0.18	$ 0.17	$ 0.16	$ 0.16	$ 0.15
5	$ 0.23	$ 0.22	$ 0.21	$ 0.20	$ 0.19	$ 0.18	$ 0.17	$ 0.17	$ 0.16
6	$ 0.26	$ 0.24	$ 0.23	$ 0.22	$ 0.21	$ 0.20	$ 0.18	$ 0.17	$ 0.16
7	$ 0.29	$ 0.27	$ 0.25	$ 0.24	$ 0.22	$ 0.21	$ 0.20	$ 0.18	$ 0.17
8	$ 0.32	$ 0.30	$ 0.28	$ 0.26	$ 0.24	$ 0.22	$ 0.21	$ 0.19	$ 0.18
9	$ 0.36	$ 0.33	$ 0.31	$ 0.28	$ 0.26	$ 0.24	$ 0.22	$ 0.20	$ 0.19
10	$ 0.40	$ 0.37	$ 0.34	$ 0.31	$ 0.28	$ 0.26	$ 0.23	$ 0.21	$ 0.19
11	$ 0.45	$ 0.41	$ 0.37	$ 0.34	$ 0.30	$ 0.27	$ 0.25	$ 0.22	$ 0.20
12	$ 0.51	$ 0.45	$ 0.41	$ 0.37	$ 0.33	$ 0.29	$ 0.26	$ 0.23	$ 0.21
13	$ 0.57	$ 0.50	$ 0.45	$ 0.40	$ 0.35	$ 0.31	$ 0.28	$ 0.25	$ 0.22
14	$ 0.64	$ 0.56	$ 0.49	$ 0.43	$ 0.38	$ 0.34	$ 0.29	$ 0.26	$ 0.23
15	$ 0.71	$ 0.62	$ 0.54	$ 0.47	$ 0.41	$ 0.36	$ 0.31	$ 0.27	$ 0.23
16	$ 0.80	$ 0.69	$ 0.60	$ 0.52	$ 0.45	$ 0.38	$ 0.33	$ 0.28	$ 0.24
17	$ 0.89	$ 0.77	$ 0.66	$ 0.56	$ 0.48	$ 0.41	$ 0.35	$ 0.30	$ 0.25
18	$ 1.00	$ 0.85	$ 0.72	$ 0.61	$ 0.52	$ 0.44	$ 0.37	$ 0.31	$ 0.26
19	$ 1.12	$ 0.94	$ 0.80	$ 0.67	$ 0.56	$ 0.47	$ 0.39	$ 0.33	$ 0.27
20	$ 1.25	$ 1.05	$ 0.87	$ 0.73	$ 0.61	$ 0.50	$ 0.42	$ 0.34	$ 0.28
21	$ 2.66	$ 2.04	$ 1.57	$ 1.22	$ 0.95	$ 0.74	$ 0.58	$ 0.46	$ 0.36
22	$ 2.98	$ 2.26	$ 1.73	$ 1.32	$ 1.02	$ 0.79	$ 0.62	$ 0.48	$ 0.38
23	$ 3.33	$ 2.51	$ 1.90	$ 1.44	$ 1.10	$ 0.85	$ 0.65	$ 0.51	$ 0.39
24	$ 3.73	$ 2.79	$ 2.09	$ 1.57	$ 1.19	$ 0.91	$ 0.69	$ 0.53	$ 0.41
25	$ 4.18	$ 3.09	$ 2.30	$ 1.72	$ 1.29	$ 0.97	$ 0.73	$ 0.56	$ 0.42

Figure 4.4. Compound interest table of a $0.13/kW/h electric rate over a period of 25 years.

Note that if a commercial entity pays federal income tax on subsidized energy financing, additional incentives will not reduce the ITC amount. For more information refer to "The SEIA Guide to Federal Tax Incentives" (PDF). Further information regarding federal income tax can also be found on the SIAE Web link.

Figure 4.4 shows a compound interest table of a $0.13/kW/h electric rate over a period of 25 years. Figure 4.5 shows a compound interest table of a $0.13/ kW/h electric rate with an 8% life cycle inflation rate.

Impact of ITC on Depreciation Calculations

For federal tax purposes, the Modified Accelerated Cost Recovery System (MACRS) program (as discussed previously) allows for accelerated depreciation over a period of 5 years. MACRS and the 30% investment tax credit are set up to make it easier to purchase a renewable energy system.

It is important to note that when calculating the depreciation on a commercial solar energy system, the *tax depreciation basis* (TDB) is distinct from the *tax credit basis* (TCB). Essentially, for the full 30% credit, the first-year depreciation value will be 30% less. For example, a solar power system costing $100,000 can only be depreciated down to $70,000 ($100K – 30% credit). For this reason, the IRS counts the full value of the credit alongside accelerated depreciation.

Furthermore, IRS rules allow companies to apply half the value of the tax credit when determining the basis on which to calculate depreciation. As such, the tax depreciation basis that a company claims for the solar energy system is reduced by 50% of the tax credit amount.

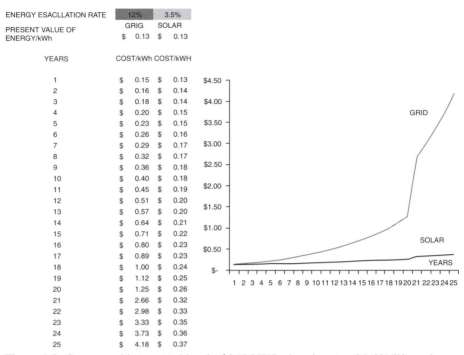

ENERGY ESACLLATION RATE	12%	3.5%
	GRIG	SOLAR
PRESENT VALUE OF ENERGY/kWh	$ 0.13	$ 0.13

YEARS	COST/kWh	COST/kWH
1	$ 0.15	$ 0.13
2	$ 0.16	$ 0.14
3	$ 0.18	$ 0.14
4	$ 0.20	$ 0.15
5	$ 0.23	$ 0.15
6	$ 0.26	$ 0.16
7	$ 0.29	$ 0.17
8	$ 0.32	$ 0.17
9	$ 0.36	$ 0.18
10	$ 0.40	$ 0.18
11	$ 0.45	$ 0.19
12	$ 0.51	$ 0.20
13	$ 0.57	$ 0.20
14	$ 0.64	$ 0.21
15	$ 0.71	$ 0.22
16	$ 0.80	$ 0.23
17	$ 0.89	$ 0.23
18	$ 1.00	$ 0.24
19	$ 1.12	$ 0.25
20	$ 1.25	$ 0.26
21	$ 2.66	$ 0.32
22	$ 2.98	$ 0.33
23	$ 3.33	$ 0.35
24	$ 3.73	$ 0.36
25	$ 4.18	$ 0.37

Figure 4.5. Compound interest table of a $0.13/kW/h electric rate with 8% life cycle inflation rate.

This rule can be illustrated by examining a hypothetical company that installs a commercial solar electric system costing $200,000. The company's tax depreciation basis will be equal to project costs minus half the allowable credit: $200,000 – (50% × $60,000) = $170,000. This example illustrates the difference between the tax credit basis and the tax deprecation basis.

Another scenario concerns earnings and profit. As a rule, gross income is reduced by depreciation to arrive at net income. As such, large allowable depreciation in a given year lowers the net income, which results in lower net income or profits. This in turn lowers the taxable dividends paid to shareholders. In such a scenario when determining profits, companies may omit the downward basis adjustments and base their calculations on the full cost basis of the system. A company could lower its tax liability over the short term while investing in a solar energy system that may produce long-term energy savings. These rules have been specifically designed to provide an incentive for companies to invest in renewable power. It should also be noted that the tax depreciation basis is used to calculate taxable gains or losses. The 30% federal credit does not affect the book depreciation basis.

Return on Investment (ROI)

Government incentives combined with recent decreases in solar equipment prices make the investment in solar power a good financial decision for businesses. Solar power systems are considered long-term, low-risk, and high-return investments. In general solar power systems may result in tax free annual return on investment of 5–11%. Therefore, solar power systems are considered to be quite competitive with

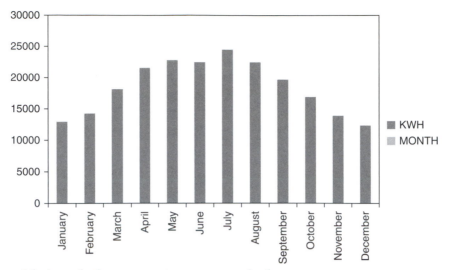

Figure 4.6. Annual solar power system energy production.

higher-risk investments such as stocks and bonds. Moreover, as utility rates increase, the annual return increases. Another attribute of solar power systems is that when installed in commercial projects such as shopping centers and office buildings they increase property values and rents and render properties environmentally responsive.

Solar Power System Salvage Value

Perhaps an important issue often overlooked when calculating return on investment of solar power systems is the dynamic salvage value of solar power technologies. Unlike any capital equipment financing scenario, solar power systems if well designed and properly maintained do not lose their power production capability and continue to produce electrical energy after expiration of the contract life cycle.

Even though from an accounting point of view they are depreciated in 5 years, solar power systems produce electrical energy over a guaranteed period of their contractual life cycle of 20–25 years and continue energy production as long as the photovoltaic cells are functional; hence, unlike mechanical equipment PV systems, they have a dynamic salvage value that is subject to their extended longevity.

Therefore, the salvage value must be considered as an important positive equity value that when applied judiciously can make a significant difference in the financial viability of solar power systems. This is a key issue that must be carefully evaluated by owners and providers of PPA organizations.

Figure 4.6 illustrates annual solar power system energy production. Figure 4.7 illustrates solar power life cycle output degradation. Figure 4.8 illustrates the importance of the dynamic depreciation of the solar power system financial profile of 1000-kW solar power economic contributions. The four spreadsheets represent the significance of the added future value of a solar power system beyond a life cycle of 25 years. Salvage values of two scenarios of a solar power system are calculated over an additional 25 years of service.

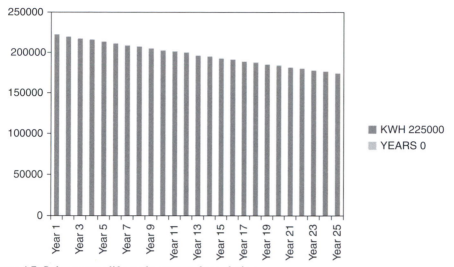

Figure 4.7. Solar power life cycle output degradation.

The spreadsheets use an identical solar power system configuration with 1000 kW capacity and 5.5 hours of nominal daily solar insolation. Figure 4.8 represent a financing calculation based on a 10% annual escalation cost at a present electrical energy rate of $0.18/kW/h. Future value calculations are based on 4% annual interest. Future value of energy contributions is calculated for the primary life cycle of 25 years and an additional extended life cycle of 25 years, which constitute the dynamic salvage value of the solar power system with an annual solar power output energy degradation rate of 1%. Figure 4.8 uses more stringent computational parameters with an annual energy cost escalation of 8% and annual energy output degradation of 1.5%.

From the economic analysis that follows, it is clear that dynamic salvage value in either of the scenarios is a significant value that must be taken into consideration when calculating the overall return on investment. Figure 4.6 depicts the graph of annual solar power system energy production. Figure 4.7 shows a graph of solar power system output power degradation over a period of 25 years. Detailed analysis of a typical solar power financing is discussed in Chapter 8.

It should be noted that key to achieving maximum return on investment is to maintain solar power systems at peak power performance efficiency during the entire life cycle of the system. Achieving peak power output performance requires careful system planning, a thorough engineering design, and proper system design maintenance and upkeep, which will be discussed later in this chapter.

Interpretation of the Preceding Computations

- The first line from the top defines daily solar power insolation as 5.5 hours. The insolation value is within the boundaries of annual solar aperture that occurs in the Southern California region such as at Los Angeles. In eastern desert areas of California such as Coachella Valley, insolation hours are about 6 hours. Daggett, California, has 8 hours of mean annual daily insolation

INSOLATION	5.5	HOURS/DAY	ENERGY ESCAL.		8%		PV = FV /(1+i)<n	
SOLAR POWER	1000	Kw	INTEREST RATE		4%		$	14,070,438
YEAR ONE	2007500	kW	YEARS		50			
DEPRECIATION	1%	PERCENT	ANNUAL		ANNUAL			
YEAR	PERFORMANCE	ANNUAL KW	NERGY COST/K\		VALUE			
26	75%	1505625	$	1.23	$	1,856,022		
27	74%	1485550	$	1.33	$	1,977,777		
28	73%	1465475	$	1.44	$	2,107,135		
29	72%	1445400	$	1.55	$	2,244,532		
30	71%	1425325	$	1.68	$	2,390,426		
31	70%	1405250	$	1.68	$	2,356,758		
32	69%	1385175	$	1.81	$	2,508,937		
33	68%	1365100	$	1.96	$	2,670,382		
34	67%	1345025	$	2.11	$	2,841,601		
35	66%	1324950	$	2.28	$	3,023,124		
36	65%	1304875	$	2.46	$	3,215,504		
37	64%	1284800	$	2.66	$	3,419,318		
38	63%	1264725	$	2.87	$	3,635,162		
39	62%	1244650	$	3.10	$	3,863,658		
40	61%	1224575	$	3.35	$	4,105,448		
41	60%	1204500	$	3.62	$	4,361,198		
42	59%	1184425	$	3.91	$	4,631,592		
43	58%	1164350	$	4.22	$	4,917,338		
44	57%	1144275	$	4.56	$	5,219,160		
45	56%	1124200	$	4.93	$	5,537,804		
46	55%	1104125	$	5.32	$	5,874,028		
47	54%	1084050	$	5.75	$	6,228,605		
48	53%	1063975	$	6.21	$	6,602,322		
49	52%	1043900	$	6.70	$	6,995,970	FUTURE VALUE	
50	51%	1023825	$	7.24	$	7,410,346	$	99,994,147
		75744180	$	3.36	AVERAGE/KW			

Figure 4.8. Economic energy output profile of a 1000-kW (1-MW) solar power system with a 25-year life cycle, 10% energy escalation, and annual 1% performance degradation.

- Also shown on the line is percent energy cost escalation. In view of considerable investment requirements to upgrade the existing transmission line infrastructure, most electrical service providers such as Southern California Edison (SCE), San Diego Gas & Electric (SDGE), and other providers, under special oversight of the California Public Utilities Commission (CPUC), have been authorized to pass on the investment expenses in the form of annual energy cost escalation over an extended period. Annual energy cost escalation experienced in the past few years has range from 10% to 12%. For the preceding example we have used two scenarios, one at 10% and the other with an 8% annual energy cost escalation.
- The leftmost of the values shows a present value (PV) formula that is used to calculate aggregated energy contribution of the two phases of solar power life cycles.

- The second line from the top shows the field value of the spreadsheet as being a 1000-kW solar system (a 1-MW solar farm). The next field to the right is the interest rate applied in the present value calculation, and finally the last field on the right is computed present value of the totalized future value shown in the lower fifth column of the tabulation.
- The third line from the top displays calculated annual energy production of a 1-MW solar power system (1000 kW × 5.5 h × 365 days). The adjacent field shows the first or second life cycle of a solar power system (each at 25 years).
- The fourth line from the top shows the percent annual solar power system energy production degradation.
- Column 1- Years of service
- Column 2 – Percent energy production or system performance
- Column 3 – Annual energy production
- Column 4 – Cost of electrical utility rate
- Column 5 – Annual value of energy contribution
- Column 6 – Accumulated value of energy during each life cycle

In Chapter 8, we will discuss the preceding computational methodology in greater detail.

Project Financing

The following financing discussion is specific to large alternative and renewable energy projects such as solar, wind, and geothermal projects that require extensive amounts of investment capital. Project financing of such large projects, similar to that of large industrial projects, involves long-term financing of capital intensive material and equipment.

Since most alternative energy projects in the United States are subject to state and federal tax incentives and rebates, project financing involves highly complex financial structures in which project debt and equity, rebate, federal and state tax incentives, and cash flow generated by grid energy power are used. In general, project lenders are given a lien on all of the project assets including property, which enables them to assume control of a project over the terms of the contract.

Since renewable energy and large industrial projects involve different levels of transactions such as equipment and material purchase, site installation, maintenance, and financing, a special purpose entity is created for each project. This shields other assets owned by a project sponsor from the detrimental effects of project failure.

As special purpose joint ventures, these types of entities have no assets other than the project. In some instances, capital contribution commitments by the owners of the project company are necessary to ensure that the project is financially sound.

Alternative energy technology project financing is often more complicated than alternative financing methods commonly used in capital intensive projects such as transportation, telecommunication, and public utility industries.

Renewable energy type projects in particular are frequently subject to a number of technical, environmental, economic, and political risks. Therefore, financial institutions and project sponsors evaluate inherent risks associated with a particular project's development and operation and determine whether it can be financed.

To minimize risk, project sponsors create special entities that consist of a number of specialist companies operating in a contractual network with each other and that allocate risk in a way that allows financing to take place.

In general a project-financing scheme involves a number of equity investors, known as sponsors, which include hedge funds as well as a syndicate of banks that provide loans for the project. The loans are most commonly nonrecourse loans, which are secured by the project itself and paid entirely from its cash flow, rebates, and tax incentives. Projects that involve large risks require limited recourse financing secured by a surety from sponsors. A complex project finance scheme also may incorporate corporate finance, securitization, options, insurance provisions, or other further measures to mitigate risk.

5 Long-Term Project Financing and Power Purchase Agreements

Introduction

A Power Purchase Agreement (PPA) is a long-term legal contract between an electricity provider and power purchaser. During the contractual period, which typically lasts 20–25 years, the purchaser agrees to buy electrical energy from the electricity provider. Under such a contract agreement the electrical energy supplier is considered to be an independent power provider (IPP). Energy sales under PPA type contracts are regulated by state or local governments. Under a PPA contract agreement, the electrical energy provider is most often the developer and owner of solar power or sustainable energy production technology. However, the seller in some instances could be an entity that buys the energy from several sources and resells it to various purchasers.

In the United States Power Purchase Agreement contracts are regulated by the Federal Energy Regulatory Commission. Under the Energy Policy Act of 2005, solar or sustainable PPA providers are considered exempt from wholesale energy producer regulations. Most prevalent PPA contracts fall into two categories, namely, solar (SPPA) or wind (WPPA).

PPAs for renewable energy projects are a class of *lease-option- to-buy* financing plans that are specifically tailored to underwrite the heavy cost burden of the project. PPAs, which are also referred to as *third party ownership* contracts, differ from conventional loans in that they require significant land or property equity, which must be tied up for the duration of the lease. PPAs have the following significant features, which make them unique as financial instruments.

Long-Term Industrial Project Financing

Unlike conventional financing of capital equipment or construction projects, long-term financing of industrial projects is based upon cash flow generated by the projects. This type of financing involves schemes in which private investors and their banks provide required loans for the project. Such loans are usually secured by revenue and income generated by the project operation as well as project assets. Furthermore, in the event of compliance with loan compliance terms and conditions, lenders are given the right to assume partial or total control of the project.

In order to avert risks and liabilities associated with failure of large-scale projects, lenders shield and protect their organizational assets by creating special purpose entity structures for each project. Since large projects are usually quite complex in nature, they may involve considerable technical, economic, and environmental challenges; as such, financing of large projects involves risk identification and feasibility studies that determine the fundamental basis for preventing unacceptable investment losses.

In large-scale programs, financing of projects is distributed in a consortium of multiple parties, so as to divide the associated risks among multiple stakeholders. In projects that involve large risks, financing of a project may necessitate so-called limited resource financing, which involves participation of *surety* or insurance companies.

Long-term industrial project financing as discussed earlier has been prevalent for several centuries and has been common in the construction of electrical power generation projects, mining, and transportation type projects worldwide. More recently the Public Utility Regulatory Policy Act of 1978, known as PURPA, was devised by the U.S. Congress to facilitate financing of utility and government type programs that have a long-term revenue stream. That policy provided the foundation for a new financing scheme, termed the Power Purchase Agreement (PPA), which encouraged domestic renewable energy development and conservation that resulted in deregulation of electric power generation and formation of private power generation entities. A subsequent policy enacted in 1994, referred to as the Public Utilities Holding Company Act, provided the foundation for lending organizations to finance alternative and sustainable energy generation projects worldwide.

Participants in Large-Scale Project Financing

Financing of large-scale projects is generally quite complex and involves numerous parties. Some of the important issues that form essentials of the projects include establishment of the contractual framework, engineering, procurement and construction documents, operation and maintenance procedures and agreements, concession deeds, a shareholder agreement, agreements between lenders and owners, loan agreement conditions, intercredit agreement among multiple lenders, and much more.

Essentially all major project financing contractual documents consist of three fundamental categories, namely, shareholder or sponsor documents, project documents, and finance documents. Figure 5.1 depicts the typical structure of long-term project financing.

Engineering Procurement Construction (EPC) and Contractual Documents

The main purposes of engineering, procurement, and construction documents as they relate to sustainable and solar power energy generation projects must include guidelines and obligations for engineers and contractors for design, construction, and delivery of turnkey projects that include predefined performance specifications.

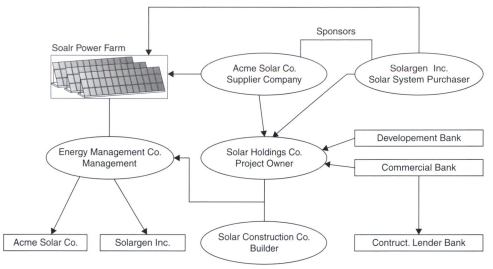

Figure 5.1. Typical structure of long-term project financing.

The core of EPC contracts includes a description of engineering design methodologies; schedule of completion of design and construction specification documentation; predetermined engineering fee schedules, construction integration, test, and acceptance criteria; system performance; and warranties. One of the most significant aspects of realizing successful completion of projects is to ensure that EPC contractors are preselected properly according to their project experience. When preparing EPC contract documents, special attention must be paid to outlining the following:

- Project specification, which must include a thorough feasibility study, which must be substantiated by EPC contractors
- Outline of expected project finance ceiling
- Payment methodology
- Completion of engineering and construction schedule
- Completion of guarantee date and liquidation damages
- System performance guarantee

System Operation and Maintenance Guarantee Agreement Document

In order to ensure peak energy production performance, solar or sustainable energy power projects must incorporate contractual agreements between the project company (owner) and the system operations and maintenance (O&M) company responsible for performance management of the project. The O&M company could either be part of the project company or i a third party subcontract operator. The O&M contractual document must include the following:

- Definition of operation and maintenance services
- Responsibilities of O&M contractor
- Nature and provision of services
- Liquidated damages
- Schedule of fees

Shareholder Agreement

When projects are financed by several financing organizations, the joint partnership is registered as a special purpose company (SPC). Under such conditions project sponsors develop contractual agreements that define special responsibilities and relationships among the parties. A contractual agreement among multiple financial partners usually includes the following:

- Percentages of capital injected funds
- Voting requirements
- Resolution of disputes and jurisdiction of law
- Shares and dividend policy
- Management of special purpose venture (SPV)
- Disposal and preemption rights

Off-Take Contractual Agreement

In the case of solar or sustainable energy projects, "off-take" is a term that defines the party who buys the energy produced or the entity that delivers the energy to the end users. In large-scale revenue producing projects such as solar power forms, off-take contractual agreements specify price and amount of energy that result in a revenue stream. The main objective of the agreement is to ensure that the principal entity responsible for the project has the financial stability and the required steady income from the project to generate funds to pay its debt obligations and cover the project's O&M expenses. The principal contents of off-take agreement contracts include the following:

- Off-taker's payment obligation for energy received
- Definition of PPA entity responsibilities
- Long-term sales contract agreement as regards the price of a unit of energy and amounts of energy produced (subject to agreed market index)
- Agreement as regards energy cost escalation, O&M cost inflation, and system power output degradation
- Variability of energy supply
- Availability and consistency of energy supply

Loan Agreement

The loan agreement contract document specifies the obligation between the project company or the borrower and the lender. The contractual document specifies agreement terms and conditions that govern the relationship between the parties and include provisions and clauses that cover specific requirements of the project. The loan agreement contract documents generally include the following:

- General loan conditions
- Conditions required for each capital drawdown
- Conditions under which the loan obligations must be paid
- Payment mechanism and methodology
- Interest payment

- Financial obligations and covenants
- Dividend restrictions
- Warranties

Creditor Agreements

Projects that are financed by multiple creditors include contractual documents that govern obligations among participating creditors. Such documents, which are commonly referred to as *intercreditor agreements,* outline governing terms and relationships among lenders vis-a-vis borrowers' obligations. Intercreditor agreements may include the following:

- General terms and conditions
- Order of drawdown of capital
- Cash flow
- Limitations and liabilities of creditors, their rights and obligations
- Voting rights
- Notification of defaults
- Order of applying the proceeds of debt recovery
- Subordination principles that may apply between senior and junior debt providers (also referred to as mezzanine debt providers)

Term Sheet

A term sheet is a document that defines agreement terms between lender and borrower. The term sheet outlines specific financing terms and conditions as applied to a project. The document defines fundamental conditions for credit approval and financing of projects that are endorsed by all lenders and borrowers as a prerequisite for credit approval.

- They take advantage of federal and state tax incentives, which may otherwise have no value for public agencies, municipalities, counties, nonprofit organizations, or businesses that do not have significant profit margins.
- Properties where renewable energy systems equipment and materials such as solar PV power support structures are installed must be leased for the entire duration of the contract agreement, which may exceed 20 years.
- Solar power or the renewable energy system must be connected to the electrical grid.
- Power generated by the renewable energy system must primarily be used by the owner.
- Depending on the lease agreement, excess power produced from the power cogeneration system is credited to the third party owner.
- Equity of the leased property must have liquidity value exceeding the value of the project.

Power Purchase Agreement

In the United States, the Solar Power Purchase Agreement (SPPA) contracts are contingent upon the federal solar investment tax credit, which under the Emergency

Economic Stabilization Act of 2008 was extended to fiscal year 2016. In general, SPPA financing benefits profitable entities that are subject to taxation, which can take advantage of the federal tax credit. To take advantage of the credit, groups of investors, solar power providers, finance and lending organizations, and solar power contractors create special purpose PPA entities that finance, design, install, monitor, and maintain solar power system installations for the duration of the PPA contractual life cycle.

The American Recovery and Reinvestment Act, which was passed in 2009, permits the solar investment tax credit to be combined with tax exempt financing, which allows significant reduction in capital investment in PPA type projects.

Power Purchase Financing

Under PPA contracts, the electrical energy provider assumes total responsibility for funding, engineering design, construction, maintenance, and monitoring of energy production. The PPA provider also assumes complete responsibility for the sale of electrical energy at an agreed contractual price for the term of the contract, which may extend for 20–25 years. In some instances, the purchaser of the electrical energy is offered various options to renew the PPA contract at the end of the contract or to purchase the solar power system at a predetermined fair market value.

Power purchase contract agreements are considered as legally binding throughout the terms of the agreement. In general, the power purchaser is obligated to buy the power generated by the solar power system; likewise the power provider agrees not to sell the power to other clients.

PPA Termination Date

As mentioned, PPA contract agreements expire upon completion of their agreement life cycle. However, under certain circumstances both parties may be allowed to terminate their contract under certain conditions, which may include system underperformance, degradation of energy delivery, or unavoidable natural disasters such as earthquake, hurricane, and flooding. PPA contracts may also include terms that allow buyers to reduce their energy use in the event that the after tax value of electricity is changed to the purchaser's disadvantage. In some instances, PPA contracts may also include terms that may allow parties to negotiate a force majeure to mitigate the issue.

In PPA type agreements electrical energy either can be delivered either on the high side of the entrance service transformer (also called a bus bar sale) or to the distribution side of the transformer.

Power Purchase Price

Power purchased under terms of a PPA agreement could be flat or be escalated over a period. Contractual agreements include clauses that guarantee annual electrical energy production and solar power system power depreciation. It should be noted that since the seller under the PPA agreement cannot sell excess energy to other clients, system design and power production estimates must be sufficiently accurate to prevent energy over- or underproduction.

As a rule, under the PPA contract agreement, purchasers have the authority to monitor power production and audit records of power production.

Power Purchase Agreement Overview for Nonprofit Organizations

As discussed previously, PPA financing is a third party ownership business model that requires the owners or providers to procure, install, and operate a solar power installation on the customer's premises for extended duration of 20–25 years. In general, PPA type contracts are designed to allow nonprofit organizations and government agencies to enter into long-term power purchase agreements, which cannot take advantage of federal tax incentive programs. In essence, the PPA system providers consist of entities that provide investment capital for financing solar power projects in return for tax benefits. Such a PPA organization can be considered a tax investment organization.

Tax investors participating in PPA projects form limited liability corporations (LLCs), which are backed by several financial institutions. PPA entities in addition to taking advantage of federal tax incentives receive state rebates, which usually amount to 50% of the investment, and benefit from long-term sales of electrical energy. In view of the significant amount of tax incentive, PPA entities that did not have the benefit of the federal tax exemption could not exist as viable financial investment entities.

Advantages of Power Purchase Agreements

The following are some benefits of PPA financing for nonprofit organizations. In general, power purchase agreements have the following significant advantages:

- Projects are financed on equity of properties, such as unused grounds or building rooftops, that otherwise have no value.
- Owners are not burdened with intensive project costa.
- Power purchase agreements guarantee owners a hedge against electrical energy escalation costs.
- There is significantly less risk of energy cost escalation associated with third party PPAs than with grid purchased electrical energy.
- The owners assume no responsibility for maintenance and upkeep of the leased equipment or grounds for the duration of the lease period.
- Upon completion of the lease agreement period, owners are offered flexible options for ownership.
- All PPAs intrinsically constitute turnkey *design build contracts,* which somewhat relieve the owners of detailed technical design.

Disadvantages of Power Purchase Agreements

Since power purchase agreements essentially constitute a contract rather than an engineering design and procurement agreement, they inherently include a number of undesirable features, which in some instances could neutralize the benefits discussed previously.

Some of the issues associated with power purchase agreements are as follows:

- PPA contracts are extremely complex and convoluted. Contract agreements drafted include legal language and clauses that strongly favor the third party provider.
- PPA contracts incorporate stiff penalties for premature contract terminations.
- PPA or third party ownerships in general involve a finance company, an intermediary such as a sales and marketing organization, a design engineering organization, a general contractor, and in some instances a maintenance contractor. Considering the fragmented responsibilities and the complexities embodying the collaborative effort of all entities and the life cycle of the contract, the owners must exercise extreme diligence in executing PPA contracts.
- The owners have no control over the quality of design or materials provided; therefore, extra measures of caution should be exercised when evaluating final ownership of the equipment.
- In general, owners who elect to enter into a power purchase agreement, such as nonprofit organizations, municipalities, city governments, or large commercial industries, seldom have experienced engineering or legal staff who have had previous exposure to PPA type contracts.
- The owners tie up the leased grounds or buildings for extended periods and assume responsibility for insuring the property against vandalism and damage due to natural causes.
- In the event of power outages, third party ownership agreement contracts penalize the owners for loss of power output generation.
- PPA contracts include yearly energy escalation costs, which represent a certain percentage of the installed cost and must be evaluated with extreme diligence and awareness, as these seemingly small inflationary costs could neutralize the main benefit, which is the hedge against energy cost escalation.
- PPAs for large renewable solar power cogeneration contracts are relatively new financial instruments. Therefore, their owners must be careful to take proper measures to prevent unexpected consequences.

Preparation for PPA Proposal

Unlike conventional capital intensive projects, power purchase agreement type contracts completely bypass proven engineering design measures, which involve project feasibility studies, preliminary design and econometric analysis, design documentation, construction documentation, design specification, and procurement evaluation (which are based on job specific criteria).

In order to ensure a measure of control and conformance to the project needs, it is recommended that the owners refer to an experienced consulting engineer or legal consultant who is familiar with power purchase agreement type projects.

In order to prevent or minimize unexpected negative consequences associated with power purchase agreements, the owners are advised to incorporate the following documents, reports, and studies in their request for a proposal (RFP):

- For ground mount installations, employ the services of qualified engineering consultants to prepare a negative environmental impact report, site grading, drainage, and soil study.

- Provide statistical power consumption and peak power analysis of present and future electrical demand loads.
- Provide a set of electrical plans, including single line diagrams, main service switchgear, and power demand calculations.
- Conduct an energy audit.
- Provide detailed data about site topology and present and future land use.
- If applicable, provide data regarding local climatic conditions, such as wind, sand, or dust accumulation conditions, and cyclic flooding conditions.
- For roof mount systems, provide aerial photographs of roof plans that show mechanical equipment, air vents, and roof hatches. Drawings must also accompany architectural drawings that show parapet heights and objects that could cause shading.
- Specifications should outline current electrical tariff agreements.
- The document should also incorporate any and all special covenants, conditions, and restrictions associated with the leased property.
- To ensure system hardware reliability for the extent of the contract, the RFP must include a generic outline of hardware and data acquisition and monitoring software requirements.
- The specifications must request the provider to disclose all issues that may cause noncompliance.
- Expected power output performance guarantees, as well as projected annual power generation requirements, must be delineated Owners are also advised to conduct preliminary renewable energy production studies that would enable them to evaluate energy production potential as well as economic analysis of possible alternatives.

Power Purchase Agreement Contract Structure for Solar Power Systems

In order to prepare a PPA request for a proposal document, the owner's legal counsel and management personnel must familiarize themselves with various elements of the contract agreement. Agreements involving third party ownership consist of two parts: legal and technical. The following are some of the most significant points of PPA type contracts that third party purchase providers must respond to and evaluate accordingly:

Contractual matters of interest

- DC output size of the PV modules in kWh
- AC or U.S. Power Test Condition (PTC) output of the photovoltaic modules in kWh
- Expected AC power output of the solar system in its first year of installation
- Expected life cycle power output in kWh DC
- Expected life cycle power output in kWh AC
- Guaranteed minimum annual power output performance in kWh AC
- Terms of contractual agreement
- Penalty or compensation for performance failure
- Price structure at the end of the contract with client paying 0% of the cost
- Price structure at the end of the contract with client paying 50% of the cost

- Price structure at the end of the contract with client paying 100% of the cost
- Expected average yearly performance during life cycle of the contract
- Expected mean yearly performance degradation during life cycle of the contract
- Assumed PPA price per kWh of electrical energy
- Initial cost of power purchase agreement
- PPA yearly escalation cost as a percentage of the initial energy rate
- Net present value over 25 years
- Proposed cost reduction measures
- Net present value of reduction measures
- Annual inflation rate
- Projected annual electricity cost escalation
- First year avoided energy cost savings
- Total life cycle energy saved in kWh
- Total life cycle energy PPA payment
- Cost of PPA buyout at the end of life cycle
- PPA expenses
- Total life cycle pretax savings
- Total project completion time in months
- Customer training
- Insurance rating

Technical matters of interest

- PV module manufacturer and type
- PV module technology
- PV module efficiency rating
- PV module DC Watts
- PV module PTC Watts as listed under California Energy Commission (CEC) equipment and product qualification listing
- Total PV module count
- Percent yearly solar power output degradation
- PV module warranty in years after formal test acceptance and commissioning
- Inverter make and model as listed under CEC equipment and product qualification
- Inverter kilowatt rating
- Number of inverters used
- Inverter performance efficiency
- Inverter basic and extended warrantees
- Solar power tracking system (if used)
- Tracking system tilt angle in degrees east and west
- Number of solar power tracker assemblies
- Kilowatts of PV modules per tracker
- Ground or pedestal area requirement per 100 kW of tracker; for large solar power farms, tracker footprint must be accounted for in acres per megawatt of land required
- Tracker or support pedestal ground penetration requirements
- Tracker above ground footing height
- Tracker below ground footing height

- Wind shear withstand capability in miles per hour
- Environmental impact during and after system installation (if applicable)
- Lightning protection scheme
- Electrical power conversion and transformation scheme and equipment platform requirements
- Equipment mounting platforms
- Underground or above ground DC or AC conduit installations
- PV module washing options, such as permanent water pressure bibs, automatic sprinklers, or mobile pressure washers
- Service options and maintenance during life cycle of the PPA

  Experience in large-scale installation

- Engineering staff's collective experience in photovoltaic design and power engineering
- In-house or subcontracted engineering
- In-house or subcontracted installation crew
- Years of experience in solar power tracker system type installations (if applicable)
- Years of collective experience in PPA contracting
- Location of management, engineering, installation, and maintenance depots
- Availability of PV modules and specific power purchase agreements with major national and international manufacturers
- Name of the primary entity assuming full contractual responsibility and project bonding
- Names of each contractor or subcontractor taking part in the PPA
- Years of collaboration with outsourced entities
- Data acquisition and monitoring system
- Data acquisition and control system (DACS) certification by the rebate agency, such as the California Solar Initiative (CSI)
- Data acquisition system has a proprietary provider or a third party certified provider?
- DACS power measurement and transmission intervals in minutes
- Monitored data, such as weather, wind speed, humidity, precipitation, and solar irradiance
- CSI certified reporting scheme
- Customer WEB access key
- On site electrical display and printing capability, and associated options
- On site integration capability with customer's data monitoring system
- Periodic data reporting format and frequency
- Presentation and visual aids, such as bar chart displays of statistical solar power monitored information and solar power array configuration displays
- On demand reporting
- Proactive solar power system diagnostic capability

Proposal Evaluation

As mentioned, long-term financing agreements such as PPAs are inherently complicated and demand extensive due diligence by the owner, legal counsel, and consulting engineers alike. In order to execute a power purchase agreement successfully, the owner must fully appreciate the importance of the collective effort of experts and the collaborative effort required in preparation of specifications and requests for proposal documents. Prior knowledge of specification and evaluation points (outlined earlier), when exercised properly, would prepare owners to evaluate comparative value advantages among PPA providers.

Conclusion

Even though the initial investment in a solar power cogeneration system requires a large capital investment, the long-term financial and ecological advantages are so significant that their deployment in the existing project should be given special consideration.

A solar power cogeneration system, if applied as per the recommendations reviewed here, will provide considerable energy expenditure savings over the life span of the recreation facility and provide a hedge against unavoidable energy cost escalation.

Special Note

In view of the depletion of existing CEC rebate funds, it is recommended that applications for the rebate program be initiated at the earliest possible time. Furthermore, because of the design integration of the solar power system with the service grid, the decision to proceed with the program must be made at the commencement of the construction design document stage.

Special Funding for Public and Charter Schools

A special amendment to the CEC mandate, enacted on February 4, 2004, established a Solar Schools Program to provide a higher level of funding for public and charter schools. This is to encourage the installation of photovoltaic generating systems at more school sites. Currently, the California Department of Finance has allocated a total of $2.25 million for this purpose. To qualify for the additional funds, the schools must meet the following criteria.

- Public or charter schools must provide instruction for kindergarten or any of the grades 1 through 12.
- The schools must have installed high-efficiency fluorescent lighting in at least 80% of classrooms.
- The schools must agree to establish a curriculum tie-in plan to educate students about the benefits of solar energy and energy conservation.

Principal Types of Municipal Lease

There are two types of municipal bonds. One type, referred to as a "tax-exempt municipal lease," has been available for many years and is used primarily for the purchase of equipment and machinery that have a life expectancy of 7 years or less. The second type is generally known as an "energy-efficiency lease" or a "power purchase agreement." It is used most often on equipment being installed for energy-efficiency purposes, in cases in which equipment has a life expectancy of more than 7 years. Most often this type of lease applies to equipment classified for use as a renewable energy cogeneration, such as solar PV and solar thermal systems. The other common type of application that can take advantage of municipal lease plans includes energy efficiency improvement of devices such as lighting fixtures, insulation, variable-frequency motors, central plants, emergency backup systems, energy management systems, and structural building retrofits.

The leases can carry a purchase option at the end of the lease period for an amount ranging from $1.00 to fair market value. They frequently have options to renew the lease at the end of the lease term for a lesser payment than the original payment.

A tax exempt municipal lease is a special kind of financial instrument that essentially allows government entities to acquire new equipment under extremely attractive terms with streamlined documentation. The lease term is usually for less than 7 years. Some of the most notable benefits are:

- Lower rates than conventional loans or commercial leases
- Lease-to-own with no residual and no buyout
- Easier application, such as same-day approvals
- No "opinion of counsel" required for amounts less than $100,000
- No underwriting costs associated with the lease

Tax Exempt Municipal Lease

Municipal leases are special financial vehicles that provide the benefit of exempting banks and investors from federal income tax. This allows for interest rates that are generally far below conventional bank financing or commercial lease rates. Most commercial leases are structured as rental agreements with either nominal or fair-market-value purchase options.

Borrowing money or using state bonds is strictly prohibited in all states, since county and municipal governments are not allowed to incur new debts that will obligate payments that extend over multiyear budget periods. As a rule, state and municipal government budgets are formally voted into law; there is no legal authority to bind the government entities to make future payments.

As a result, most governmental entities are not allowed to sign municipal lease agreements without the inclusion of nonappropriation language. Most governments, when using municipal lease instruments, consider obligations as current expenses and do not characterize them as long-term debt obligations.

The only exceptions are bond issues or general obligations, which are the primary vehicles used to bind government entities to a stream of future payments. General obligation bonds are contractual commitments to make repayments. The government bond issuer guarantees to make funds available for repayment, including raising taxes if necessary. In the event that adequate sums are not available in the general fund, "revenue" bond repayments are tied directly to specific streams of tax revenue. Bond issues are very complicated legal documents that are expensive and time consuming and in general have a direct impact on the taxpayers and require voter approval. Hence, bonds are exclusively used for very large building projects such as creating infrastructure such as sewers and roads.

Municipal leases can be prepaid at any time without a prepayment penalty. In general, a lease amortization table included with a lease contract shows the interest principal and payoff amount for each period of the lease. There is no contractual penalty and a payoff schedule that can be prepared in advance. It should also be noted that equipment and installation can be leased.

Lease payments are structured to provide a permanent reduction in utility costs when used for the acquisition of renewable energy or cogeneration systems. A flexible leasing structure allows the municipal borrower to level out capital expenditures from year-to-year. Competitive leasing rates of up to 100% financing are available with structured payments to meet revenues. This could allow the municipality to acquire the equipment without having current fund appropriation.

The advantages of a municipal lease program include the following:

- Enhanced cash flow financing allows municipalities or districts to spread the cost of an acquisition over several fiscal periods, leaving more cash on hand.
- A lease program is a hedge against inflation since the cost of purchased equipment is figured at the time of the lease and the equipment can be acquired at current prices.
- Flexible lease terms structured over the useful life span of the equipment can allow financing of as much as 100% of the acquisition.
- Low-rate interest on a municipal lease contract is exempt from federal taxation. They have no fees and have rates often comparable to bond rates.
- Full ownership at the end of the lease most often includes an optional purchase clause of $1.00 for complete ownership.

Because of budgetary shortfalls, leasing is becoming a standard way for cities, counties, states, schools, and other municipal entities to get the equipment they need today without spending their entire annual budget to acquire it. It should be noted that municipal leases are different from standard commercial leases because of the mandatory nonappropriation clause. This states that the entity is only committing to funds through the end of the current fiscal year, even when signing a multiyear contract.

6 Solar Power Rebates, Financing, and Feed-In Tariffs Programs

Introduction

In this chapter, the reader will be introduced to financial topics that affect the viability of large-scale solar power system programs. As with any other capital investment projects, the proper appreciation of financial issues associated with large-scale solar power projects becomes a prerequisite for justifying the validity of a project. As such, the overall financial assessment becomes as important as designing the system. The rebate and tariff programs discussed in the following reflect the specifics of California's programs, which have set the standards for the United States. Therefore, the readers are advised to inquire about the status of the programs that prevail in other states.

California Solar Initiative Rebate Program

This chapter is a summary of the California Solar Incentive (CSI) and Feed-in Tariff programs, which are essential to understand when dealing with large-scale solar power systems. It should be noted that the California Energy Commission (CEC) has developed all regulatory policy and solar power system equipment certification standards that have been used throughout the United States and abroad. The CSI and feed-in tariffs discussed in this chapter are applicable to all states within the country.

On January 1, 2007, the state of California introduced solar rebate funding for the installation of photovoltaic power cogeneration, which was authorized by the California Public Utilities Commission (CPUC) and the Senate. The bill referenced as SB1 has allotted a budget of $2.167 billion, which will be used over a 10-year period.

The rebate funding program known as the CSI is a program that awards incentive plans on the basis of performance, unlike earlier programs, which based rebates on calculated projections of system energy output. The new rebate award system categorizes solar power installations into two incentive groups. An incentive program referred to as the Performance Based Incentive (PBI) addresses photovoltaic installations of 100 kilowatts or larger and provides rebate dollars based on the solar power cogeneration's actual output over a 5-year period. Another rebate, referred to as Expected Performance Based Buy Down (EPBB), is a one-time lump sum

Table 6.1. *CSI Program budget by administrator*

Utility	Total budget %
Pacific Gas & Electric	43.7%
Southern California Edison	46.0%
San Diego Gas & Electric	10.3%

incentive payment program for solar power systems with performance capacities of less than 100 kilowatts. This is a payment based on the system's expected future performance.

The distribution and administration of the CSI funds are delegated through three major utility providers, which service various state territories. The three main service providers that administer the program in California are

- Pacific Gas and Electric (PG&E), which serves Northern California
- Southern California Edison (SCE), which serves Central California
- San Diego Regional Energy Office (SDREO)/San Diego Gas and Electric (SD&G), which serves Southern California

It should be noted that municipal electric utility customers are not eligible to receive CSI funds from the three administrator agencies.

All of the service providers administering the CSI program have Web pages that enable clients to access online registration databases, which provide program handbooks, reservation forms, contract agreements, and all other forms required by the CSI program. All CSI application and reservation forms are available on www.csi.com.

The principal object of the CSI program is to ensure that 3,000 megawatts of new solar energy facilities are installed throughout California by the year 2017.

CSI Fund Distribution

The CSI's fund distribution is administered by the three main agencies in California and has a specific budget allotment, which is proportioned according to the demographics of power demand and distribution. CSI budget allotment values are shown in Table 6.1.

The CSI budget shown in the table is divided into two customer segments: residential and nonresidential. Table 6.2 shows relative allocations of CSI solar power generation by customer sector.

CSI Power Generation Targets

In order to offset the high installation costs associated with solar power installation and promote PV industry development, the CSI incentive program has devised a plan that encourages customer sectors to take immediate advantage of rebate initiatives, which are intended to last for a limited duration of 10 years. The incentive program is currently planned to be reduced automatically over the duration of the 3,000 MW of solar power reservation, in 10 step-down trigger levels that gradually distribute

Table 6.2. *CSI solar power production targets for residential, commercial, and government sectors*

Administrator	Customer class*	Current step	Internal MW in step	Unused MW from previous steps	Revised total MW in steps	Issued conditional reservation letters (MW)	MW remaining	MW under review
PGE	Residential	8	36.10	3.67	39.77	25.77	14.00	2.06
	Nonresidential	8	73.20	8.53	81.73	45.12	36.62	0.00
SCE	Residential	6	28.80	0.72	29.52	21.15	8.37	0.87
	Nonresidential	8	77.10	14.56	91.66	32.58	59.08	1.32
CCSE	Residential	8	8.50	1.24	9.74	6.45	3.29	0.62
	Nonresidential	8	17.30	3.54	20.84	0.27	20.57	8.17

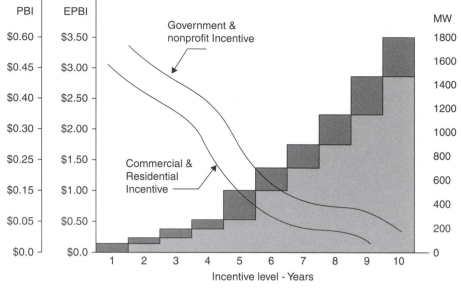

Figure 6.1. CSI solar power rebate incentive step level graph.

the power generation over 10 allotted steps. CSI MW power production targets are proportioned among the administrative agencies by residential and nonresidential customer sectors (see Table 6.2).

In each of the 10 steps, CSI applications are limited to the trigger levels. Table 6.2 shows set trigger stages for SCE and PG&E client sectors. Once the trigger level allotments are complete, the reservation process is halted and restarted at the next trigger level. In the event of trigger level power surplus, excess energy allotment is transferred forward to the next trigger level. Figure 6.1 represents a graph of the CSI solar power rebate incentive step levels.

The CSI power production targets shown in Tables 6.1 and 6.2 are based on the premises that solar power industry production output and client sector awareness will gradually increase within the next decade, and that the incentive program will eventually promote a viable industry that will be capable of providing a tangible source of a renewable energy base in California.

Incentive Payment Structure

As mentioned, the CSI offers PBI and EPBB incentive programs, both of which are based on verifiable PV system output performance. EPBB incentive output characteristics are determined by factors such as location of solar platforms, system size, shading conditions; tilt angle, and all of the other factors discussed in previous chapters. On the other hand, PBI incentives are strictly based on predetermined flat rate per kWh output payments over a 5-year period. Incentive payment levels have been devised to be reduced automatically over the duration of the program, namely, in 10 steps that are directly proportional to the MW volume reservation, as shown in Table 6.2.

As seen from the incentive distribution table, the rebate payments diminish as the targeted solar power program reaches its 3,000-MW energy output. The main

Table 6.3. *CSI solar power rebate incentives and step level: Production targets for residential, commercial, and government*

Step	Statewide MW in step	EPBB Payments (per watt)			PBI Payments (per kWh)		
		Residential	Nonresidential		Residential	Nonresidential	
			Commercial	Government/ Nonprofit		Commercial	Government/ Nonprofit
1	50	n/a	n/a	n/a	n/a	n/a	n/a
2	70	$2.50	$2.50	$3.25	$0.39	$0.39	$0.50
3	100	$2.20	$2.20	$2.95	$0.34	$0.34	$0.46
4	130	$1.90	$1.90	$2.65	$0.26	$0.26	$0.37
5	160	$1.55	$1.55	$2.30	$0.22	$0.22	$0.32
6	190	$1.10	$1.10	$1.85	$0.15	$0.15	$0.26
7	215	$0.65	$0.65	$1.40	$0.09	$0.09	$0.19
8	250	$0.35	$0.35	$1.10	$0.05	$0.05	$0.15
9	285	$0.25	$0.25	$0.90	$0.03	$0.03	$0.12
10	350	$0.20	$0.20	$0.70	$0.03	$0.03	$0.10

reasoning behind downscaling the incentive is based on the presumption that the solar power manufacturers will, within the next decade, be in a position to produce larger quantities of more efficient and less expensive photovoltaic modules. As a result of economies of scale, the state will no longer be required to extend special incentives to promote the photovoltaic industry by using public funds.

Expected Performance Buy-down

As mentioned earlier, EPBB is a one-time, initial incentive that is based upon a photovoltaic power cogeneration's estimated or predicted future performance. This program is targeted to minimize program administration works for relatively small systems that do not exceed 100 kWh. Factors that affect the computations of the estimated power performance are relatively simple and take into consideration such factors as panel count, PV module certified specifications, location of the solar platform, insolation, PV panel orientation, tilt angle, and shading losses. These are all entered into a predetermined equation that results in a buy-down incentive rate.

The EPBB program applies to all new projects, excluding systems that have Building Integrated Photovoltaics (BIPVs). The EPBB one-time incentive payment calculation is based on the following formula:

EPBB Incentive Payment = Incentive Rate × System Rating (kW) × Design Factor
System Rating (kW) = (Number of PV modules × CEC PTS value × CEC inverter listed efficiency)/(1000) for kilowatt conversion

Imposed special design requirements are the following:

- All PV modules must be oriented between 180 and 270 degrees
- The optimal tilt for each compass direction shall be within the range of 180 and x degrees for optimized summer power output efficiency

- Derating factors associated with weather and shading analysis must be taken into account
- Systems must be on an optimal reference and location
- PV tilt must correspond to the local latitude

It should be noted that all residential solar power installations are also subject to EPBB incentive payment formulations.

Performance Based Incentive

As of January 1, 2007, the PBI incentive applied to solar power system installation that was equal to or exceeded 100 kilowatts. As of January 1, 2008, the base power output reference was reduced to 50 kW. By January 1, 2010, it was reduced to 30 kW. Each of the BPI payments is limited to a duration of 5 years following the completion of the system acceptance test. Also included in the plan are custom-made building integrated photovoltaic systems (BIPVs).

Host Customer

All beneficiaries of CSI programs are referred to as the "host customer," which includes not only the electric utility customers but also retail electric distribution organizations such as PG&E, SCE, and SDG&E. Under the rules of the CSI program, all entities that apply for an incentive are referred to as applicants, hosts, or system owners.

In general, host customers must have an outstanding account with a utility provider at the location of solar power cogeneration. In other words, the project in California must be located within the service territory of one of the three listed program administrators.

Upon approval of the reservation, the host customer is considered the system owner and retains sole rights to the reservation. The reservation is a payment guarantee by CSI that cannot be transferred by the owner. However, the system installer can be designated to act on behalf of the owner.

To proceed with the solar power program, the applicant or the owner must receive a written confirmation letter from the administering agency then apply for authorization for grid connectivity. In the event of project delays beyond the permitted period of fund reservation, the customer must reapply for another rebate to obtain authorization.

According to SCI regulations, there are several categories of customers who do not qualify to receive the incentive. Customers exempted from the program are organizations in the business of power generation and distribution, publicly owned gas and electricity distribution utilities, and any entity that purchases electricity or natural gas for wholesale or retail purposes.

As a rule, the customer assumes full ownership upon reception of the incentive payment and technically becomes responsible for the operation and maintenance of the overall solar power system.

It should be noted that a CSI applicant is recognized as the entity that completes and submits the reservation forms and becomes the main contact person who must communicate with the program administrator throughout the duration of the project. However, the applicant may also designate an engineering organization or a system integrator, an equipment distributor, or even an equipment lessor to act as the designated applicant.

Solar Power Contractors and Equipment Sellers

Contractors from California who specialize in solar power installation must hold an appropriate State of California Contractors State License. In order to be qualified as an installer by the program administrator, the solar power system integrator must provide the following information:

- Business name and address
- Principal name or contact
- Business registration or license number
- Contractor's license number
- Contractor's bond (if applicable) and corporate limited liability entities
- Reseller's license number (if applicable)

All equipment such as PV modules, inverters, and meters sold by equipment sellers must be Underwriters Laboratory (UL) approved and certified by the California Energy Commission. All equipment provided must be new and must have been tested for a period of 1 year. The use of refurbished equipment is not permitted. It should be noted that experimental, field-demonstrated, or proof-of-concept operation type equipment and materials are not approved and do not qualify for the rebate incentive. All equipment used must therefore have UL certification and performance specifications that would allow program administrators to evaluate equipment performance.

According to CEC certification criteria, all grid connected PV systems must carry a 10-year warranty and meet the following certification requirements:

- All PV modules must be certified to UL 1703 standards
- All grid connected solar watt hour meters for systems less than 10 kW must have an accuracy of ±5%. Watt hour meters greater than 10 kW systems must have a measurement accuracy of ±2%.
- All inverters must be certified to UL 1741 standards.

PV System Sizing Requirements

It should be noted that the primary objective of solar power cogeneration is to produce a certain amount of electricity to offset a certain portion of the electrical demand load. Therefore, the power production of PV systems is set in a manner so as not to exceed the actual energy consumption during the previous 12 months. The formula applied for establishing maximum system capacity is

Maximum system power output (kW) = 12 months of previous energy used (kWh)/(0.18 × 8760 hours/year)

The factor of 0.18 × 8760 = 1577 hours/year can be translated into an average of 4.32 hours/day of solar power production, which essentially includes system performance and derating indexes applied in CEC photovoltaic system energy output calculations.

The maximum PV system under the present CSI incentive program is limited to 1000 kW, or 1 MW. However, if the preceding calculation limits permit, customers are allowed to install grid connected systems of up to 5 megawatts, for which only 1 megawatt will be considered for receiving the incentive.

For new construction where the project has no history of previous energy consumption, an applicant must substantiate system power demand requirements by engineering system demand load calculations that will include present and future load growth projections. All calculations must be substantiated by corresponding equipment specifications, panel schedules, single line diagrams, and building energy simulation programs such as eQUEST, EnergyPro, or DOE-2.

Energy Efficiency Audit

Recent rules enacted in January 2007 require that all existing residential and commercial customers will be obligated to provide a certified energy efficiency audit for their existing building when applying for a CSI rebate. The audit certification, along with the solar PV rebate application forms, must be provided to the program administrator for evaluation purposes.

Energy audits can be conducted either by calling an auditor or by accessing a special Web page provided by each administrative entity. In some instances, energy audits can be waived if the applicants can provide a copy of an audit conducted in the past 3 years or provide proof of a California Title 24 Energy Certificate of Compliance, which is usually calculated by mechanical engineers. Projects that have a national LEED™ certification are also exempt from energy audit.

Warranty and Performance Permanency Requirements

As mentioned previously, all major system components are required to have a minimum 10 years of warranty by manufacturers and installers alike. All equipment, including PV modules and inverters, is required to be replaced at no cost to the client in the event of malfunction. System power output performance and electrical equipment breakdown must include 15% power output degradation from the original rated projected performance for a period of 10 years.

To be eligible for CSI, all solar power system installations must be permanently attached or secured to their platforms. PV modules supported by quick disconnect means and installed on wheeled platforms or trailers are not considered as legitimate stationary installations.

During the course of project installation, the owner or its designated representative must maintain continuous communication with the program administrator and provide all required information regarding equipment specification, warranties, platform configuration, design revisions and system modifications, updated construction schedules, and construction status on a regular basis.

In the event the location of PV panels is changed and panels are removed or relocated within the same project perimeters or service territory, the owner must inform the CSI administrator and establish a revised PBI payment period.

Insurance

Currently, the owners or host customers of all systems equal to or greater than 30 kW receiving CSI are required to carry a minimum level of general liability insurance. Installers must also carry workman's compensation and business auto insurance coverage. Since U.S. government entities are self-insured, the program administrators require only proof of coverage.

Grid Interconnection and Metering Requirements

The main criterion for grid system integration is that the solar power cogeneration system must be permanently connected to the main electrical service network. As such, portable power generators are not considered eligible. In order to receive the incentive payment, the administrator must receive proof of the grid interconnection record. In order to receive additional incentives, customers whose power demand coincides with California's peak electricity demand become eligible to apply for Time-of-Use (TOU) tariffs, which could increase their energy payback.

All installed meters must be easily located to allow an administrator's authorized agents to have easy access for tests or inspections.

Inspection

All systems rated from 30 to 100 kW that have not adopted a PBI will be inspected by specially designated inspectors. In order to receive the incentive payment, the inspectors must verify system operational performance, installation conformance to applications, and eligibility criteria and grid interconnection.

System owners who have opted for EPBB incentives must install the PV panels in proper orientation and produce power that is reflected in the incentive application.

In the event of inspection failure, the owner will be advised by the administrator about shortcomings regarding material or compliance, which must be mitigated within 60 days. Failure to correct the problem could result in cancellation of the application and a strike against the installer, applicant, seller, or any party deemed responsible.

If entities identified as responsible for mitigating the problem fail three times, they will be disqualified from participating in CSI programs for a period of 1 year.

CSI Incentive limitations

The prerequisites for processing CSI programs are based upon the premise that a project's total installed out-of-pocket expenses by the owner do not exceed the eligible costs. For this reason, the owner or the applicant must prepare a detailed project cost breakdown that will highlight only relative embedded costs of the solar power system. A worksheet designed for this purpose is available for the California Solar Incentive. For additional detail refer to the Web page, www.csi.com.

It is important to note that clients are not permitted to receive incentives under other sources. In the event a project may be qualified to receive an additional incentive from another source for the same power cogenerating system, the first incentive amount will be discounted by the amount of the second incentive received.

Nevertheless, the overall combined incentive amount must be assured not to exceed total eligibility costs.

At any time during project construction, administrators reserve the right to conduct periodic spot checks and random audits to make certain that all payments received were made in accordance with CSI rules and regulations.

CSI Reservation Steps

The following is a summary of steps for an EPBB application:

- Reservation forms must be completed and submitted with the owner's or applicant's original signature.
- Proof of electric utility service or account number for the project site must be shown on the application. In the case of a new project, the owner must procure a tentative service account number.
- System description worksheets, available on the CSI Web page, must be completed.
- Electrical system sizing documents, as discussed, must be attached to the application form.
- If the project is subject to tax exemption, form AB1407 Compliance for Government and Non-profit Organizations must be attached to the application.
- For existing projects, energy efficiency audit or Title 24 calculations must be submitted as well.
- To calculate the EPBB, use the CSI Web page calculator (www.csi-epbb.com).
- Attach a copy of the executed purchase agreement from the solar system contractor or provider.
- Attach a copy of the executed contract agreement if system ownership is given to another party.
- Attach a copy of the grid interconnection agreement if available. Otherwise, inform the administrator about steps taken to secure the agreement. To submit payment claims, provide the following documents to the administrator:

  - Wet signed claim form available on the CSI Web page.
  - Proof of authorization for grid integration.
  - Copy of the building permit and final inspection sign-off.
  - Proof of warranty from the installer and equipment suppliers.
  - Final project cost breakdown.
  - Final project cost affidavit.

For projects categorized as BPI or nonresidential systems 10 kW or larger, the owner must follow these steps:

- Reservation form must be completed and submitted with owner's or applicant's wet signature.
- Proof of electric utility service or account number for the project site must be shown on the application. In case of a new project, the owner must procure a tentative service account number.

- The system description worksheet, available on the CSI Web page, must be completed.
- Electrical system sizing documents, as discussed previously, must be attached to the application form.
- Attach an application fee (1% of the requested CSI incentive amount).
- If the project is subject to tax exemption, form AB1407 Compliance for Government and Non-profit Organizations must be attached to the application.
- For existing projects, energy efficiency audit or Title 24 calculations must be submitted as well.
- Forward the printout of calculated PBI. Use the CSI Web page calculator.
- Attach a copy of the executed purchase agreement from the solar system contractor or provider.
- Attach a copy of the executed contract agreement if system ownership is given to another party.
- Attach a copy of the grid interconnection agreement if available. Otherwise, inform the administrator about steps taken to secure the agreement.
- Complete proof of project milestone.
- Provide host customer certificate of insurance.
- Provide system owner's certificate of insurance (if different from the host's).
- Provide a copy of the project cost breakdown worksheet.
- Provide a copy of alternative system ownership, such as lease/buy agreement.
- Provide a copy of RFP or solicitation document if the customer is a government, a nonprofit organization, or a public entity.

To submit the claim for the incentive documents to the administrator, the owner or the contractor must provide the following:

- Wet signed claim form available on the CSI Web page
- Proof of authorization for grid integration
- Copy of the building permit and final inspection sign-off
- Proof of warranty from the installer and equipment suppliers
- Final project cost breakdown
- Final project cost affidavit

In the event of incomplete document submission, the administrators will allow the applicant to provide missing documentation or information within 20 days. Information provided must be in written form mailed by the U.S. postal system. Faxes or hand-delivered systems are not allowed.

All changes to the reservation must be undertaken by a formal letter that explains legitimate delay justification. Requests to extend the reservation expiration date are capped to a maximum of 180 calendar days. Written time extension requests must explicitly highlight circumstances that were beyond the control of the reservation holder, such as the permitting process, manufacturing delays, the extended delivery of PV modules or critical equipment, or acts of nature. All correspondence associated with the delay must be transmitted by letter.

Incentive Payments

Upon the completion of final field acceptance and submission of the referenced documents (for EPBB projects), the program administrator will within a period of 30 days issue a complete payment. For PBI programs, the first incentive payment commences and is issued within 30 days of the first scheduled performance reading from the watt meter. All payments are made to the host customer or the designated agent.

In some instances, a host can request that the administrator assign the entire payment to a third party. For payment reassignment, the host must complete a special set of forms provided by the administrator.

The EPBB one-time lump sum payment calculation is based on the following formula:

$$\text{EPBB Incentive Payment} = \text{Final CSI System Size} \times \text{Reserved EPBB Incentive Rate}$$

PBI payments for PV systems of 100 kW or greater are made on a monthly basis over a period of 5 years. The payments are based on the actual electric energy output of the photovoltaic system. If chosen by the owner, systems less than 100 kW can also be paid on the PBI incentive basis.

The PBI payment calculation is based on the following formula:

$$\text{Monthly PBI Incentive Payment} = \text{Reserved Incentive Rate} \times \text{Measured kWh Output}$$

In the event of PV system size change, the original reservation request forms must be updated and incentive amount recalculated.

An Example of the Procedure for Calculating the CSI Rebate

The following example is provided to assist the reader with details of CSI reservation calculation requirements. Reservation forms in addition to completing host customer and project site information, require a set of calculations to satisfy system projected performance output requirements. Regardless of system size and classification as EPBB or PBI, supportive documentation for calculating the solar power system cogeneration remains identical.

In order to commence reservation calculations, the designer must undertake the following preliminary design measures:

Outline the solar power cogeneration system net unobstructed platform area.

- Use a rule of thumb to determine watts per square feet of a particular PV module intended for use. For example, a PV module area of 2.5 square feet × 5 square feet = 12.5 square feet, which produces 158 watts PTC, will have an approximate power output of about 14 watts per square foot.
- By dividing the available PV platform area by 12.5 square feet, we can determine the number of panels required.

In order to complete the CSI reservation forms referenced earlier, the designer must also determine type, model, quantity, and efficiency of the CEC approved inverter.

For example, let us assume that we are planning for a ground mount solar farm with an output capacity of 1 MW. The area available for the project is 6 acres, which is adequate for installing a single axis solar tracking system. Our solar power module and inverter are chosen from approved CEC listed equipment.

Solar power system components used are as follow:

- PV module – SolarWorld, Powermax 175p, unit DC watts = 175, PTC = 158.3 watts AC, number of panels required 6680 units.
- Inverter – Xantrex Technology – PV225S-480P, 225 kW, efficiency 94.5%

Prior to completing the CSI reservation form, the designer must use the CSI EPBB calculator (available at www.csi-epbb.com) to determine the rebate for systems that are smaller than 100 kW or larger. Even though BPI calculation is automatically determined by the CSI reservation form spreadsheet, EPBB calculation determines a design factor number required by the form.

EPBB Calculation Procedure

To perform the EPBB calculation, the designer must enter the following data in the blank field areas:

- Project area zip code (for example, 92596)
- Project address information
- Customer type, such as residential, commercial, or government or nonprofit
- PV module manufacture, model, associated module DC and PTC rating, and unit count
- Inverter manufacturer, model, output rating, and percent efficiency
- Shading information, such as minimal
- Array tilt, such as 30 degrees; for maximum efficiency, the tilt angle should be close to latitude
- Array azimuth in degrees, which determines orientation; for north facing south, use 180 degrees

After entering the data, the CSI calculator will output the following results:

- Optimal tilt angle at proposed azimuth
- Annual kWh output at optimal tilt facing south
- Summer month output from May to October
- CEC-AC rating – a comment will indicate whether the system is greater than 100 kW
- Design correction factor – required for calculating the CSI reservation form
- Geographic correction
- Design factor
- Incentive rate – dollars/watt
- Rebate incentive cash amount if system is qualified as EPBB

California Solar Initiative Reservation Form Calculations

As is the EPBB, the CSI reservation is a Web page spreadsheet that can be accessed at www.csi.com. The data required to complete the solar equipment are the same as the ones used for EPBB, except that EPBB design factors derived from the preceding calculation must be inserted into the project incentive calculations.

The California Solar Initiative Program Reservation form consists of the following six major information input data fields:

1. Host customer – information required includes customer name, business class and company information, taxpayer identification, contact person's name, title, mailing address, telephone, fax, and e-mail
2. Applicant information if system procurer in not the host customer
3. System owner information
4. Project site information, same as EPBB; system platform information such as available building or ground area must be specified; in this field the designer must also provide the electrical utility service account and meter numbers if available; for new projects there should be a letter attached to the reservation form indicating the account procurement status
5. PV and inverter hardware information identical to that used in calculating EPBB
6. Project incentive calculation; n this field the designer must enter the system rating in kW (CEC) and design factor data obtained from the EPBB calculation

After completing the entries, the CSI reservation spreadsheet will automatically calculate the project system power output size in watts. In a field designated Total Eligible Project Cost, the designer must insert the projected cost of the system. CSI automatically produces per watt installed cost, CSI rebate amount, and the system owner's out-of-pocket expenses.

The following CSI reservation request calculation is based on the same hardware information used in the previous EPBB calculation.

Data entry information and calculation steps are as follows:

- Platform – single axis tracker
- Shading – none
- Insolation for zip code 92596, San Bernardino, California = 5.63 average hour/day
- PV module – SolarWorld model SW175 mono/P
- DC watts – 175
- PTC – 158.2
- PV count – 6680
- Total power output – 1057 kW AC
- Calculated CSI system size by the spreadsheet – 1032 kW AC
- Inverter – Xantrex Technology Model PV225S-480P
- Power output capacity – 225 kW AC
- Efficiency – 94.5%
- Resulting output = $1057 \times 94.5\% = 999$ kW
- CSI – EPBB design factor = 0.975
- CSI system size = $999 \times 0.975 = 975$ kWh
- PV system daily output = 957×5.63 (insolation) = 5491 kWh/day
- Annual system output = 5491×365 (days) = 2,004,196 kWh per year
- Assuming an incentive class for a government or a nonprofit organization for the year 2007, allocated performance output per kWh = \$0.513

- Total incentive over 5 years = 2,004,196 × \$0.513 = \$1,028,152.00Projected installed cost = \$8,500,000.00 (\$ 8.50/watt)
- System owner out-of-pocket cost = \$3,489,510.00
- Application fee of 1% of incentive amount = \$50,105.00

Equipment Distributors

Eligible ERP manufacturers and companies who sell system equipment must provide the CEC with the following information on the equipment seller information form (CEC-1038 R4). – -

- Business name, address, phone, fax, and e-mail address
- Owner or principal contact
- Business license number
- Contractor license number (if applicable)
- Proof of good standing in the records of the California Secretary of State, as required for corporate and limited liability entities
- Reseller's license number

Special Funding for Affordable Housing Projects

California Assembly Bill 58 mandates that the CEC establish an additional rebate for systems installed on affordable housing projects. These projects are entitled to qualify for an extra 25% rebate above the standard rebate level, provided that the total amount rebated does not exceed 75% of the system cost. The eligibility criteria for qualifying are as follows.

The affordable housing project must adhere to California health and safety codes. The property must expressly limit residency to extremely low, very low, lower, or moderate income persons and must be regulated by the California Department of Housing and Community Development.

Each residential unit (apartments, multi-family homes, etc.) must have individual electric utility meters. The housing project must be at least 10% more energy-efficient than current standards specified.

Special Funding for Public and Charter Schools

A special amendment to the CEC mandate enacted established a Solar Schools Program to provide a higher level of funding for public and charter schools to encourage the installation of photovoltaic generating systems at more school sites. To qualify for the additional funds, the schools must meet the following criteria:

- Public or charter schools must provide instruction for kindergarten or any of the grades 1 through 12.
- The schools must have installed high-efficiency fluorescent lighting in at least 80% of classrooms, and they must agree to establish a curriculum tie-in plan to educate students about the benefits of solar energy and energy conservation.

Principal Types of Municipal Lease

There are two types of municipal bonds. One type is referred to as a "tax exempt municipal lease," which has been available for many years and is used primarily for the purchase of equipment and machinery that have a life expectancy of 7 years or less. The second type is generally known as an "energy efficiency lease" or a "power purchase agreement" and is used most often on equipment being installed for energy efficiency purposes. It is used where the equipment has a life expectancy of greater than 7 years. Most often this type of lease applies to equipment classified for use in renewable energy cogeneration, such as solar PV and solar thermal systems. The other common type of application that can take advantage of municipal lease plans includes the energy efficiency improvement of devices such as lighting fixtures, insulation, variable-frequency motors, central plants, emergency backup systems, energy management systems, and structural building retrofits.

The leases can carry a purchase option at the end of the lease period for an amount ranging from $1.00 to fair market value and frequently have options to renew the lease at the end of the lease term for a lesser payment over the original payment.

Tax Exempt Municipal Lease

A tax exempt municipal lease is a special kind of financial instrument that essentially allows government entities to acquire new equipment under extremely attractive terms with streamlined documentation. The lease term is usually for less than 7 years. Some of the most notable benefits are:

- Lower rates than conventional loans or commercial leases
- Lease-to-own: there is no residual and no buyout
- Easier application, such as same-day approvals
- No "opinion of counsel" required for amounts less than $100,000
- No underwriting costs associated with the lease

Entities That Qualify for a Municipal Lease

Virtually any state, county, or city municipal government and its agencies (such as law enforcement, public safety, fire, rescue, emergency medical services, water port authorities, school districts, community colleges, state universities, hospitals, and 501 other organizations) qualify for municipal leases. Equipment that can be leased under a municipal lease include essential use equipment and remediation equipment such as vehicles, land, or buildings. Some specific examples are listed here:

- Renewable energy systems
- Cogeneration systems
- Emergency backup systems
- Microcomputers and mainframe computers
- Police vehicles

- Networks and communication equipment
- Fire trucks
- Emergency management service equipment
- Rescue construction equipment such as aircraft helicopters
- Training simulators
- Asphalt paving equipment
- Jail and court computer-aided design (CAD) software
- All-terrain vehicles
- Energy management and solid waste disposal equipment
- Turf management and golf course maintenance equipment
- School buses
- Water treatment systems
- Modular classrooms, portable building systems, and school furniture such as copiers, fax machines, and closed circuit television surveillance equipment
- Snow and ice removal equipment
- Sewer maintenance

The transaction must be statutorily permissible under local, state, and federal laws and must involve something essential to the operation of the project.

The Difference between a Tax Exempt Municipal Lease and a Commercial Lease

Municipal leases are special financial vehicles that provide the benefit of exempting banks and investors from federal income tax, allowing for interest rates that are generally far below conventional bank financing or commercial lease rates. Most commercial leases are structured as rental agreements with either nominal or fair-market-value purchase options.

Borrowing money or using state bonds is strictly prohibited in all states, since county and municipal governments are not allowed to incur new debts that will obligate payments that extend over multiyear budget periods. As a rule, state and municipal government budgets are formally voted into law. Therefore, there is no legal authority to bind government entities to make future payments.

As a result, most governmental entities are not allowed to sign municipal lease agreements without the inclusion of nonappropriation language. Most governments, when using municipal lease instruments, consider obligations as current expenses and do not characterize them as long-term debt obligations.

The only exceptions are bond issues or general obligations, which are the primary vehicles used to bind government entities to a stream of future payments. General obligation bonds are contractual commitments to make repayments. The government bond issuer guarantees to make funds available for repayment, including raising taxes if necessary. In the event that adequate sums are not available in the general fund, "revenue" bond repayments are tied directly to specific streams of tax revenue. Bond issues are very complicated legal documents that are expensive and time consuming and in general have a direct impact on the taxpayers and require voter approval. Hence, bonds are exclusively used for very large building projects, such as creating infrastructure including sewers and roads.

Municipal leases automatically include a nonappropriation clause; therefore, they are readily approved without counsel. Nonappropriation language effectively relieves the government entity of its obligation in the event that funds are not appropriated in any subsequent period for any legal reason.

Municipal leases can be prepaid at any time without a prepayment penalty. In general, a lease amortization table included with a lease contract shows the interest principal and payoff amount for each period of the lease. There is no contractual penalty, and a payoff schedule can be prepared in advance. It should also be noted that equipment and installation can be leased.

Lease payments are structured to provide a permanent reduction in utility costs when used for the acquisition of renewable energy or cogeneration systems. A flexible leasing structure allows the municipal borrower to level out capital expenditures from year to year. Competitive leasing rates of up to 100% financing are available with structured payments to meet revenues that could allow the municipality to acquire the equipment without having current fund appropriation. The advantages of a municipal lease program include the following:

- Enhanced cash flow financing allows municipalities or districts to spread the cost of an acquisition over several fiscal periods, leaving more cash on hand.
- A lease program is a hedge against inflation since the cost of purchased equipment is figured at the time of the lease and the equipment can be acquired at current prices.
- Flexible lease terms structured over the useful life span of the equipment can allow financing of as much as 100% of the acquisition.
- Low-rate interest on a municipal lease contract is exempt from federal taxation, has no fees, and has rates often comparable to bond rates.
- Full ownership at the end of the lease most often includes an optional purchase clause of $1.00 for complete ownership.

Because of budgetary shortfalls, leasing is becoming a standard way for cities, counties, states, schools, and other municipal entities to get the equipment they need today without spending their entire annual budget to acquire it.

Municipal leases are different from standard commercial leases because of the mandatory nonappropriation clause, which states that the entity is only committing to funds through the end of the current fiscal year (even if they are signing a multiyear contract).

Solar Power Project Financing

The following financing discussion is specific to large alternative and renewable energy projects such as solar, wind, and geothermal projects that require extensive amounts of investment capital. Project financing of such large projects, similar to that of large industrial projects, involves long-term financing of capital intensive material and equipment.

Since most alternative energy projects in the United States are subject to state and federal tax incentives and rebates, project financing involves highly complex financial structures. Project debt and equity, rebate, federal and state tax incentives, and cash flow generated by grid energy power are used to finance projects. In general, project lenders are given a lien on all of the project assets, including property, which enables them to assume control of a project over the terms of the contract.

Since renewable energy and large industrial projects involve different levels of transactions such as equipment and material purchase, site installation, maintenance, and financing, a special purpose entity is created for each project. This shields other assets owned by a project sponsor from the detrimental effects of project failure.

As special purpose joint ventures, these types of entities have no assets other than the project. In some instances, capital contribution commitments by the owners of the project company are necessary to ensure that the project is financially sound.

In particular, alternative energy technology project financing is often more complicated than alternative financing methods commonly used in capital intensive projects found in the transportation, telecommunication, and public utility industries.

Renewable energy type projects in particular are frequently subject to a number of technical, environmental, economic, and political risks. Therefore, financial institutions and project sponsors evaluate inherent risks associated with a particular project's development and operation and determine whether projects are financeable.

To minimize risk, project sponsors create special entities, which consist of a number of specialist companies operating in a contractual network with each other. They allocate risk in a way that allows financing to take place.

In general, a project financing scheme involves a number of equity investors (known as sponsors), which include hedge funds and a syndicate of banks that provide loans for the project. The loans are most commonly nonrecourse loans, which are secured by the project itself and paid for entirely from its cash flow, rebates, and tax incentives. Projects that involve large risks require limited recourse financing secured by a surety from sponsors. A complex project finance scheme may also incorporate corporate finance, securitization, options, insurance provisions, or other further measures to mitigate risk.

California Feed-in Tariffs

California Energy Commission Report on Feed-In Tariffs

The following feed-in tariff discussion is a summary of a 2008 workshop report that was published by the California Energy Commission. California has a Renewables Portfolio Standard (RPS) that requires the state's investor-owned utilities, energy service providers, and community choice aggregators to serve 20% of retail sales with renewable resources by 2010. Publicly owned utilities are required to develop RPS programs as well. As indicated in the *2007 Integrated Energy Policy Report* (IEPR), California set a 20% green energy target that was to be met by 2010 (to date, this has not been realized). California has also set a renewable energy target of 33% by 2020 and is expected to need new policy tools to meet this aggressive target.

The RPS referenced previously explores the potential approaches to expanding the use of feed-in tariffs as a mechanism to aid in making California's renewable generation objectives a reality. There are a great variety of potential feed-in tariff policy design options and policy paths. The report examines feed-in tariff options for design issues such as appropriate tariff structure, eligibility, and pricing. The report also considers policy goals and objectives, stakeholder comments on materials presented in the Energy Commission's feed-in tariff design issues, the options workshop held on June 30, 2008, as well as data accumulated from feed-in tariff experiences in Spain and Germany. The report identifies six representative policies for consideration. The pros and cons of each of the six policies are explored and analyzed in detail. In addition, the report explores the potential interaction of the policy paths, examines the interaction of feed-in tariff policies with other related policies, and discusses issues related to the potential for establishing a specific program for the California feed-in tariff program. The six policy paths that are examined in the report span a range of directions, timing, and scope.

Policy Path 1

This policy is designed to be similar to the feed-in tariff system currently in place in Germany but is conditional in that it will be triggered only if California's 20% renewable energy goal is not met by 2010. Under this option, tariffs would become available in the 2012–13 time frame to ensure that the 33% renewables target would be met by 2020. The policy does not mandate any restrictions on generator size, and all contracts are fixed-price and long-term. The tariffs intended would be differentiated by technology and project size. Fundamentally, the program is intended to be cost based, and the preliminary price settings would be set competitively and not administratively. The use of emerging resources would be capped, so as to limit ratepayer impacts. In addition, the use of long-term contract and technology differentiation would provide a degree of price stability to investors while promoting a diversity of renewable resources.

Policy Path 2

The second policy is essentially a pilot program that would involve a single utility that could generate more than 20 megawatts (MWs) of electrical energy. It would go into effect immediately without any sort of trigger mechanism. Long-term fixed-price contracts would be available for projects coming on-line within a 3-year window, after which the policy would be reevaluated. There would be no limit to the quantity of generation eligible to use this tariff, as the limited duration would serve to constrain its overall use. Tariff payments under this option would be value based, with payments differentiated only by production profiles. These include times of production, contribution to peak, and other factors (this also takes into account environmental issues). The value-based payments are intended to alleviate some ratepayer concerns relative to the cost-based alternatives. However, this path may not promote the resource diversity of Policy Path 1.

Policy Path 3

This policy would be triggered by the establishment of a Competitive Renewable Energy Zone (CREZ) designated for feed-in tariff procurement in the 2010–11 timeframe, allowing generation within the CREZ to proceed aggressively with development once transmission expansion is committed (without being constrained by the timing and risk of an RPS competitive solicitation). The policy is essentially a cost-based system. However, tariff prices would be set administratively rather than through the use of competitive benchmarks. This option would be limited geographically by the CREZ footprint, and the quantity eligible to take the feed-in tariff price would be capped at the CREZ transmission limit. This option would also target generators greater than 1.5 MW. In terms of the renewable resource potential and available/planned transmission in the CREZ, this option would help alleviate worries of undersubscription of new transmission lines and support a diverse mix of renewable resources.

Policy Path 4 This policy is supposed to be a solar-only pilot feed-in tariff. It will include elements of Policy Paths 1 and 3 that are cost based, using competitive benchmark rates. Rather than being limited to a specific window of time, however, the pilot scale for the tariff would be accomplished by limiting long-term contract availability to a single utility territory. Eligibility would be limited to solar installations larger than the net metering limit of 1 MW. This policy would also have a capacity cap option. Although this option could provide incentives for larger systems, as solar energy is above market, it may not provide enough renewable energy and diversity for the state to meet its goals. This option could be established independently or in concert with another policy path.

Policy Path 5 This policy is limited to sustainable biomass. Tariffs would be cost based and differentiated by size and biomass fuel feedstock. Unlike the solar-only option, the biomass path would be available in every market, rather than on a pilot scale in a single utility, and would not be capped. Finally, unlike all of the other policy paths that would incorporate long-term contracts or price guarantees, the contract term would be either short or medium term in acknowledgment of the fuel price risk that longer-term contracts would place on biomass developers and investors. As discussed later, this option could be established independently or in concert with another policy path.

Policy Path 6 This policy is intended to establish feed-in tariffs that will be available statewide to generators up to 20 MW in size without any conditions, to address a perceived gap in the current RPS solicitation process. This policy would offer cost-based, long-term prices differentiated by size and technology. Unlike in Policy Path 1, however, prices would not be based on a competitive benchmark, and the tariff quantity would be uncapped. It is not limited to one technology and therefore might be helpful in enabling the state to meet its diversity goals. Nevertheless, the California feed-in tariff policy report is intended to stimulate varied stakeholder input on which feed-in tariffs options could best help California meet its renewable energy objectives.

As discussed previously, a feed-in tariff is an offering of a fixed-price contract over a specified term with specified operating conditions to eligible renewable energy generators. Feed-in tariffs can be either an all-inclusive rate or a fixed premium payment on top of the prevailing spot market price for power. The tariff price paid will represent estimates of either the cost or the value of renewable generation. In the future, feed-in tariffs will be offered by the interconnecting utilities, which will set standing prices for each category of eligible renewable generators; the prices will be available to all eligible generators. Tariffs will essentially be based upon technology type, resource quality, and project size and may decline on a set schedule over time.

Benefits and Limitations of Feed-in Tariffs

As with other policies, feed-in tariffs provide benefits and limitations, a number of which depend upon the design of the tariff system. From the generator's perspective, the benefits of a feed-in tariff include the availability of a guaranteed price, buyer, and long-term revenue stream without the cost of solicitation. Market access is enhanced by feed-in tariffs, as project timing is not constrained by periodic scheduled solicitations. In addition, completion dates may not be constrained by contractual requirements, quantities are often uncapped, and interconnection is typically guaranteed. Together, these characteristics can help to reduce or alleviate generator revenue uncertainty, project risk, and associated financing concerns. Feed-in tariffs reduce transaction costs for both buyer and seller and are more transparent to administer than the current system. Because responding to standing tariffs is likely to be less costly and less complex than competitive solicitations, feed-in tariffs may also increase the ability of smaller projects or developers to help the state meet its Renewables Portfolio Standard (RPS) and greenhouse gas emission reduction goals. Policy makers can target feed-in tariffs to encourage specific types of projects and technologies if so desired. However, there are limitations to how a feed-in tariff might function in California. Total feed-in tariff costs cannot be predicted accurately because, despite the predetermined payments, the quantity of generation responding to a feed-in tariff is not typically predetermined.

One key concern is how the tariff fits in a deregulated market structure, including questions of who pays, how payments are distributed, what portion of rates would be used to recover tariff costs, and how to integrate electric production purchased through feed-in tariffs into utility power supplies. Another question specific to California is whether feed-in tariffs would work in concert with California's existing RPS law or would require changes in that law.

Getting the price right can be challenging. If the price is set too high, the tariff introduces the risk of overpaying and overstimulating the market. This risk may be exacerbated when the tariff is open to large projects in regions with ample resource potential. On the other hand, if the tariff is set too low to provide adequate returns to eligible projects, it may have little effect on stimulating the development of new renewable energy generation. A range of approaches for setting the price are discussed in the six options considered in this report.

Recommended Steps for a Successful Solar Power Project

The following discussion is intended to be a comprehensive guide that provides a step-by-step methodology for the procurement and management of large-scale solar power projects. The same guidelines could also be applied to any alternative energy system project, such as wind or geothermal. The main purpose of the steps outlined later is to allow investors to analyze their alternative of solar power energy needs, to assess system costs, and to evaluate savings that could be realized by the installation of appropriately sized solar power systems.

In view of the complexities of utility tariffs, the solar power interconnectivity to grid, and solar power system design and coordination, it is recommended that organizations interested in solar power projects should work with highly competent and experienced consultants who can assist with the project and represent the owner's interest in system feasibility study, design, contract negotiation, and project supervision. The following are a summarized recommendation and step-by-step processes that must be followed to achieve proper results. Similar recommendations are also suggested by the DOE and National Renewable Energy Laboratories.

Step No. 1 – Preliminary Project Evaluation

Prior to proceeding with any solar power system project, owners must engage the services of expert solar power engineering consultants to conduct a comprehensive investigation of solar power system requirements. Some of the critical issues to be reviewed and considered are as follows:

- How would a solar power project meet the long-term objectives of the electrical energy cogeneration of the organization?
- Could the organization commit a solar power platform for a period of 20 to 25 years?
- What are the implications of future business expansion and extended electrical energy requirements?
- What are the present and future projections of the electrical demand load profile?
- Does the electrical utility support solar power grid interconnectivity?
- What credits do electrical utility providers supply?

Responses to all of the queries could be provided by expert electrical/solar power engineering consultants, who in turn must do the following:

- Conduct a detailed feasibility study
- Identify present and future energy demand issues and propose solutions
- Provide an engineering design and economic analytical report
- Follow up with detail design and project supervision

Step No. 2 – Energy Use and Efficiency Cost Evaluation

The next step in implementing a solar power project is analyzing existing and future electrical energy consumption profiles and exploring ways and means for reducing electricity consumption efficiency and conservation in the following areas:

- Indoor and area lighting
- Building envelope and insulation
- Efficiency of heating and cooling systems
- All energy intense equipment, such as pumps or devices that operate around the clock

Note that energy conservation and energy audit are critical to any solar power system installation and can result in considerable savings.

Step No. 3 – Identifying Solar Power Platform(s)

As mentioned earlier, the first process for undertaking a solar power project is to conduct a thorough feasibility study. An experienced solar power system engineering consultant and the owners must consider this step to be perhaps the most critical to a solar power project. A well-studied feasibility report, if executed proficiently, will provide responses to all questions in steps 1 and 2 and provide a clear projection of the anticipated energy production potential as well as the financial metrics of the project. Moreover, a feasibility study will clearly delineate the potential power production capacities of the available platforms and associated footprints as well as the type of technologies best suited to achieve the objectives of the project.

Step No. 4 – Finding Experienced Solar Power/Electrical Engineering Consultants

It is suggested that in order to achieve the best technical and financial results, owners must avoid direct negotiations or contract agreements with solar power integrators or PPA organizations. It is also imperative to ensure that competitive bidding is based on the foundations of a thorough technical specification that is specifically prepared to address the unique requirements of a solar power project.

The criteria for selecting a solar power consultant should primarily be based on the following: consulting engineers qualified to undertake the design and implementation of large-scale systems must first and foremost have an electrical engineering degree, must hold state registration as a professional engineer (PE), and must have the following experience in electrical power and solar power system design. The following are some of the required qualifications that a large-scale solar power consultant must possess:

- Employ independent consultants that have significant familiarity with photovoltaic system technologies but have absolutely no involvement with solar power integrators
- A minimum of 5 to 10 years design experience in electrical power engineering
- At least 5 years experience in medium (200 to 500 kW) to large-scale (multimegawatt) solar power system design
- Solar power design experience that must include roof mount, ground mount, carport, and building integrated photovoltaic power systems
- Must have a minimum of 5 years experience with medium and large-scale solar power design and integration
- Must be well versed in solar power system econometric analysis and have familiarity with solar power economic profiling software
- Should have a track record of accomplishment with the specific type of solar power system
- Personal references that support experience in similar types of projects
- Scope of service provided should include

  1. Conduct a solar power system feasibility study
  2. Design solar power system
  3. Develop specific solar/electrical power system technical specification (request for proposal [RFP])

4. Assist clients with grid interconnectivity and rebate process paperwork
5. Assist clients with contract award negotiation
6. Coordinate with solar power integrator's project development activity
7. Conduct periodic construction supervision
8. Provide clients periodic project progress reports
9. Verify and approve solar power contractor's submittal and shop drawings
10. Verify and approve solar power contractor's field test and integration specifications
11. Monitor solar power system test and measurement documents
12. Actively take part in solar power system final test and commissioning
13. Evaluate customer training curriculum and take active part in customer training

Step No. 5 – Criteria for Engaging Solar Power Integrators

Upon completion of engineering design, evaluation of RFP, and award of the solar power contract (whether privately financed or third party financed), the client and the engineering consultant must ensure that solar power integrators meet the following standards:

- They have a track record of accomplishment with the specific type of solar power system.
- They have personal references that support experience in similar types of projects.
- Contractors are fully bondable and responsible in financing construction in financing the construction.
- They carry appropriate insurance.
- They have totally evaluated the solar power system specifications included in the RFP and are in full conformance with the letter and the intent of the RFP.
- They have studied issues associated with the logistics of the construction site; have taken into consideration environmental disturbances, material storage, and assembly areas; and have evaluated the safety and security of the project site.
- They have financial partners with the experience and sophistication to follow through the project or project life cycle.
- The integrator has total expertise in and knowledge of large-scale solar power systems.
- Ensure that in the event of third party ownership, the solar power purchase provider (SPPA) works with experienced solar power integrators and engineers.
- In SPPA type contract agreements, verify specific terms and conditions, study the power production calculations and assumptions, and evaluate the econometrics and validity of the calculation criterion used.
- Validate whether the contractor has studied environmental impact issues that may be associated with a specific geographic location.
- Contractor proposals must include references to project-specific support structure requirements.
- Evaluate integrator's contractual documents and their flexibility in accommodating minor changes.
- Evaluate solar power integrator's project management methodology and project control procedures.

- Pay careful attention to solar power equipment and materials provided. Be aware of shortcuts and the use of untried or uncertified equipment. Be prepared to invest more in a high-quality product.
- Quantify terms and conditions of optional or incremental long-term operation and maintenance cost items.

Step No. 6 – Solar Power Contract Negotiation

Owners must ensure that contract negotiations are supported and represented by their organization members and professionals that can support the owner's interest. Issues to be taken into consideration at minimum should include

- Present and future electric utility rates and escalations
- Financial incentives
- Contractual terms and conditions
- In the case of SPPA, site lease agreement and exit strategies
- End of contract terms and preterm options
- Econometric profiling methodologies used to calculate solar power system salvage value

Step No. 7 – Owner, Consultant, and Contractor Collaboration

On large-scale solar power projects, solar power systems study, engineering design contract negotiation, and the contract award process may take several months, depending on the number of contract participants and the type of contract, whether self-financed or SPPA.

In the case of SPPA type projects, a third solar power service provider, before committing itself to the program, has to secure federal, state, and rebate incentives. In addition, the commencement of solar power projects requires engineering document plan check approval and permits. To expedite project development, in some instances solar power construction incentives and rebates can be secured by the owners.

To facilitate project execution further, owners and their consulting engineers must jointly explore and understand local permitting processes and grid connectivity issues as related to large-scale solar electrical power generation.

Step No. 8 – Long-Term Solar Power System Operation and Maintenance

In order to assume successful solar power projects, regardless of the type of contract, owners must consider full or partial participation in solar power system monitoring and maintenance.

In general, SPPA provider contracts in some instances stipulate maintenance as an optional cost during the life cycle of the contract. With a comprehensive data acquisition and monitoring system and a small staff of maintenance personnel, considerable cost savings could be realized.

The following is a summary of a typical Solar Power Purchase Agreement (SPPA) contract time line:

- Upon receiving a letter of intent from a customer, the SPPA provider must conduct a site survey and a feasibility study.
- The PPA agreement is signed by the customer.
- Design documents and rebate applications are cosigned and filed by the SPPA provider.
- The project investment process is initiated.
- The design timeline is established and approved by the client.
- The developer or solar power system integrator must complete detailed solar power system engineering and submit shop drawings to the customer's engineering consultant for review and approval.
- The permitting and construction process is initiated.
- The customer's solar power consultant conducts periodic supervision of the project.
- Upon partial or full completion, solar power system arrays are tested and commissioned.
- The SPPA provider assumes responsibility for system operation for the duration of the solar power project contractual life cycle.

Solar Power System Ownership

Solar power ownership involves the direct purchase of solar power systems and equipment maintenance during the life cycle of the operation. In order to make the project more cost effective, the solar power systems must be connected to local grid utility companies that support net metering and provide retail credit for surplus electrical energy sent to the grid. The financing of large-scale solar power systems may involve cash purchases; financing through loans, bonds, or grants; or lease-to-own options.

The following are some of the ownership options and issues that must be taken into consideration.

Cash Investment Option

- Requires high initial investment
- Has the advantage of a tax write-off for owners who have tax liability
- Can take 100% of rebate incentive plans
- Electricity cost from the solar power system could be a significant advantage in reducing or totally eliminating peak power penalties or offseting peak power utility tariffs
- There are no monthly payments for the use of generated solar power electricity
- There are no financial balloon payments
- There is no final purchase payment
- Of invested capital 100% belongs to the owner
- Property value is appreciated to the full extent of the investment
- Owner assumes full system maintenance responsibility, which may also be subcontracted to an outside source
- Of the clean power attribute (Solar Renewable Energy Certificate value) 100% is owned

Debt or Loan Option

- Initial payment is low or regular
- Owners can take advantage of tax write-off on interest payment and apply ownership accelerated depreciation
- Electricity cost from the solar power system could be a significant advantage in reducing or totally eliminating peak power penalties or offsetting peak power utility tariffs
- There are no monthly payments for the use of generated solar power electricity
- There are no financial balloon payments
- There is no final purchase payment
- Monthly loan payments are known and fixed
- Balloon payments could be negotiated
- Allowance for ability to prepay debt
- Final debt payment could be scheduled
- Of invested capital 100% belongs to the owner
- Property value is appreciated to the full extent of the investment
- Owner assumes full system maintenance responsibility, which may also be subcontracted to an outside source
- Of clean power attribute (Solar Renewable Energy Certificate value) 100% is owned

Operating Lease Option

- Initial payment is considered regular
- Owner can write off the entire payment
- Owner cannot use depreciation
- All tax incentives are passed on to the lessor
- Electricity cost from the solar power system could be a significant advantage in reducing or totally eliminating peak power penalties or offsetting peak power utility tariffs
- Monthly lease payment to the owner is negotiable
- Prepayment penalties could be negotiated
- Final purchase payment may involve the following scenarios

  1. No prepayment
  2. Option to buy or return the system to the lessor
  3. Final purchase price defined as "fair market value" determined by the Internal Revenue Service (IRS); no allowance for a $1.00 purchase buy-down

- Capital appreciation can only apply after purchase at the end of the lease, which may not be a financially viable option
- System owners must contract system maintenance and incur monthly costs
- Clean power attributes are usually owned by the system owner. As such, the customer does not own the GREEN value of the Solar Renewable Energy Certificate (SRECs) until the system is purchased.

Suggestions Regarding Purchase Options

The following are several suggestions that should be taken into consideration when planning to purchase a lease-option to make a solar power system investment.

Step No. 1 – Conduct thorough research of a solar power project's financial profile. Always commence solar power system projects with an engineering/ financial feasibility study, as discussed previously.

Step No. 2 – Evaluate the energy efficiency measures required to reduce electrical energy consumption and its associated costs.

Step No. 3 – From the solar power feasibility report, identify the best solar power platforms and locations that could yield the largest amount of electrical energy.

Step No. 4 – Evaluate the estimated cost of the solar power investment as reflected in the feasibility study report. Feasibility study reports, as discussed earlier, should provide a relatively accurate estimate of the system installed cost per watt, which will include federal and state rebates, incentives, and depreciations.

Step No. 5 – Upon completion of solar power system preliminary design and system specification, seek proposals from several well-established solar power contractors based on the selection criteria discussed earlier. It is recommended that solar power contractors carry a certificate from the National American Board of Certified Energy Practitioners (NABCEP).

Step No. 6 – The engineering system design documentation and specifications must indicate a "turnkey" installation, whereby the installers must assume total responsibility for providing detailed engineering design documents, shopping drawings, obtaining all needed permits, and in some instances accepting the incentive payments on the customer's behalf.

Step No. 7 – Ensure that the installer will at all times coordinate solar power system installation and integration with the owner's engineering consultant.

Step No. 8 – Ensure that owner's maintenance personnel are merged and involved during all phases of solar power system engineering design and construction and that they fully take part in system test and acceptance procedures.

Advantages and Responsibilities of Ownership

The following are advantages and responsibilities of solar power systems:

- Building value increases in direct proportion to the solar power system investment.
- Solar-generated electricity adds a special multiplier to the building value since it offers a considerable hedge against escalating energy costs.
- It provides owners with the ability to use the system to meet environmental policy goals and requirements.
- It provides owners with perpetually known energy costs, which are derived from a free solar feedstock.
- Financing is simple and straightforward and does not involve complexities.
- To take advantage of peak power production, owners must properly maintain the system.

- A comprehensive monitoring system must be deployed to provide a precise overview of solar power system performance.
- Owners must sign a contract with critical solar power component manufacturers, such as manufacturers of inverters, and purchase extended warranties.
- Owners are advised to hire a broker to sell their SRECs for additional income.

Finally, it should be noted that the life span of existing mono-silicon solar PV modules can readily extend beyond 30 years. Solar power system installations deployed 40 to 50 years ago, though somewhat derated, are functioning properly. Therefore, self-financing must always be given priority consideration.

A detailed financial analysis of a large-scale solar power system, discussed further, exemplifies the validity of solar power investment.

Lease-to-Own Agreement

In lease-to-own agreement contracts, the host customer generally pays little or none of the system cost at the outset of the contract and then makes fixed payments over a period to the lessor. Essentially, in a lease agreement the owner does not purchase the system at the outset of the project. However, the owner does so over time by making regular payments to the lessor.

Unlike the Solar Power Purchase Agreement (SPPA), under which one buys the power that solar power produces, in a lease financing agreement, the payments are fixed and do not fluctuate with the amount of energy produced and used.

It should be noted that local and state government entities in general have access to special financing programs and have resources for solar electricity production that make ownership much easier. It is recommended that municipal and local government entities read a publication by the National Renewable Energy Laboratories (NREL) entitled *Solar Photovoltaic Financing: Deployment on Public Property by Local and State Governments.* The document provides a detailed analysis and funding resources for financing SPPA.

It should also be noted that only an SPPA contract can specify how much solar power generated electricity will cost. With an SPPA contract, users only buy the electricity that the system produces. In this type of contract, the solar power system owners are usually subject to production fluctuations that may result from climatic and maintenance conditions. Under any PPA type agreement, customers are not subject to ownership risks.

Green Energy Pricing Program

Under various state utility commission programs, electrical energy service providers and utilities are allowed to charge a small premium on their energy bill in order to buy a portion of their electricity from green sources.

Within the United States, 800 utilities have green energy programs that add green energy premiums to their tariffs. Premiums paid to the utilities go toward the purchase of installations of solar or other types of green energy that may be generated by wind turbines or geothermal systems. It should be noted that all rebates and incentives provided for solar power projects are a result of the green energy premiums that are passed on to all electrical power users.

Advantages of SPPA

SPPA financing for various organizations can be advantageous under the following conditions:

- The organization has limited financing options.
- The organization uses a lot of energy during peak power hours (as with hospitals, water treatment plants, multiple shift manufacturing plants, etc.).
- The organization prefers not to make a large investment and opts to pay for solar electricity through the normal operating budget.

In short, solar power electricity is a viable choice for organizations whether they wish to own or buy it from a utility company. By opting for an SPPA contract, electrical power users are guaranteed that they can purchase electrical energy without a major capital investment and associated maintenance costs, which will have a predicted cost escalation for a period of 20 to 25 years.

On the other hand, if an organization can secure system financing through cash, debt, a bond, or lease option-to-buy financing, system ownership can offer superior benefits.

Feed-in Tariffs

A feed-in tariff (FIT) is referred to as governmental or municipal financial assistance and is offered to promote renewable energy tariffs or payments. The main objective of feed-in tariff policies instituted by many countries worldwide is to accelerate private investment and financing in renewable energy technologies. In general, special financial assistance in the form of tax incentives or low-interest rates is offered to long-term renewable energy producers. In general, favorable investment incentives are based on the cost of energy generation and the type of technology. For instance, wind power energy generation is awarded a lower per kilowatt hour (kWh) price than solar photovoltaic generated power, which costs more to produce.

As a rule, worldwide feed-in tariffs often include what is referred to as tariff digression, a mechanism by which the set price of energy per kilowatt is lowered over a period of several years. Tariff regression is a mechanism devised to encourage technological cost reductions. The principal purpose of feed-in tariffs is to compensate and encourage renewable energy producers by providing them with price certainty and financial assistance for long-term renewable energy investments.

Under a feed-in tariff, renewable electricity producers, which may include residential, business, farmers, as well as private power generation projects, and owners are paid special tariffs for producing renewable electricity that can be connected to the electrical grid. Such measures promote generation of diverse technologies such as wind, solar, and biogas by private customers by providing a reasonable return on their investments.

As of 2011, feed-in tariff policies had been adopted and were active in Algeria, Australia, Austria, Belgium, Brazil, Canada, China, Cyprus, the Czech Republic, Denmark, Estonia, France, Germany, Greece, Hungary, Iran, the Republic of Ireland, Israel, Italy, Kenya, the Republic of Korea, Lithuania, Luxembourg, the Netherlands, Portugal, South Africa, Spain, Switzerland, Tanzania, Thailand, and Turkey.

Origins of Feed-in Tariff

The first feed-in tariff policy was implemented in the United States in 1978 under President Jimmy Carter, who endorsed the National Energy Act (NEA). The act consisted of five separate segments, one of which was the Public Utility Regulatory Policies Act, better known as PURPA. The purpose of the NEA was to promote energy conservation and to develop new renewable or alternative energy resources, such as wind, solar, or geothermal power.

The NEA encompassed a provision that required utilities to purchase electricity generated from independent power producers at rates that did not exceed their avoided cost. The so-called avoided costs consisted of expenses that a utility company would incur to provide that same electrical generation. Another provision included in the law was that utilities were not permitted to own more than 50% of projects. This clause was introduced to encourage private or public investors to enter the electricity generation industry.

On the basis of those laws, the state of California established the Public Utility Commission, which developed special Standard Offer Contracts, including Standard Offer No. 4 (SO4). No. 4 (SO4) makes the use of the use of fixed prices that took into consideration the expected long-run cost of power generation. At the time of the law's enactment, it was believed that in the long run, the cost of electricity would continue to increase as a result of oil and gas prices escalation (this holds true today). On the basis of the assumption, CPU established an escalating schedule of fixed purchase prices, which was designed to project avoided costs of long-term electrical energy generation. As a result of the law in the early 1990s, a number of power producers invested private funds and installed 1,700 MW of wind capacity, mostly in California's Coachella Valley, which is still in service today. In the same period, Florida and Maine adopted a similar law, which also led to significant deployment of renewable energy power generation. The NEA was implemented to encourage nonutility power generation. Ever since adoption of the law there has been a significant lobby of opposition from utility companies. This has unfortunately hindered the advancement of solar photovoltaic power generation in the United States. However, as of the mid-1990s, as a result of the encouragement of nonutility generation, most public utility commissions throughout the United States had enacted special state and municipal measures toward increasing competition in the U.S. electricity industry.

Effects of Feed-In Tariff on Electricity Rates

It is a well-known fact that in countries that have promoted feed-in tariff laws, grid tied electrical power under a feed-in tariff has been shown to add a small annual increase in the price of electricity per customer. This is a result of the fact that electricity generated from renewable energy sources is typically somewhat more expensive than electricity generated from sources such as coal or nuclear power. However, in some instances it has been shown that the incremental cost increase in the electrical tariff can be offset by what is called the price-dampening effect. This involves the composite pricing of large amounts of lower-cost renewable energy sources such as wind and solar photovoltaic power that can lower spot market renewable energy prices. Renewable energy price dumping is currently prevalent in Spain, Denmark, and Germany. The lowering of the renewable energy cost by price dumping is contingent upon the extent of the overall mix of renewable energy sources.

Grid Parity

In the recent past, rapid deployment of renewable energy under feed-in tariffs in European countries such as Germany, Denmark, and Spain has resulted in the reduction of renewable energy costs. It has also resulted in an accelerated expansion of technologies, a trend that has been lacking in the United States.

The large-scale deployment of renewable energy technology results in an advance toward grid parity. Achieving grid parity pricing, because of numerous varying factors such as the amount of power generation during day- and nighttime, climatic variations over the course of years, and climatic and geographic conditions, at best becomes a moving target. As a result, the price of electricity generated from renewable energy sources in the United States varies significantly from high cost in states like Hawaii and California, to lower cost in states such as Wyoming and Idaho. Likewise, in islands (where the prime sources of electrical power generation depend on diesel generators) the cost of renewable electrical power generation can compete favorably with that of grid electrical power, with time-of-use pricing during the peak power demand hours of 11 am to 8 pm. "Time-of-use" refers to the electrical energy rate charged during special hours of the day. For instance, under time-of-use pricing the electrical energy costs more between the hours of 11 am and 5 pm, referred to as "peak power energy use."

Nowadays, grid parity is being achieved in certain states as a result of higher-efficiency photovoltaic solar power systems and tracking and concentrator type technologies.

Quota-Based Renewable Energy Policies

In general certain renewable energy power generation policies, such as quota-based systems (also referred to as Renewable Portfolio Standards [RPS]) and subsidies, result in limitations for renewable energy. In such subsidized energy production environments, the supply of renewable energy is achieved by mandating renewable energy generation providers to deliver a portion of their electricity from renewable energy sources to their customers. Upon meeting their mandate, power generation companies are allowed to collect green electricity certificates, which are a tradable financial instrument. According to the policy concept, the green electricity certificates are considered to promote lowering prices paid to renewable energy developers. Such policies make an assumption that market forces will eventually result in competition and lowering of renewable energy prices. In such a scenario, no single buyer or seller will have sufficiently large market share to influence renewable energy prices significantly. However, the fundamental problems with the quota system thus far have been that there are no assurances that in the long term there is nothing to prevent the renewable energy producers from becoming noncompetitive with companies that produce electrical power from coal fired power plants, which may result in demise of their businesses. A basic problem could be a shortage of basic material used in production of solar photovoltaic cells, which may include silicon crystal and rare earth metals.

As a result there is great reluctance among investors to be involved with quota-based policies. In general, investors who favor quota policies are multinational electric utilities that have financial resources to deploy large-scale solar power systems and have the capability to produce integrated electrical power from solar, wind, and conventional electrical power generation facilities.

Feed-In Tariffs across the World

The following prevalent feed-in tariffs data can be accessed through the Internet. The information provided is intended to allow the reader a prompt overview of varying feed-in tariffs and policies that are currently promoted by fifty countries across the world. The equivalent values of feed-in tariffs in various denomination have been shown in native currencies.

Algeria

To cover the additional costs of producing electricity from renewables and the costs of diversification, producers of electricity from renewables will get a bonus for each kilowatt hour produced, marketed, or consumed. For electricity generated from solar or radiant heat only, the bonus is 300% of the price per kWh of electricity produced by the market operator defined by law, when the minimum contribution of solar energy represents 25% of all primary energy. For electricity generated from facilities using solar thermal solar-gas hybrid systems, the bonus is 200% of the price per kWh.

For contributions of solar energy below 25%, the said bonus is paid in the following conditions:

Solar Share	Percent of Bonus
25%	200%
20% to 25%	1805%
15% to 20%	160%
10% to 15%	140%
5% to 10%	100%
0% to 5%	0

Furthermore, the feed-in tariff provides bonuses for electricity generated by cogeneration. The bonus will be 160% with thermal energy use of 20% of all primary energy used. The bonuses for solar generated electricity and cogeneration are cumulative. Remuneration of the generated electricity will be guaranteed over the whole plant lifetime.

Australia

Feed-in tariffs were introduced by the states, in 2008 in South Australia and Queensland, 2009 in the Australian Capital Territory and Victoria, and 2010 in New

South Wales, Tasmania, and Western Australia. In the Northern Territory there are only local feed-in tariff schemes. A uniform federal scheme to supersede all state schemes has been proposed by the Tasmanian Greens Senator Christine Milne, but not enacted.

Canada

Ontario introduced a feed-in tariff in 2006 and revised it in 2009; a draft proposal increases the tariff from 42¢/kWh to 80.2¢/kWh for microscale (≤10 kW) grid-tied photovoltaic projects. Ontario's FIT program also includes a tariff schedule for larger projects up to and including 10-MW solar farms at a reduced rate. As of April 2010, several hundred projects had been approved, including 184 large-scale projects, worth $8 billion altogether.

China

China has set a fixed feed-in tariff for new onshore wind power plants in a move that will help struggling project operators realize profits. The National Development and Reform Commission (NDRC), the country's economic planning agency, announced at the weekend four categories of onshore wind projects. According to the region, these projects will be able to apply for the tariffs. Areas with better wind resources will have lower feed-in tariffs, while those with lower outputs will be able to access more generous tariffs.

The tariffs per kilowatt hour are set at 0.51 yuan (US 0.075, GBP 0.05), 0.54 yuan, 0.58 yuan, and 0.61 yuan. These represent a significant premium on the average rate of 0.34 yuan per kilowatt hour paid to coal-fired electricity generators.

Czech Republic

The Czech Republic introduced a feed-in tariff by Act of Law No. 180/2005 for a wide range of renewable sources in 2005. The tariff is guaranteed for 15–30 years (depending on the source). Supported sources are small hydropower (up to 10 MW), biomass, biogas, wind, and photovoltaics. As of 2010, the highest tariff was 12.25 CZK/kWh for small photovoltaic.

Germany

First introduced in 2000, the Erneuerbare-Energien-Gesetz (EEG) law is reviewed on a regular basis and the 2010 version is currently in force. Its predecessor was the 1991 Stromeinspeisegesetz. As of May 2008, the cost of the program added about €1.01 (USD1.69) to each monthly residential electric bill. In 2011 the additional costs were expected to rise to €0.035/kWh. Feed-in tariff rates vary depending on the size and locations of the systems and are shown in the table. Since 2009, there have been additional tariffs for electricity immediately consumed rather than supplied to the grid with increasing returns if more than 30% of overall production is consumed on site. This is to incentivize demand side management and to help develop solutions to the intermittency of solar power.

Peak Power Dependent FIT in cents/kWh

Type		2004	2005	2006	2007	2008	2009	2010	Jul 2010	Oct 2010	2011
Rooftop mounted	Up to 30 kW	57.4	54.53	51.80	49.21	46.75	43.01	39.14	34.05	33.03	28.74
	Between 30 kW and 100 kW	54.6	51.87	49.28	46.82	44.48	40.91	37.23	32.39	31.42	27.34
	Above 100 kW	54.0	51.30	48.74	46.30	43.99	39.58	35.23	30.65	29.73	25.87
	Above 1000 kW	54.0	51.30	48.74	46.30	43.99	33.00	29.37	25.55	24.79	21.57
Ground mounted	Contaminated grounds	45.7	43.4	40.6	37.96	35.49	31.94	28.43	26.16	25.37	22.07
	Agricultural fields	45.7	43.4	40.6	37.96	35.49	31.94	28.43	–	–	–
	Other	45.7	43.4	40.6	37.96	35.49	31.94	28.43	25.02	24.26	21.11

India

India inaugurated its most ambitious solar power program to date on January 9, 2010. The Jawaharlal Nehru National Solar Mission (JNNSM) was officially announced by the prime minister of India on January 12, 2010. This program aims to install 20,000 MW of solar power by 2022. The first phase of this program aims to install 1000 MW by paying a tariff fixed by the Central Electricity Regulatory Commission (CERC) of India. While in spirit this is a feed-in tariff, there are several conditions on project size and commissioning date. The tariff for solar PV projects is fixed at Rs. 17.90 (USD 0.397/kWh). The tariff for solar thermal projects is fixed at Rs. 15.40 (USD 0.342/kWh). The tariff will be reviewed periodically by the CERC.

Iran

SANA introduced FIT in 2008 for purchasing renewable energy from investors. A price of 1300 Rials/kWh was set for renewable electricity. For 4 hours in the middle of the night, the price is 900 Rials.

Republic of Ireland

From 2006 to 2011 the sitting ministers for communications energy and natural resources, Noel Dempsey and Eamon Ryan, at various times announced feed-in tariffs for renewable electricity in Ireland. However, they have never actually established the tariffs. This has seriously hindered the growth of renewable electricity in Ireland. In 2011 The Renewable Energy Federal Incentive Tax (REFIT) was being negotiated between DCENR and the EU, and the industry was hopeful that REFIT would be available. However, much of the investment funds was lost to oversea projects projects at this stage. Industry associations (such as the Irish Bio-Energy Association) are strongly lobbying government and government departments to make feed-in tariffs available to the renewable energy sector.

Israel

On June 2, 2008, the Israeli Public Utility Authority approved a feed-in tariff for solar plants. The tariff is limited to a total installation of 50 MW during 7 years, whichever is reached first, with a maximum of 15 kW installation for residential and 50 kW for commercial. Bank Hapoalim offered 10-year loans for the installation of solar panels. The National Infrastructures Ministry announced that it would expand the feed-in tariff scheme to include medium-sized solar power stations ranging from 50 kilowatts to 5 megawatts. The new tariff scheme caused the solar company Sunday Solar Energy to announce that it would invest $133 million to install photovoltaic solar arrays on kibbutzim, which are social communities that divide revenues among their members.

South Africa

South Africa's National Energy Regulator (NERSA) announced 31 March 2009 the introduction of a system of feed-in tariffs designed to produce 10 TWh of electricity per year by 2013. The feed-in tariffs announced were substantially higher than those in NERSA's original proposal. The tariffs, differentiated by technology, will be paid for a period of 20 years.

NERSA said in its release that the tariffs were based, as in most European countries, on the cost of generation plus a reasonable profit. The tariffs for wind energy and concentrating solar power are among the most attractive worldwide.

The tariff for wind energy, 1.25 ZAR/kWh (€0.104/kWh, $0.14 USD/kWh, $0.17 CAD/kWh) is greater than that offered in Germany (€0.092/kWh), and more than that proposed in Ontario, Canada ($0.135 CAD/kWh).

The tariff for concentrating solar, 2.10 ZAR/kWh (€0.175/kWh), is less than that in Spain (€0.278/kWh), but offers great promise in the bright sunlight of South Africa. NERSA's revised program followed extensive public consultation.

Stefan Gsänger, Secretary General of the World Wind Energy Association, said in a release that "South Africa is the first African country to introduce a feed-in tariff for wind energy. Many small and big investors will now be able to contribute to the take-off of the wind industry in the country. Such decentralised investment will enable South Africa to overcome its current energy crisis. It will also help many South African communities to invest in wind farms and generate electricity, new jobs and new income. We are especially pleased as this decision comes shortly after the first North American feed-in law has been proposed by the Government of the Canadian Province of Ontario.

Spain

The current Spanish feed-in legislation are Royal Decree 1578/2008 (Real Decreto 1578/2008) for photovoltaic installations and Royal Decree 661/2007 for other renewable technologies injecting electricity into the public grid. Originally under 661/2007, photovoltaic feed-in tariffs have recently (September 2008) been developed under a separate specific law frame because of the rapid growth experienced by this technology since the release of the original scheme.

The current Photovoltaic Decree 1578/2008 categorizes installations in two main groups with differentiated tariffs:

I) Building Integrated installations: with 34 c€/kWh in systems up to 20 kW of nominal power, and for systems above 20 kW with a limit of nominal power of 2 MW tariff of 31 c€/kWh.

II) Non integrated installations: 32 c€/kWh for systems up to 10 MW of nominal power.

For other technologies, Decree 661/2007 sets up the following:

Switzerland

Energy Source	Feed-in Tariff
Cogeneration systems	Maximum FIT of 13.29 c€/kWh during lifetime of system
Solar thermoelectric	26.94 c€/kWh for the first 25 years
Wind systems	Up to 7.32 c€/kWh for the first 20 years
Geothermal, wave, tidal, and sea-thermal	6.89 c€/kWh for the first 20 years
Hydroelectric	7.8 c€/kWh for the first 25 years
Biomass and biogas	Up to 13.06 c€/kWh for the first 15 years
Waste combustion	Up to 12.57 c€/kWh for the first 15 years

Switzerland introduced the so-called cost-covering remuneration for feed-in to the electricity grid. The Cost Recovery Factor (CRF) applies to hydropower (up to 10 megawatts), photovoltaics, wind energy, geothermal energy, biomass, and waste material from biomass. It will be applicable for a period of between 20 and 25 years, depending on the technology. The implementation is done through the national grid operator SWISSGRID.

Thailand

In 2006, the Thai government enacted a feed-in tariff that provides an adder (incentive) paid on top of utility avoided costs (differentiated by technology type and generator size) and guaranteed for 7–10 years. Solar receives the highest, 8 baht/kWh (about U.S. cents 27/kWh). Large biomass projects receive the lowest at 0.3 baht/kWh (at about 1 U.S. cent per kWh). Additional per-kWh subsidies are provided for projects that offset diesel use in remote areas. Under the FIT program, as of March 2010, 1364 MW of private sector Renewable Energy (RE) was online with an additional 4104 MW in the pipeline with signed PPAs. Biomass makes up the bulk of this capacity: 1292 MW (online) and 2119 MW (PPA only). Solar electricity is second but rapidly catching up, with 78 MW online and signed PPAs for an additional 1759 MW.

Ukraine

Ukraine introduced the On Feed-In Tariff law on September 25, 2008. The law guarantees grid access for renewable energy producers (small hydro up to 10 MW, wind, biomass, photovoltaic, and geothermal). The feed-in tariffs for renewable power producers are set by the national regulator. As of July 2011, the following tariffs per kWh were applied: biomass – UAH 1.3446 (EUR 0.13), wind – UAH 1.23 (EUR 0.12), small hydro – UAH 0.8418 (EUR 0.08), solar – UAH 5.0509 (EUR 0.46). In case of significant fluctuations of the national currency against the Euro, the feed-in tariff is adjusted to reflect the changes.

United Kingdom

Feed-in tariffs in the United Kingdom were first announced in October 2008 by the UK secretary of state for energy and climate change. It was announced that Britain would implement a scheme by 2010, in addition to its current renewable energy quota scheme (ROCS). In July 2009, the secretary presented details of the scheme, which began in early April 2010.

In March 2011, less than a year into the scheme, the new coalition government announced that support for large-scale photovoltaic installations (greater than 50 kW) would be cut. This was in response to European speculators' lining up to establish huge solar farms in the West Country, which would have absorbed disproportionate amounts of the fund.

On June 9, 2011, the tariffs would be cut for solar PV systems above 50 Kw after August 1, 2011. Many were disappointed with the decision of the DECC, especially after long discussion with public. Still some have a positive opinion about the new feed-in tariffs, from the report published by PVinsights. It is believed that the total subsidies for the solar PV industry are not changed, but the UK government tries to trim the tariffs for large systems in order to benefit smaller systems. Since the fast track review is based on the long-term plan to reach an annual installation of 1.9G W in 2020, There is no doubt that the UK market will still be a growing market in coming years.

United States

On April 2011, U.S. state legislatures were considering adopting a FIT as a complement to their renewable electricity mandates.

"While feed-in tariffs are most closely associated with solar photovoltaic panels, utilities managing the programs in Vermont and California will also pay a set price for electricity generated from other renewable sources, like wind."

California's Sacramento Municipal Utility District (SMUD) program is open to homeowners who are not participating in another program (called net metering) that allows people whose system is producing more electricity than they need to sell the excess back to the utility, thus reducing their electric bill. But once their bill falls to zero, the homeowners receive no more money from the system.

As long as they are not part of the net-metering program and not seeking the $1.90- to $2.20-per-watt ratepayer subsidy for their new panels under the state's "Million Solar Roofs" program, small generators can sell their power to SMUD. The rates would depend on the time of day the power is generated, ranging from a low of 5 or 6 cents a kilowatt hour to 30 cents on a hot summer afternoon; the size of eligible systems is capped at 5 megawatts (and the program overall has a 100-megawatt cap).

The Vermont law caps the size of individual systems at 2.2 megawatts. Solar energy fetches a fixed price of 30 cents a kilowatt-hour, and other forms of renewables fetch lower rates."

California

The California Public Utilities Commission (CPUC) approved a feed-in tariff on January 31, 2008, effective immediately. As of January 1, 2010, state laws have allowed homeowners who produce more than they use over the course of the year to opt in to be able to sell the excess power to the utility at a currently undisclosed rate (the rate decision had a January 1, 2011 deadline). Previously, the homeowner would get no credit for overproduction over the course of the year. In order to get the California Solar Initiative (CSI) rebate the customer is not allowed to install a system that inherently overproduces, thereby encouraging efficiency measures to be installed after solar installation. This overproduction credit is not available to certain municipal utility customers, namely, Los Angeles Water and Power.

Hawaii

In September 2009, the Hawaii Public Utilities Commission required Hawaiian Electric Company (HECO & MECO & HELCO) to pay above-market prices for renewable energy fed into the electric grid. The policy offers the project a set price and standard 20-year contract. The PUC plans to review the initial feed-in tariff 2 years after the program starts and every 3 years thereafter. Table 6.3 represents CSI solar power rebate incentive step levels and production targets for residential, commercial, and government.

7 Importance of Solar Power System Peak Power Performance and Solar Power System Hazard Mitigation

Introduction

This chapter covers the significance of power output performance of all types of solar power system installations. Regardless of type of technology, whether solar photovoltaic power, concentrator type solar power systems, or solar thermal technologies, system upkeep and maintenance are of great importance. To take advantage of maximum return on investment, solar power systems, though quite robust and reliable from a physical and operation point of view, in order to harvest maximum amount of energy production, must at all times be kept at peak performance. This chapter discusses the significance of upkeep of solar power systems.

Effects of Dust on Solar Power System Energy Output Performance

As discussed in Chapter 1, solar power output performance is directly proportional to the amount of solar irradiance that impacts the photovoltaic modules. Therefore, any obstacle that blocks or refracts the solar rays will prevent photons from being absorbed and converted into electrical current. Most common of all such obstacles are airborne particulates such as dust and water particulate in the air that gradually settle on the surface of PV panels.

In urban areas in addition to dust and moisture, oil particulates emitted from vehicle exhausts and ash from chimneys and smokestacks can affect power output performance of solar power systems. Specifically in desert areas, solar power systems frequently are subjected to dust storms, which can in a short span of time considerably reduce power output performance of solar photovoltaic and solar thermal electrical energy output.

The effect of power output reduction resulting from precipitants is also dependent on the tilt angle of the PV system installation as well as the type of support platforms. The larger the tilt angle, the lesser the particulate precipitation. For instance, a flat mounted PV platform at 5-degree tilt will accumulate dust somewhat more rapidly than one that is mounted at a 25-degree angle.

Another factor that affects solar power energy production depends on the solar PV module frame assembly and the clearance band of the solar cells from the frame edges. In general under rainy conditions the accumulated dust particles on the surface of PV modules are washed downward and gradually accumulate around the edges of

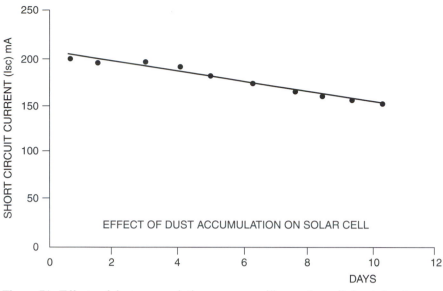

Figure 7.1. Effects of dust accumulation on mono-silicon solar cell short circuit current Isc.

the frame. In some instances when solar panels are not washed periodically, gradual accumulation of dust called edge shading can completely block small segments of lower cells, resulting in total blockage of electrical current flow. Since solar cells within PV modules are connected in series, blockage of a single cell can render a solar module totally dysfunctional. In turn, since solar strings are configured from tandem interconnection of series of PV modules, serious edge shading, depending on the location of a single panel, can obstruct current flow through the entire string.

Experiments conducted on solar cell dusting reported in the *Journal of Basic and Applied Scientific Research* (www.textroad.com by A. Ibrahim, 2010) indicate that solar photovoltaic modules open circuit voltage (Voc), short circuit current (Isc), and therefore can substantially degrade power output by dust and grime accumulation. Figure 7.1 is a graph that shows the effects of dust accumulation on mono-silicon solar cell short circuit current Isc. Figure 7.2 shows the effect of dust accumulation on mono-silicon cell open circuit voltage (Voc). Figure 7.3 depicts dust accumulation effects on power output.

The conclusion of the experiment indicates that accumulation of dust in a short span of time can result in a reduction of power output performance of a solar cell by as much as 40%. Similar studies cited in the report indicated that dust accumulations on glass plates with different tilt angles indicate that the measures of transmittance of PV plates under different climatic conditions over a period of 1 month can cause degradation by an amount ranging from 17% to 64% for tilt angles from 0 to 60 degrees. The formula that follows is used to calculate the Isc current degradation; it is based on a silicon solar cell of 100 cm^2 that produces 211.85 milliampere (mA) of current when completely clean (at the start of the test) and measures loss of -5.5612 mA per day:

$$Isc = -5.5612t + 211.85$$

Likewise, the Voc degradation formula used in measuring the effect of dust accumulation on voltage output is based on cell voltage output of 2.2 volts under

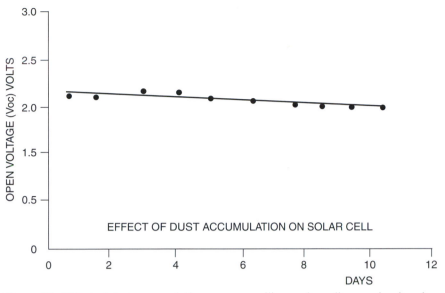

Figure 7.2. Effect of dust accumulation on mono-silicon solar cell open circuit voltage Voc.

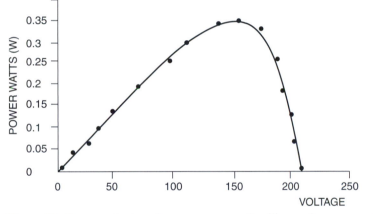

Figure 7.3. Power output performance curve of a silicon solar power cell showing power output in watts versus millivolts.

clean conditions (start of test) and measured daily degradation of – 0.0192 volt of daily voltage drop.

$$\text{Voc} = -0.0192t + 2.2$$

In both of the formulas, t stands for test per daytime insolation.

From the preceding experiment it has been discovered that under deposited dust conditions solar cell power degradation in a short period can result in significant loss of power output Therefore, in order to maintain peak power output performance, solar power systems must be kept clean throughout their project life cycle.

It should also be kept in mind that dust accumulation can be a serious impediment in maintaining full functionality of large-scale solar power installations and if not attended to properly can have considerable negative impacts on the return on investment (ROI) of a project.

Figure 7.4. Example of edge shading of solar power system installation.
Source: Photo courtesy of Vector Delta Design Group, Inc.

At a median cost of \$0.18/kWh present value and an electrical energy cost escalation of 10%, the median cost of electrical energy computed with simple linear extrapolation yields about \$0.8/kWh. Loss of income or investment due to dusting may therefore yield an accumulated loss of \$ 4,150,000.00. Net present value calculated over half of the project life cycle (median of 12.5 years), when using the present value calculation formula, $PV = FV/(1 + i)n$, results in the following:

FV (future values) = \$ 4,150,000.00 I (interest rate) = 4%

From the preceding computation, it is obvious that maintaining solar power at peak performance can help prevent substantial losses. At this juncture it is of great importance to realize that maintenance of the 5000 PV modules in a 1-megawatt system can pose a significant challenge. Yet of greater concern is the maintenance of 20, 50, or larger photovoltaic systems that will involve deployment of many thousands of PV modules. Such a deployment will impose a significant challenge to maintain peak power performance. Figure 7.4 is a photograph of edge shading that resulted from PV module soil accumulation.

Operating such a vast system at peak power in a cost efficient manner requires the use of a granular data acquisition system that can monitor and report power output performance of every module of the solar power system, however large. Such a system must deploy a microprocessor-based data acquisition system that is an integral part of each solar photovoltaic panel and will allow real time measurement of power output performance parameters such as current and voltage. This data should be transmitted to a central data monitoring system by means of wireless communications technology.

Another research project conducted by the National Renewable Energy Laboratories reflected in a conference paper titled "Partial Shading Operation of a Grid-Tied PV System" (NREL/CP-520–46001, published in June 2009) concludes that partial shading of a PV installation has a disproportionate impact on its power production. The paper presents the background and experimental results from a single string grid-tied PV system, operated under a variety of shading conditions.

The shading study presented in the paper applies not only to shading of solar photovoltaic systems by buildings, trees, and other obstacles that can cast a shadow on PV modules, but also to dirt and dust accumulation, which presents substantially more impairment to power output production. The report underlines that a partial shadows on a PV module can represent a reduction in power more than 30 times its physical size. The full report can be accessed at the NREL Web site www.nrel.gov/docs/fy09osti/46001.pdf.

> There is continued interest among photovoltaic installers, regulators and owners to obtain accurate information on photovoltaic (PV) systems operating under shaded or mismatched conditions. It is best to avoid shade where possible; while partially shaded installations can still produce useful power for a portion of the day, shading will generally result in a significant reduction in power. This is particularly true for building-integrated PV which often requires the integration of modules with existing structures in sometimes crowded urban environments. In the interest of expanding the number of PV installations worldwide and providing maximum benefit from these systems, it is useful to consider in more detail the power loss from partially shaded PV systems.

> Regulatory agencies such as the California Energy Commission have a particular interest in obtaining information on shaded PV operation in order to accurately calculate rebate incentives for shaded systems. For incentive programs focused on installed capacity rather than actual kWh production, modeling the PV system and its access to the solar resource is required to obtain the expected performance of the system. Because shading of PV systems disproportionately reduces their power output, the rebate incentives need to be adjusted accordingly.

> Prior experiments have been conducted investigating the effect of shade on various PV systems, many of which were cited in a comprehensive literature review by Woyte et al. Other recent works include simulations of partially shaded PV cells, experimental results of different maximum power point tracking algorithms under shaded conditions and the effect of shade on PV system performance.

Module Shading Effect

Typically, a crystalline silicon module will contain bypass diodes to prevent damage from reverse bias on partially shaded cells. These diodes are placed across 12–18 cells in what will be termed a "group" of cells here. The bypass diode across this group of cells will begin conducting before the power dissipated into the shaded cell is enough to involve damaging temperatures. While the main purpose of a bypass diode is not necessarily to improve module performance under shaded conditions, this is a useful by-product. The bypass diode allows current from nonshaded parts of the module to pass by the shaded part and limits the effect of shading to only the neighboring group of cells protected by the same bypass diode. When a bypass diode begins conducting, the module voltage will drop by an amount corresponding to the sum of cell voltages protected by the bypass diode plus the diode forward voltage, but current from surrounding unshaded groups of cells continues around the group of shaded cells. In Figure 7.5, N modules are shown in series, each consisting of three groups of 18 cells. Each group of cells is connected to a bypass diode that begins conducting if shading causes a cell to go far enough into reverse bias. In this example, shade causes diode D1 to short out its group of cells, reducing module 1's voltage by one-third.

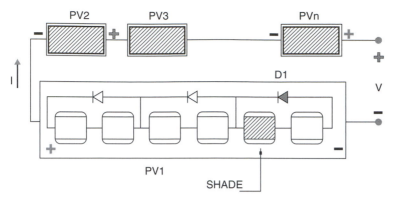

N modules are shown in series, each consisting of 3 groups of 18 cells. Each group of cells is connected to a bypass diode that begins conducting if shading causes a cell to go far enough into reverse bias. In the above example, shade causes diode D1 to short out its group of cells, reducing

Figure 7.5. Effect of partial solar PV module shading.

Figure 7.5 is a circuit diagram showing that the effect of shade on power output of typical PV installations is nonlinear in that a small amount of shade on a portion of the array can cause a large reduction in output power.

Solar Power System Physical Configuration and Economics

In order to maximize return on investment, owners and investors alike must ensure that solar power systems operate at their optimized power output performance for an extended period of 25 to 50 years. To achieve optimal energy output performance, solar power system equipment configuration and engineering design must be conducted by an expert, with appropriate care taken to protect and ensure maximum return on investment. To accomplish this, solar power system design must incorporate the following design measures.

Solar Power Systems Configuration

Engineering and system design topology must be such that it ensures the highest system operational availability. In other words, solar power systems must be designed in a fashion such that failure of a component or equipment will not result in significant deterioration of system power production. Reliable engineering design must address the following considerations.

To achieve maximum system availability, solar power design configuration must be based on modular or distributed array design, where failure of a solar string or an array would not have a significant adverse effect on the overall electrical energy output.

The solar power system must incorporate an advanced real time (near instantaneous) energy output monitoring and data acquisition system that can provide alerts for solar power system component failure or underperformance such as inverters, combiner box fuse failure, and PV module failure. During the design phase, engineering design consultants should conduct trade-off analyses about system availability and system cost and return on investment for the duration of a solar power system life cycle. This means that the engineering consultant must be proficient in both engineering design and economic analysis as related to large-scale solar power systems. For instance, choice of solar topology design may include evaluation of centralized or distributed solar power DC to AC conversion equipment (inverter) configuration, where the failure of one inverter will not result in loss of more than 5–10% of the overall solar power system power output.

Critical equipment used must be specifically selected to meet objectives of a project. For instance, inverter technologies deployed in addition to optimum efficiency must have a long track record of performance. In all instances caution should be exercised in use of devices that modify power output performance of photovoltaic modules. Such technologies can reduce overall system reliability. Examples of such devices include peak power tracking DC to DC converters, string inverters, and microinverters, which are devices that are designed to improve energy production efficiency of photovoltaic modules, yet as na result of complex electronics can reduce system reliability. In particular, use of unproven technologies that make claims of energy improvement must be evaluated carefully since their deployment as a serial component attachment to a PV module may indeed lower reliability of the overall solar power system over the expected life cycle of a solar power system. In view of evolving solar power energy output conditioning devices such as microinverters and peak power tracking devices, it should be noted that claims of 25-year warranties associated with the technology are not founded on field tested statistical data, which is usually associated with mono-silicon PV modules; therefore, deployment of the devices in series configuration may have unknown consequences with respect to system operational reliability and system maintenance. The significance of the use of micropower conversion technologies will be discussed further later in this chapter.

Solar power system design must take into consideration significant issues related to maintenance and operational cost, since the burden of upkeep can have an adverse effect on return on investment over the life cycle of a solar power system.

Large-scale solar power systems, whether privately financed or leased, ideally have an automatic energy monitoring and computerized troubleshooting and diagnostic system that can alert system operators to rectify system performance abnormalities with specificity that will minimize troubleshooting and component or equipment replacement time.

Operational issues associated with peak power output performance include shading, edge shading, PV module reverse diode failure, and module mortality. Other factors affecting solar power output performance may include dust and dirt accumulation as well as particulate precipitation.

It should be noted that large-scale solar power systems may include thousands of photovoltaic modules, hundreds of solar strings, and numerous solar arrays and subarrays that may cover many acres of land. In such configurations, even a small percentage of solar power energy loss can translate into a significant loss of investment return, which must be prevented.

When designing large solar power systems, care must be exercised to conduct return on investment analysis with regard to system availability and system reliability. Any short-term savings with respect to the deployment of centralized power conversion devices that may result in cost savings must be evaluated in the larger context of peak power energy output performance of the overall system that would have the least amount of system downtime and best operational reliability as well as maintainability.

Solar Power System Reliability and Economics

In the context of solar power generation there is a direct nexus between return on investment and solar power system availability or reliability. Simply stated, every hour of solar power system downtime equals loss of energy production or loss of income. To illustrate further the significance of reliability, investors must have an appreciation for technical terminology such as "mean time between failures" (MTBF) and "system availability", and their correlation with the equipment warranty.

"Mean time between failure" is the time in days or years that a piece of equipment or a device is supposed to operate before it fails. In other words, MTBF is a figure of merit that indicates life expectancy of a product. In general, equipment warranties offered by manufacturers are based on statistical data related to the life span of a product.

"Mean time to repair," also referred to as MTTR, is time required to repair equipment. In general, MTTR is not a quantifiable figure since it implies availability of components, complexity of equipment design and repair, procurement time of the replacement components, skill and competency of technicians undertaking the repair, time or travel, and many other factors that may affect time required to render the equipment functional. For instance, replacement of a particular type of a fuse that requires a few minutes because of lack of availability of the specific model may involve several days for parts to be ordered and delivered. This scenario could result in several days of equipment downtime.

Another term that defines system performance is "system availability," which measures the percentage of the time a system can be relied upon to perform without failure. For instance, electrical outage of a grid system in various parts of a distribution system in critical zones can have an availability of 99.98% and in other regions on the same grid such as in a remote location have availability of 99.91%.

The relationship of MTTR and MTBF and availability is calculated with the following formula:

$$\text{Availability} = \text{MTBF} / (\text{MTBF} + \text{MTTR})$$

When configuring any system regardless of the technology, system reliability becomes subject to parallel and series deployment of various components or system equipment that is either used in series or tandem or in parallel. Figure 7.6 depicts series and parallel system configuration.

In a series configuration integrity of system availability and reliability becomes directly dependent on the least reliable equipment or device in the chain. Therefore, the resulting reliability of a chained system configuration of series combination becomes equal to the product of all reliability figures of the entire chain.

For instance, if we designate "R" as device or equipment and "a" as availability, then availability of a series system will be defined by the following equation:

$$As = R1a \times R2a \times R3a \times Rna$$

In the case of parallel system configuration, probability of availability formulation is based upon the definition of percent unavailability (Ua) and system percent outage (So) and percent system availability Ap by the following equations:

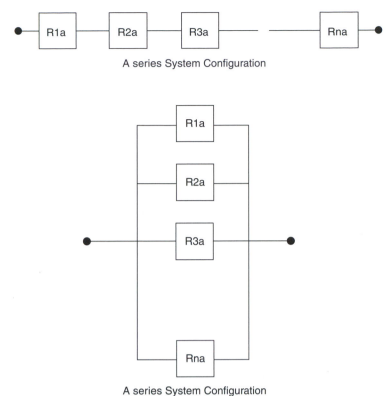

A series System Configuration

Figure 7.6. Series and parallel system configuration.

Unavailability = 1 – Availability or Ua = 1 – Ap Outage = Unavailability × Hours/ hours per year Availability = 1 – (U1a) × (U2a) × (U3a) × (Una) or Ap = 1 – (1 – R1a) × (1 – R2a) × (1 – R3a) × (Rna)

On the basis of the preceding reliability or system availability calculation it could be concluded that in order to have a more reliable solar power system, solar array configuration topologies must be designed in a distributed or parallel fashion to effect greater reliability, which translates into smaller solar array systems and a larger number of inverters. Hence, economic evaluation must take into account initial investment cost versus system operational downtime for the entire life cycle of the investment period.

Distributed solar power array systems greatly enhance maintenance and repair downtime as well as direct replacement of complete inverters, as opposed to major component replacement, which can substantially increase the MTTR time. Figure 7.6 is a block diagram of series and parallel system configuration.

Deployment of Micropower Inverters and DC to DC Peak Power Conversion Devices

In order to appreciate significant trade-offs of parallel and serial power conditioning and conversion devices it is important to understand the principal purpose and functional attributes of microinverters, string inverters, and DC to DC peak power conversion equipment and technologies that have recently been introduced into the photovoltaic solar power system industry.

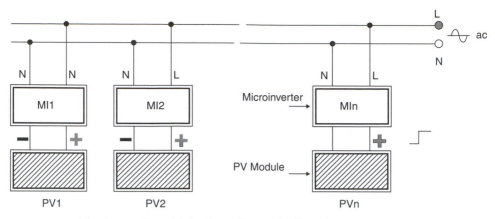

Microinverter Based Solar Power System Configuration

Figure 7.7. A microinverter solar power system diagram.

Solar Microinverters and String Inverters

Solar micro-nverters, commonly referred to as microinverters, are electronic devices that convert DC current generated from a single PV module into alternating or AC current.

As shown in Figure 7.7, above the electrical AC power generated from a series of microinverters is combined on an electrical bus bar and accumulated into AC combiners or accumulators that are sent directly to an electrical load.

Figure 7.8 shows a system configuration of a string inverter device, which is used to convert summation of DC current of a solar power string, which may consist of a series connection of 10–14 PV modules.

The main advantage of microinverters over string inverters is that they reduce losses associated with shading or malfunction of any PV module within the chain forming the solar string. The shading or malfunctions otherwise will disproportionately reduce power output of the entire array. With microinverters being connected to the electrical buss bar in parallel, the event of shading or failure of a single PV module will have a minimal effect on the solar power system power output that will be proportional to the power output capacity of single PV module.

Power Point Tracking System

Maximum power point tracking (MPPT) in the context of solar power panels implies electronic power output management electronic circuitry that automatically adjusts the electrical load to achieve the greatest possible power output. Variations in electric load result during solar irradiance variations that may be caused by passing of transitional clouds, temperature variations, shading that causes instantaneous variations in voltage, and current characteristics of solar cells.

Because of specific physical characteristics of solar cells, variation in ambient conditions always results in complex variations in power output performance that result in a nonlinear output current and voltage relationship. Figure 7.9 depicts the voltage and current I-V (current-voltage) curve of a typical photovoltaic solar cell, which results from the presence and absence of solar irradiance.

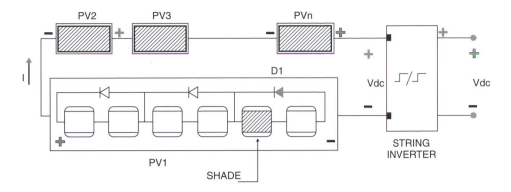

String-Inverter Based Solar Power System Configuration

Figure 7.8. A solar string inverter system configuration.

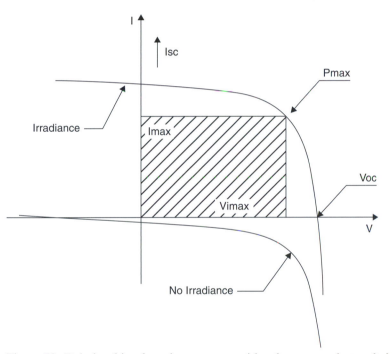

Figure 7.9. Relationship of maximum power with reference to photovoltaic current-voltage characteristic output under solar irradiance conditions. Shaded area indicates maximum power output zone.

The main purpose of an MPPT circuit is to sample output of each solar photovoltaic module and apply the power to a load in the form of resistance to achieve maximum power for given instantaneous environmental variations that affect solar power output performance. The MPPT circuit essentially uses electronic circuitry to maintain a power equation relationship between current and voltage defined by the following Ohm's law formulas:

$$P \text{ (power)} = V \text{ (voltage)} \times I \text{ (current)}, V = I \times R, \text{ and by substitution } P = I2 \times R \text{ or } P = V2/R$$

With reference to the V-I curve shown in Figure 7.9, it is evident that photovoltaic cells under varying solar irradiance conditions have a single operating point where the values of voltage and current are at maximum (Vmax and Imax) that result in a maximum of power output (Pmax). The Pmax value is achieved when the rate of change of power dP divided by the rate of change of voltage dV equals zero dP/dV = 0. At this point the characteristic resistance of the load connected to the PV module and the intrinsic resistance of the module become equal. Figure 7.9 shows the relationship of maximum power with reference to photovoltaic current-voltage characteristic output under solar irradiance conditions where the shaded area indicates the maximum power output zone.

Maximum power point tracking is achieved by use of specific iterative logic circuits that constantly search for the Pmax point that allows PV modules to deliver maximum power.

In general, all conventional inverters incorporate the MPPT circuit for an entire solar array, where the current flow to the inverter is the aggregated sum of all solar strings connected to the inverter. In view of the fact that all panels within the solar strings may have different current and voltage characteristics (as a result of partial shading or manufacturing tolerances), the collective maximum power point of the solar array will be expected to be lower when modules are connected in tandem. Therefore, use of microinverters offers an advantage that eliminates losses that may incur in string circuits.

Incidentally, the same MPPT technology is used in solar power charge regulators where a bank of batteries are charged during solar irradiance hours. In such applications the battery charger monitors and regulates the battery bank voltage when fully charged. Charging of batteries by the solar power system in the morning hours usually starts at a voltage that is below the solar array peak power point. However, at noon, the time when the voltage builds up, the regulator maintains the peak power point tracking by directing the excess power to an external load up until such time as power production exactly matches the demand. The same MPPT technology is used to drive electrical motors and pumps by solar power systems.

Advantages and Disadvantages of Microinverters and String Inverters

As discussed, the main advantage of use of microinverter systems is to mitigate transitional or seasonal shading, inadvertent accumulation of dirt, and side shading, as well as failure of an internal bypass diode or module infant mortality that may occur in rare instances.

It is also important to note that operational characteristics that give certain operational advantage to such equipment is somewhat irrelevant in the context of large-scale solar power system design. As referenced in previous chapters, among the fundamental principles when designing a large-scale solar power system are evaluation of the adequacy of the solar power platform and evaluation of shading, which are specifically devised to avoid shading. Therefore, if solar power systems are studied and designed properly, there should be no requirement to add complexity to systems by deploying thousands of electronic devices that may indeed not be capable of enhancing solar power system output performance and yet could diminish the overall system reliability and result in loss of return on investment.

Some of the notable disadvantages of microinverters are that they generally include complex electronics, they are more expensive than conventional inverters, and most of all the complexity of circuitry always diminishes system reliability. Even though microinverters may enhance the individual PV module power performance profile under shading or transitional or adverse environmental conditions, use of the devices in series with solar PV modules diminishes system reliability over the entire life cycle of a project.

Likewise, string inverters, in view of their complex electronics when connected in series with solar power strings, diminish system reliability and availability and may impose serious burdens on maintenance and operation of large-scale power systems that are configured from thousands of PV modules.

The following calculation based on the reliability calculations is intended to demonstrate inclusion of serially connected microinverters and string inverters in the context of a typical 1-megawatt system. It should be noted that reliability calculations studied later are not intended to diminish some of the advantages of the micropower conversion technologies; rather, the main purpose of the exercise is to demonstrate consequences of deployment of sequential combinations of various devices and the resulting increase in probability of longer system downtime and consequent loss of revenue.

Example of Effect of Microinverters and String Inverter Deployment on Solar Power System Reliability

The following exercise is intended to demonstrate the effect and consequence of deployment of microinverter, string inverter, DC to DC converter, or any energy enhancement circuitry in *series* with solar PV modules. Even though microenergy conversion devices have recently entered the market, equipment manufacturers claim a warranty of 20 to 25 years for technologies that have recently entered the industrial trial phase. Regardless of the viability of claims, the reliability calculations that follow are intended to demonstrate the importance of solar power system equipment deployment and methodologies that must be used to ensure adequate system reliability.

In the following example reliability or availability calculations are applied to the deployment of a microinverter and string inverter in a hypothetical 1000-kW solar power system located in a region that that has a mean daily solar insolation of 5.5 hours. It should be noted that in a microinverter-based solar power system configuration each PV module is connected to a microinverter, which means that a 1-megawatt solar power system consisting of a 5000 DC PV module, with PVWatts conversion efficiency rate of 20% will require 5000 microinverters. On the other hand, the same 1-megawatt solar power system using string inverters with strings configured with 10 PV modules will deploy 500 string inverters.

With reference to the preceding hypothetical solar power systems and assumption, the calculation reflected in Figure 7.1 (a) and (b) yields results of the downtime of each system configuration. Loss of solar power system or availability in a day, if translated into insolation hours, can be applied to mean electrical utility rate values, which could be interpreted as an economic or investment loss. Once again, the reader must take note that deployment of microconversion technologies discussed earlier has specific benefits that in certain situations can be economically

```
MICROINVERTER SYSTEM MTBF CALCULATIONS
AVAILABILITY = MTBF/(MTBF + MDT)
UNAVAILABILITY = 1 - AVAILABILITY%
YEARLY OUTAGE = AVAILABILIT x HOURS/YEAR
OUTAGE = UNAVAILABILITY X HOURS/YEAR
SERIAL CONNECTED SYSTEMS = R1a x R2a x r3a x ...Ran
PARALEL CONNECTED SYSTEMS =          1-(1-R1a)(1-R2a)(1-R3a)...(1-Rna)
POWER OUTPUT CAPACITY - KW            1000
ASSUMED DAILY SOLAR INSOLATION HOURS
MEAN TIME BETWEEN FAILIUR IN HOURS
OF OPERATION - MTBF                  5.5
```

MTTR		MEAN TIME TO REPAIR IN HOURS	
YEARLY OUTAGE		DOWN TIME PER YEAR IN HOURS	
EQUIPMENT	PV - SYSTEM	MICRO INV.	DC-DC CONVERT.
RELIABILITY BLOCK ID	Pva	Mia	DC/DC
MTBF -YEARS	20	20	20
HOURS PER YEAR	8760	8760	8760
HOURS	175200	175200	175200
MTTR (MEAN TIME TO REPAIR) HRS	72	72	72
AVAILABILITY	99.96%	99.96%	99.96%

RELIABILITY COMPUTATION MICROINVERTERS			% RELIABILITY
1- MICROINVERTER IN SERIES WITH SOLAR PV		R1 = PV x MI	99.92%
DIFFERENTIAL REDUCTION OF AVAILABILITY		Ami = Pva - IMa	0.04%
RELIABLE DOWN TIME HOURS PER YEAR			3.60
SOLAR PV AVAILABILITY MINUS DIFFERENTIAL REDUCTION IN AVAILABILITY			
RELATIVE DOWN TIME HOURS PER 20 YEARS			**71.94**
LOSS OF AVAILABILITY IN SOLAR DAYS **DURING LIFECYCLE**			**13.08**
MODULE WATTS	250		
NUMBER OF MODULES AT 80% CONV. EFFIC.	5000		
TOTAL LOSS IN SOLAR DAYS DURING LIFE			**65,401**

Figure 7.10. Effect of microinverter and string inverter deployment and their effect on solar power system reliability.

justified. Figure 7.10 shows computational details of the effect of deployment of a microinverter and string inverter and their effect on solar power system reliability.

Introduction to Wireless Data Acquisition and Control

The solution to the output performance challenges described is provided by Solar Analytic Solutions Inc. (SAS), with patent pending granular solar PV module Wireless Intelligent Solar Power Reader (WISPR) remote data acquisition and control unit and the Master Data Acquisition and Control System called O-DACS

RELIABILITY CALCULATION FOR DC-DC STRINC CONVERTERS

2 - ASSUME DC-DC CONV. IN STRING OF 12 PV MODULES

AVAILABILITY Adc/dc= (Pva1 x Pva2 .....Pva12) x (DC/DC1 x DC/DC2 x...DC/DC12)

	99.51%	x	99.51%
		= 99.02%	

DIFFERENTIAL REDUCTION IN AVAILABILIITY	0.489%
RELATIVE DOWNTIMEHOURS PER YEAR	42.87
SOLAR STRING AVAILABILITY MINUS DIFFERENTIAL REDUCTION IN AVAILABILITY RELATIVE DOWN TIME SOLAR STRING HOURS PER 20 YEARS	857.46
RELIABILITY CALCULATION FOR DC-DC STRING CONVERTER DURING LIFECYCLE	**155.90**

MODULE WATTS	250	
PV COUNT PER STRING	12	
NUMBER OF STRINGS AT 80% CONV. EFFIC.	500	
TOTAL LOSS IN SOLAR DAYS DURING LIFE		77951

Figure 7.10. (continued)

state-of-the-art wireless communication and control system. In addition to granular data acquisition, the system described in the following incorporates electronic circuitry that provides vital life safety and system shut-down mechanisms that can prevent significant financial losses that may result from inherent hazards associated with solar power photovoltaic systems.

The following is a generalized overview of a solar power installation whether a roof or ground mount system.

Solar Power Farms are very large solar power installations where literally thousands of photovoltaic panels are used to provide grid type (very large current) power.

Solar power system configurations consist of a number of PV modules connected in series (and sometimes in parallel) just like tandem connected batteries. The analogy of a PV module is that of a battery that produces power only when exposed to solar rays. When PVs are connected in series, the power output is accumulated in proportion to the number of connected modules. A series of connected PV modules (usually 8 to 14) is referred to as a PV string.

A PV string, depending on the type of solar photovoltaic module, produces a nominal current (I) and voltage (V), which is a measure of the electrical DC power generated.

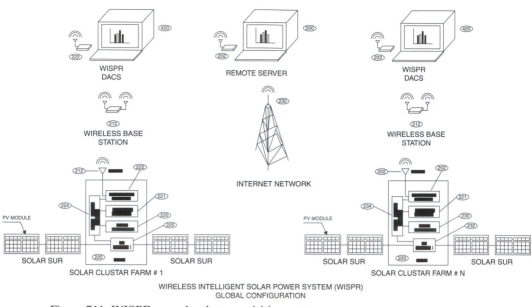

Figure 7.11. WISPR granular data acquisition system.
Source: Graphics courtesy of Solar Analytic Solution.

DC power produced by a PV string is converted to AC power by an electrical/ electronic apparatus referred to as an inverter. Depending on the power conversion capacity requirements, inverters are designed to accept several PV strings (which may range from 4 to 126).

A solar array consists of a group of strings (i.e., 20, 30, or larger number) that are connected to an inverter. A subarray is simply a number of strings within an array.

Large groups of solar arrays consisting of PV strings and inverters form a solar power system. As mentioned, a solar farm is a very large number of solar power subsystems which may be installed on many acres of land. Figure 7.11 depicts the Solar Analytic Solutions Inc. patented WISPR granular data acquisition system diagrams.

Wireless Intelligent Solar Power Reader (WISPR)

WISPR is a device that is attached to each individual solar PV module that is capable of reading the power output from each unit. The measured information is broadcast via wireless communication to a central communications hub that in turn forwards the information to a centrally located Data Acquisition and Control System (DACS). Special communication radio devices embedded within the WISPR unit transmit and receive data from the DACS. A solar farm, for instance, depending on its physical configuration, may have thousands of PV units and an equivalent number of WISPR modules dedicated or attached to each PV unit.

One of the significant functions of the WISPR/DACS system is to retrieve power output data from each individual PV unit, in a sequential addressing or scanning method. Data received from WISPR is then processed by the DACS software, which in turn is used to generate power output performance reports, in the form of statistical information (either printed or displayed). Having a unique ability to monitor and acquire data from each PV module is one of the significant features of the system, since data acquired provide the ultimate accuracy in solar power measurement.

However, the most important feature of the WISPR-based solar power system is that every individual PV module output or the entire solar power array and subarray systems can be physically disabled (shorted) by the O-DACS.

The disabling of the solar power output is of great significance in the case of fire, earthquake, or even maintenance security, since in the case of an emergency, very high voltages generated by solar power strings (600 volt DC) could be a serious hazard to firefighters and maintenance personnel, which could be totally prevented by automatic or manual intervention at the O-DACS.

Each WISPR unit communicates with the central data acquisition control by means of long range radio via a single WAN gateway installed in each solar power system configuration or a solar power farm. In turn, each WISPR unit communicates with other units by embedded frequency hopping radio modules. WISPR has no relevance to AC electric metering. Figure 7.11 shows a block diagram of the Wireless Intelligent Solar Power Reader (WISPR) patented technology, which is designed to provide granular data acquisition.

Analytic Solutions

To clarify the last point, WISPR is a measurement and control device. The term "WISPR DC" has no special meaning, except that the power measured from each PV is in the form of a direct current.

Product Summary Overview

The following discussion is intended to inform the reader about inherent safety hazard and operational dangers associated with solar power systems and solutions that are offered by WISPR.

All categories of solar power system installations whether residential, commercial roof mounted solar systems, parking solar canopies, or large solar farms, while exposed to solar irradiation generate high-voltage DC power that may cause severe electrical shock and fire hazard, thereby causing life safety and property damage.

Photovoltaic modules, regardless of technology, whether silicon or film technology, when connected in series strings, under exposure to solar irradiance produce 300- to 600-volt DC voltages, which under certain circumstances may cause severe burns, fire, and even fatality.

Solar power string cables or feeder cables from DC combiner boxes, even operating at very low tension of 100 volts, if shorted, can cause sparks and under some circumstances cause cable insulation melt- down. In installations where solar power systems are within close vicinity of flammable or combustible materials, such as shrubs or vegetation, in the event of tremor or accidental damage to current carrying cables they may cause sparks that may result in fires.

Even though solar power systems have been around for some time, the technology deployment is still nascent. Even though solar power system deployments in the past couple of decades have proliferated dramatically, as yet the industry as a whole has not addressed the life safety and fire hazard issues that are inherent in solar power systems.

A conundrum associated with the solar power system is that manufacturers of photovoltaic modules are seldom involved with solar power system design or deployment. The engineers who are responsible for the overall system design have to contend with intrinsic characteristics of PV modules described earlier. It should be noted that photovoltaic modules are essentially miniature electrical power generators, and as long as they are exposed to solar rays, they generate electrical power that cannot be turned off or shut down. At present, the only way to stop PV modules from producing power under solar irradiance is either to cover them with a canvas or to short-circuit the output leads of every module, which is a functionally impracticable option.

Since solar power systems are configured from hundreds and thousands of tandem or series connected photovoltaic strings, covering them with a canvas is not a practical option. For example, a 1-megawatt solar power farm may have as many as 5000 PV modules and could cover an area of approximately 5–7 acres; therefore, use of a cover becomes impossible. The same holds true for a moderate 250-kW roof mount solar power system, which may be configured by stringing several thousand PV modules together and may cover several thousand square feet of roof space.

Technology Solution

Prior to discussing hazard mitigation measures for solar power systems, it would be useful to explore circumstances and causes of solar power system hazards.

In general, medium- and large-scale solar power systems fall within three categories, namely, roof mount solar systems, carport or trellis solar systems, and ground mount solar systems.

Each of the three categories has some unique and common features associated with life safety and hazard.

Roof Mount Solar Power System Hazards

Scenario 1 – Building Fire In the event of building fire during daylight hours, firefighters must have the means to shut down the main service electrical power. Under normal settings the main switchgear breaker is turned off. However, since roof mount solar power systems produce uninterrupted power under solar irradiance they cannot be turned off during daytime conditions. Since quenching fire usually involves breaking roofs with steel axes under extremely wet conditions, firefighters face imminent danger of serious electrical shock hazards. Figure 7.12(a) and (b) as well as Figure 7.13 are photographs before and after a fire of a roof-mount solar power system. The source of the photograph is a public domain report that can be accessed at the Web address http://www.nfpa.org/foundation.

Scenario 2 – Earthquake Under moderate or high earthquake tremor conditions, solar power support structures or arrays may be subject to breakage, which could result in voltage spark activity that could result in fire or life safety hazard for the maintenance personnel.

It should be noted that even though most solar power systems are configured with a Data Acquisition and Monitoring System (DAMS), such systems are primarily

Figure 7.12 (a). Roof mount solar power system before fire.
Source: Bakersfield, California, Fire Department. (b). Roof mount solar power system after fire.

designed to monitor and report power output performance of the solar power systems; however, they do not have the capability or the means to detect system malfunction. As a result, under such circumstances, physical system damage that may require manual service intervention may expose maintenance personnel to high-voltage parts of the system that could cause serious life safety hazard.

Scenario 3 – Maintenance and Repair When maintaining a solar power system, during daylight hours, technicians are constantly in danger of exposure to high DC

Exhibit H

Figure 7.13. DC current of a solar power system causing electrical metallic conduit meltdown.

voltage. In order to detect a malfunctioning solar array subsystem or a PV module, within a solar array string, maintenance personnel must constantly be vigilant about taking extreme safety measures to prevent electrical shock when inspecting or repairing solar power system equipment or components.

Carport or Trellis Mount Solar Power System Hazards

Carport systems in general involve the configuration of photovoltaic modules mounted on metallic structures that are exposed to the public. Such solar power systems are configured with identical devices, components, and wiring methodology to roof mount systems, except that the main power carrying cables and conduits, referred to as feeders, are usually installed underground.

In the event of earthquake, shorting of solar PV strings to the canopy structure could result in public exposure to electrical shock. Likewise underground conduits and conductors that carry solar power to the DC accumulators could be subject to damage, which could also result in electrical shock hazard to maintenance personnel. Similar to roof mount solar power installations, during daylight hours, the solar power system shall be fully operational, generating high-voltage electrical power, which otherwise could not be shut down; therefore, maintenance and repair may result in life safety hazard.

Ground Mount Solar Power Farms

In general large-scale solar power systems cover substantial ground surface areas. For instance, a 1-megawatt solar power installation may cover 6 to 8 acres of land configuration, which may require stringing of thousands of photovoltaic modules (about 5000 PV modules per megawatt). Such large solar power system configurations include a wide array of underground feeder cables, conduits, step-up transformers, and overhead or underground electrical power transmission, which are constantly operational during daylight hours and may also be exposed to the following hazards.

Scenario 1 – Ground Fire hazard In the event of natural ground fires, which occur frequently throughout the western and southern states and western provinces of Canada, during daylight solar irradiance solar power systems could pose real and imminent hazards. Even though the main grid connectivity equipment and power conversion equipment such as inverters could be disengaged from main service, thousands of solar strings exposed to fire inevitably pose unpredictable damage and significant safety hazards for maintenance and repair personnel.

Scenario 2 – Earthquake In the event of an earthquake, photovoltaic module support platforms (whether fixed angle, single, or dual axis tracking systems) shall be subject to dislocation and breakage. In such circumstances, disruption of high-voltage-carrying cables will be inevitable, and when these cables contact metallic structures or moist ground may generate sparks that could result in serious ground fires. In situations where grounds are treated with vegetation suppression chemicals, mere accumulation of dry plant material could still create a fire hazard.

Scenario 3 – Flood In the event of flooding, exposure or immersion of high-voltage wires and conduit could create electrical shorts that could be detrimental to maintenance personnel and could cause substantial damage to equipment.

Wireless Intelligent Solar Power Reader (WISPR)

To mitigate safety and fire hazards, for each of the preceding scenarios, all of the solar power system configurations must incorporate an adjunct safety device on photovoltaic strings that would allow systematic shutdown of every PV module that can be activated from a central data acquisition and control system.

WISPR is a patented device designed to resolve all of the safety hazards discussed. WISPR is a wireless microprocessor device that embeds hardware and firmware components that communicate with a central data acquisition and control system that enables activation and deactivation of each PV module within a solar power system. The module, which is housed in a small weatherproof compartment, is designed of be an integral part of a PV module. In addition to the PV activation or deactivation control from the central data acquisition system, the unit is design to collect real time solar power performance parameters.

WISPR's internally embedded hard contact relay is activated or deactivated by either reception of its unit address or a global broadcast shutdown message, initiated from the central data acquisition and control system.

A duplex communication transceiver within WISPR allows each device to transmit and receive secured data to and from the central data acquisition and control system. The message can be an ON/OFF actuation message or a request for module power output measurement parameter scan data.

The microprocessor-based central data acquisition system incorporates a number of digital, analog, and dry contact inputs that allow interface with a variety of external sensors.

Upon sensing a hazard signal from an input port, the master data acquisition and control software broadcasts a global shutdown to all WISPR modules within the solar power system, which initiate latching of the crowbar relays that remain in *fail safe* position, which effectively shunt the positive and negative output terminals of each PV unit.

Upon restoration of normalcy, the central control broadcasts a global activation message, whereby the latched shunting of the PV module is removed.

Upon actuation of the relay and positioning of the PV module in either ON or OFF state, WISPR transmits confirmation of relay position to the central data acquisition system.

In the event of repair or maintenance, the data acquisition system operator transmits a control message(s) to WISPR units to activate or deactivate an individual or a selected group of PV modules within a solar array string.

A hand-held microprocessor-based device such as a PDA device could be used to perform limited data acquisition and control similar to the central data acquisition and control system.

Hazard input signal detection by the central data acquisition and control system could be from either a local hardwired contact or a remotely located device that could transmit signals via land line or by means of wireless communication.

Hazard signals that could activate a global shutdown of the solar power system could be as follows:

Dry contact input from a central or a local (pull station) fire alarm system
Dry contact or digital signal from an earthquake detection mechanism
Dry contact from a fire or CO_2 detection device
Dry contact from a water level float
Dry contact from an emergency pushbutton

WISPR Special Features

Some of the unique features of WISPR include the following:

Individual data acquisition and control from each individual PV module within a solar system
Real time monitoring of voltage, current, and unit temperature of each PV module frequency hopping communication system
Adaptability of the module as digital and analog data sensor and actuator

The unit only reacquires a few microamperes of its operational power from the PV module during daylight. Each module has a long-lasting lithium ion battery backup system intended to maintain module operation during extended cloudy periods when PV modules are dormant.

WISPR operation is completely transparent to PV module functional performance. The module input and output connectivity are designed to match standard U.S. PV module positive and negative IN/OUT pigtails. The unit physical package is configured to allow retrofit installation on existing solar power systems.

WISPR electronics could also be manufactured in a compact hybrid circuit form for sale as an Original Equipment Manufacturer (OEM) product that could be embedded within the PV module junction boxes during the manufacturing process.

Figure 7.8 is a diagram of a solar string inverter system configuration. Figure 7.9 is graphic representation of maximum power output with reference to photovoltaic current and voltage characteristic output under solar irradiance conditions. The shaded area of the graph represents the maximum power output zone.

8 Solar Power System Econometric and Analytical Software Solution

Introduction

The following article is a summary description of analytical software that provides instantaneous solar power system integration cost calculation and financial econometrics for large-scale grid connected solar power systems.

The program is intended to provide private owners as well as financial entities that are involved in solar power system construction funding lease purchase agreements or Power Purchase Agreements.

The main objective of the software program is to provide a software tool that allows users to insert numerous variables associated with labor, material, engineering design, solar power performance characteristics, energy cost escalation algorithms, and other factors required to perform detailed financial calculations.

Econometric Analytic Software System

A significant feature of the Solar Power Econometric Analysis is that the costing analytical methodology, unlike conventional accounting methodology used for capital equipment depreciation and return on investment, deploys specific computational algorithms that conform to *real life* solar power system integration costing, photovoltaic system performance characteristic parameters under various climatic conditions, solar platform venues, PV system intrinsic degradation, and much more.

The following are solar power system economic and analytical software solutions (SPSEAS) software subsystems.

Solar Power System Construction and Integration Costing

This software performs costing computations for the following:

A – Engineering Cost Component Analysis

- Engineering site survey and feasibility study
- Engineering feasibility report for various solar power system alternative technologies and support platforms
- Shading analysis

- Detailed engineering design documentation
- Solar system integration, test, and commissioning specification
- Assistance to owners of client in procuring rebates
- Project coordination
- Periodic construction supervision
- Field test and acceptance oversight
- Project coordination
- Solar power platform design
- Negative environmental impact report
- Power transmission and grid integration
- Civil and structural design
- Construction and performance bonding expenses
- Construction financing

B – Solar Power System Output Performance Analysis This software block performs analysis of solar power performance, which is influenced by PV module specification and various output power depreciation factors imposed by the California Solar Initiative or National Renewable Energy Laboratories (NREL) PV Watts V.2 computation.

Parameters influencing solar power performance taken into account include the following:

- Solar platform topology power output potential analysis, which includes principal hardware components such as PV module performance specification
- Type of solar module support structure system such as fixed angle flat PV configuration, single or dual axis solar power tracking system
- Topical solar irradiance and per diem power potential
- Solar power system life cycle power output vis-a-vis the expected system degradation
- Typical monthly system solar power output under varying insolation conditions

C – Solar Power System Financial Analysis The following software provides detailed analysis of the solar power system for the entire contractual life cycle period, which includes step by step solar power energy cost values for each year. Parameters that influence the cost such as system operation performance, dynamic power degradation, present or contractual unit energy cost/kWh, project grid electrical energy cost escalation, rebate profile for Performance Based Incentive (PBI), initial cost investment, salvage value, and many additional factors provide a year-to-year solar power return on investment throughout the life cycle of the contract.

The preceding computations constitute a most significant financial analysis of PPA financial costing that has not been not available to date. The analytical engine of the SPEAS program allows users to optimize critical kWh unit sales, percent annual cost escalation, and maintenance cost component instantaneously. Costing parameters when entered in various fields perform the following computations:

- Yearly average AC power output and solar power energy value computation
- Dynamic extrapolation of projected unit energy cost escalation for the entire life cycle of the contract

- Dynamic depreciation of solar power output for system operational life cycle
- Life cycle power output potential
- Progressive rebate accumulation for the PBI period based on the available unity energy fund availability
- Cumulative electrical energy cost income from the end of the rebate period up until the end of the system contractual life cycle
- Integrated accumulated income form based on annual cost escalation factor
- Savings resulting from annual maintenance

D – Comparative Analysis of Grid Power Energy Expenditure versus Solar Power System Energy Output Cost This component of the software performs cumulative energy costing as well as costing for grid connected PPA based solar power systems, for the solar project life cycle.

E – Power Purchase Agreement (PPA) Accounting In the final step of computation the software uses all of the previous results to perform financial analysis for PPA type contracts; they include the following:

- Totalized energy income from life operation
- Present value of the projected system salvage amount
- Federal tax incentive income
- State tax incentive income
- Salvage value at the end of the system life cycle
- Maintenance cost savings
- Recurring annual maintenance cost
- Net income value over the life cycle of operation

Solar Power Economics Analytical Software

The solar power economic system described in the following is based upon extensive experience in large-scale solar power system design, construction estimating, test and integration, as well as economic analysis and methodologies of financing of photovoltaic systems. Fundamentals of the software described are based upon integrated econometric analysis of solar photovoltaic power system cost components that contribute to the totality of the economics. In essence, the analytical software encompasses all issues as well as causes and effects that impact the system bottom line. To achieve the objective, the economic analysis commences with system design topology, equipment and material parameters that affect energy output performance, grid electrical energy tariff escalations, state rebates, federal tax incentives, as well as numerous factors that impact the economics of solar power systems.

Discussion of the economics of solar power systems also highlights unique intrinsic characteristics of mono-silicon photovoltaic technologies, their extended life cycle, and the resulting impact on solar power system financing.

Economic Analytical Software Components

The solar analytical software presented in the following consists of 12 analytical engines that perform various engineering and economic computations that aggregate into a single software. It should be noted that basic computational principles and accounting methodologies used in development of the analytical engines are in full compliance with the California Solar Initiative PV Watts V.2 program.

Analytical engines presented and discussed are in spreadsheet form. Basic computation algorithms include the following:

1- Photovoltaic module power output normalization
2- Solar power system loss computation
3- Solar power system design configuration
4- Solar power system power output computation
5- Solar power system construction costing
6- First-year power output performance and economic contribution
7- Solar power system first 25-year life cycle economic contribution
8- Solar power system second 25-year life cycle economic contribution
9- Solar power system first 25-year life cycle net future value economic contribution
10- Solar power system second 25-year life cycle net future value economic contribution
11- Solar power system economics summary overview
12- Effect of solar power system on CO_2 abatement

Photovoltaic Module Power Output Normalization

The spreadsheets presented in this segment represent three main types of solar power system platforms, namely, fixed angle, single axis tracking, and dual axis tracking platforms. The first set of spreadsheet and associated graphics represent a hypothetical fixed axis solar power system; it is followed by single and dual axis solar power system economic analysis.

The spreadsheet shown in Figure 8.1 is an analytical engine that is used to adjust power output performance of various types of solar power panels under various environmental temperature conditions. In general photovoltaic module specifications provided by manufacturers, in addition to specifying power output performance characteristics of the modules, include specific temperature coefficients or multipliers that affect power output performance under varying temperature conditions.

It should be noted that PV module power output parameters included in manufacturers' specifications are based on test measurements conducted in laboratories under ideal environmental conditions, under maximum terrestrial solar irradiance. Test measurements that are conducted at 25 °C are referred to as standard test conditions (STC). PV modules, when operating in environments that deviate from the STC, exhibit different power output characteristics. As such when designing solar power systems, designers must pay specific attention to adjusting or normalizing the expected power output performance of PV modules.

In the preceding specification, lines 6, 7, and 8 represent temperature coefficients that must be applied to determine power output performance under maximum and minimum temperature conditions. When designing solar power systems, special consideration must be given to ensure that solar string lengths and their aggregated Voc voltages do not exceed the maximum DC input voltage tracking capability of the inverter. By applying the VToc temperature coefficient to the temperature

PV Module Specification	Manufacturer	SANYO -HIT 210
1 - Module Rated Power (Pmax)	210	Watts
2 - Maximum Power Voltage (Vpm)	41.3	Watts
3 - Maximum Power Current (Imp)	5.09	Amperes
4 - Open Circuit Voltage (Voc) @ 25 Centigrade	50.9	Volts
5 - Short Circuit Current (Isc)	5.57	Amperes
6 - Temperature Coefficient (Pmax)	-0.336	Watts/Deg.C
7 - Temperature Coefficient (Voc)	-0.142	Volts Deg.C
8 - Temperature Coefficient (Isc)	1.92	mA/Deg.C
9 - Normal Operating Temperature (NOCT)	46	Centigrade
10 - Lowest Operational Ambient Temperature	-10	Centigrade
11 - STC Temperature	25	Centigrade
12 - Record Low Temperature - Rlt	-35	Centigrade
13 - Voltage temperature Correction Factor of VTco	-0.142	Volts/C
14 - Temperature Correction Voctf = Rlt X VTco	4.97	volts
15 - Nameplate of PV Module Per String	10	
16 - Corrected String Voc = (# strings X Voc) + Voctf	513.97	Volts
17 - CEC PTC	194.8	Watts
18 - Cell Efficiency	18.9%	
19 - Module Efficiency	16.7%	
20 - Watts Per Square Foot	15.48	Watts
21 - Maximum System Voltage	600	Volts
22 - Series Fuse Rating	15	Amps
23 - Warranted Tolerance	0 to + 10%	
24 - STRING WATTS	1948	WATTS

Figure 8.1. Photovoltaic module specification and power output normalization.

correction equation (shown on line 14), at –35 °C the corrected Voc voltage of a solar power string formed by tandem connection of 10 modules must be elevated by 4.97 volts. According to the PV module specification, under STC conditions the string Voc would have been aggregated to 509 volts (50.9×10); instead the normalized or adjusted Voc when adjusted for minimal temperature equates to 513.97 volts.

If we consider DC swing voltage of conventional inverters to be around 300–660 VDC, the temperature compensation may have little effect on the design. However, if the designer decides to lay out strings composed of 12 PV modules with Voc of 50.9 volts DC, with identical temperature coefficient, then the aggregated voltage will be quite close to the upper boundary of the inverter DC tracking voltage.

Solar Power System Loss Computation

The spreadsheet shown in Figure 8.2 represents an algorithm that computes solar power system degradation when operating under specific environmental conditions.

The first line of the upper spreadsheet shows U.S. Power Test Condition (PTC) of the PV module. The PTC value is based on test certification of the PV module established by the California Energy commission. All solar power equipment and components used in the United States that are subject to federal or state financial incentives must be listed under the CEC approved list of solar power equipment. In the preceding case, the SANYO-HIT-210 module, which is rated by the manufacturer to have 210-watt output performance capacity, is down rated to a PTC of 194.8 watts.

SOLAR POWER SYSTEM OVERALL SPECIFICATION - FIXED AXIS

1 - PV MODULE DC WATTS - PTC	**194.8**	
2 - PV MODULE DC KW - PTC	0.1948	
3 - INVERT. - IN1	DC INPUT SWING	300-600
4 - MAX STRING VOLTS	514	
5 - PV MODULE - Voc	52	
6 - STRING LENGTH	10	
7 - SOLAR INVERTER - N1		
10 - STRING PER COMBINER BOX	**10**	

SOLAR POWER SYSTEM LOSSES PER PVWATTS V.2

	ACTUAL	ALLOWED RANGE
PV MODULE NAMEPLATE DC RATING	0.98	0.8 - 1.05
INVERTER AND TRANSFORMER LOSSES	**0.97**	0.88 - 0.98
PV STRING MISMATCH	0.98	0.97 - 0.995
PV CELL MODULE DIODE LOSSES	0.995	0.97 - 0.997
DC WIRING LOSSES	0.99	0.97 - 0.99
AC WIRING LOSSES	0.99	0.98 - 0.993
SOILING LOSSES	0.99	0.3 - 0.995
SYSTEM AVAILABILITY	**0.98**	0.00 - 0.995
SHADING LOSSES	**0.98**	0.00 - 1.00
SUN TRACKING LOSSES	**96.9%**	0.95 - 1.00
SYSTEM AGING AND DEGRADATOIN LOSSES	1	0.7 - 1.00

COMBINED LOSSES	**0.84**	

Figure 8.2. Solar power system loss computation.

The third line of the upper spreadsheet specifies the DC tracking limits of the inverter used. The fourth line represents the normalized Voc voltage, the fifth line the solar string configuration lengths, and the tenth line specifies the number of solar strings that can be accommodated by a specific combiner box.

The lower spreadsheet consists of three columns. The left-most column outlines derating factors that could affect power output performance of the solar power system. The second column specifies derate multiples specific to a particular project and the right-hand column are limits of derate factors established by the CSI PV Watts V.2 calculator.

The first five figures in the second column are standard losses, some of which are associated with equipment nameplate ratings. The soiling losses, the system availability, and the shading losses shown in the table are subject to system design and engineering analysis of a particular solar power project. The sun tracking multiplier is subject to the type of solar power platform that PV modules are attached to (fixed axis, single or dual axis tracking systems); finally, aging and degradation are figures of merits that are specified by PV module manufacturers. The lower line of the middle column represents the mean computed value of the degradation multiplier that is used to calculate electrical energy output production capacity of a solar power system.

Solar Power System Design Configuration

Figure 8.3 is a spreadsheet representation of an algorithm that computes solar string, combiner box, and recombiner box DC power generated by a specific PV module

GLOBAL TAGGING IDENTIFICATION	INVERTER IDENTIFIC.	RECOMBINER IDENTIFIC.	COMBINER IDENTIFIC.	SOLAR ARRAY IDENTIFIC.	IN-STRING PV STRING IDENTIFIC.	PV MODULE IDENTIFIC.	KWD-DC
IN1RC2CB5SA1S1PV1–10	IN1	RC2	CB5	SA1	S1	PV1–10	1.948
IN1RC2CB5SA1S2PV1–10	IN1	RC2	CB5	SA1	S2	PV1–10	1.948
IN1RC2CB5SA1S4PV1–10	IN1	RC2	CB5	SA1	S3	PV1–10	1.948
IN1RC2CB5SA1S4PV1–10	IN1	RC2	CB5	SA1	S4	PV1–10	1.948
IN1RC2CB5SA1S5PV1–10	IN1	RC2	CB5	SA1	S5	PV1–10	1.948
IN1RC2CB5SA1S6PV1–10	IN1	RC2	CB5	SA1	S6	PV1–10	1.948
IN1RC2CB5SA1S7PV1–10	IN1	RC2	CB5	SA1	S7	PV1–10	1.948
IN1RC2CB5SA1S8PV1–10	IN1	RC2	CB5	SA1	S8	PV1–10	1.948
IN1RC2CB5A1S9PV1–10	IN1	RC2	CBS	SA1	S9	PV1–10	1.948
IN1RC1CB5SA1S10PV1–10	IN1	RC2	CB5	SA1	S10	PV1–10	1.948
IN1RC1CB5SA1S11PV1–10	IN1	RC2	CB5	SA1	S11	PV1–10	1.948
IN1RC1CB5SA1S12PV1–10	IN1	RC2	CB5	SA1	S12	PV1–10	1.948

COMBINER BOX IN1CB5-TOTAL KW-DC   ● 23.376

IN1RC2CB6SA1S1PV1–10	IN1	RC2	CB6	SA1	S1	PV1–10	1.948
IN1RC2CB6SA1S2PV1–10	IN1	RC2	CB6	SA1	S2	PV1–10	1.948
IN1RC2CB6SA1S4PV1–10	IN1	RC2	CB6	SA1	S3	PV1–10	1.948
IN1RC2CB6SA1S4PV1–10	IN1	RC2	CB6	SA1	S4	PV1–10	1.948
IN1RC2CB6SA1S5PV1–10	IN1	RC2	CB6	SA1	S5	PV1–10	1.948
IN1RC2CB6SA1S6PV1–10	IN1	RC2	CB6	SA1	S6	PV1–10	1.948
IN1RC2CB6SA1S7PV1–10	IN1	RC2	CB6	SA1	S7	PV1–10	1.948
IN1RC2CB6SA1S8PV1–10	IN1	RC2	CB6	SA1	S8	PV1–10	1.948
IN1RC2CB6SA1S9PV1–10	IN1	RC2	CB6	SA1	S9	PV1–10	1.948
IN1RC2CB6SA1S10PV1–10	IN1	RC2	CB6	SA1	S10	PV1–10	1.948
IN1RC2CB6SA1S11PV1–10	IN1	RC2	CB6	SA1	S11	PV1–10	1.948
IN1RC2CB6SA1S12PV1–10	IN1	RC2	CB6	SA1	S12	PV1–10	1.948

COMBINER BOX - IN1CB6 TOTAL KW-DC   23.376
SUM TOTAL   ● 46.752

Figure 8.3. Solar power system design table showing array configuration.

configuration. In addition to computing the individual and aggregated power generated by solar power strings, subarrays, and arrays, the software defines the overall nomenclature or tagging of all components and equipment used within the solar power system. As such, each individual PV module, a solar string, a combiner or recombiner box, as well as inverters, are assigned unique identification tags that are used to cross-reference various elements of the solar power system. The identification and tagging system as reflected in the spreadsheet is used throughout engineering design drawings and equipment identification plates as well as the solar power cabling system.

The system configuration spreadsheets consist of nine columns. The first column represents unique identification tags that encompass all elements that constitute the solar power.

IN1 – The second column from the left identifies the inverter to which the elements of the solar array are connected.

RC2 – Identifies the recombiner box to which the various combiner boxes (CB5 & CB6, etc.) are connected.

CB5 – Identifies combiner box no. 5, which is within the solar array SA1 configuration.

S1–S12 – Identify solar strings connected to the combiner boxes.

PV-1–PV10 – Identify the location of the PV modules within each solar string.

IN1RC2CB5SA1S1PV1-PV10 represents the principle tagging identification number for the solar power system subset.

FIXED AXIS

WATTS PER STRIN	1948	WATTS
WATTS PER COMBINER	23376	WATTS
TOTAL RECOMBINER POWER	93504	WATTS

PV MODULES PER STRING	10
STRINGS PER COMBINER	12
RECOMBINER INPUT POINTS	4
RECOMBINER BOXES	12
PV MODULE COUNT	**5760**

	W - DC		W - DC
COMBINER-1	● 93504	COMBINER-7	● 93504
COMBINER-2	● 93504	COMBINER-8	● 93504
COMBINER-3	● 93504	COMBINER-9	● 93504
COMBINER-4	● 93504	COMBINER-10	● 93504
COMBINER-5	● 93504	COMBINER-11	● 93504
COMBINER-6	● 93504	COMBINER-12	● 93504
TOTAL DC KW	◉ 561024		◉ 561024
COMBINED TOTAL DC WATTS	**1122048**		
AXIS POWER EFFICIENCY MULTIPLIER	**100%**		
ADJUSTED DC POWER OUTPUT	**1122048**		
DC TO AC DERATE MULTIPLIER	1		
AC POWER WATTS	**939183**		
TILT ANGLE DERATE MULTIPLIER			
LATTITUDE - DEGREES-a	34.8		
PV PLATFORM TILT ANGLE - DEGREES-b	**25**		
%LOSS AT 0 DEGREES -c	11%		
PERCENT EFFICIENCY	96.9%	$(1-c) + c*(b/a)$	
NET AC POWER OUTPUT kW	910090		
NET AC POWER OUTPUT WATTS		**910**	

Figure 8.4. Solar power output calculation showing summed up combiner box outputs and power derating with a fixed axis tracking system.

The left-most column shows totalized DC power capacity of each of the combiner boxes (23.376 kW DC) as well as the sum total of combined power produced by the solar subarray. Finally, figures beside the dots represent recombiner box accumulated power outputs. The following spreadsheet shown in Figures 8.3 accounts for two combiner boxes with a total accumulated 46.75 kW of DC power.

Solar Power System Power Output Computation

Figure 8.4 is a spreadsheet that represents the computation steps for determining solar power output performance potential. The first column of the upper spreadsheet represents recombiner boxes that form the overall solar power system. In this particular exercise we use the 5760 SANYO HIT 240 PV module to construct an approximate 1000-kW DC solar power system. The second column to the left represents calculated DC power output capacities of the 12 recombiner boxes, which are in turn summed to yield 112 kWD DC power. By applying the derate factor calculated previously (0.83), net solar power output becomes 939 kW of AC power.

MATERIALS

	COUNT	DESCRPTION	RATING
INVERTER	2	SATCON	500 KW
RECOMBINER BOXES - 4 POINT	4		
COMBINER BOXES - 12 POINT	48	SANYO N210	210 W
NUMBER OF STRINGS/COMBINER		12	
PV STRING LENGTH		12	
PV MODULES	**6912**		

SOLAR POWER INTEGRATION/INSTALLATION COST BREAKDOWN - FIXED AXIX COMMERCIAL

	MULIPLIER	UNIT COST PER WATT	WATTS	TOTAL 1KW
PV MODULE		$ **1.40**	1000	$ 1,400.00
DUAL AXIS TRACKER SYSTEM		$ –	1000	$ –
ROOF MOUNT SUPPORT		$ 0.35	1000	$ 350.00
INVERTER + COMBINER BOX		$ 0.35	1000	$ 350.00
ELECTRICAL MATERIALS		$ 0.35	1000	$ 350.00
LABOR		$ 0.30	1000	$ 300.00
PERMITS AND TAX		$ 0.15	1000	$ 150.00
TRANSPORTATION AND STORAGE		$ 0.20	1000	$ 200.00
MISCELL. FIELD WORKS		$ 0.20	1000	$ 200.00
CONSTRUCTION LOAN		$ 0.15	1000	$ 150.00
CONSTRUCTION BOND		$ 0.10	1000	$ 100.00
SUB TOTAL		$ 3.55	1000	$ 3,550.00
OVERHEAD AND PROFIT	15%	$ 0.53	1000	$ 532.50
INSTALLED COST	**8.0%**	$ 0.28	1000	$ 284.00
ENGINEERING SUPPORT				
ELECT-STRUCTURAL				
INSTALLED COST		$ 7.39		**$ 4,366.50**

Figure 8.5. Construction costing estimate for a fixed axis solar power system.

The lower part of the spreadsheet specifies a 25° fixed tilt angle solar PV support platform and farther down are computational algorithms that establish associated losses resulting from not tilting the installation angle at the maximum 34.8 latitude that could have resulted in yielding maximum power production capacity. Formulas adjacent to the second column in the left-most column are geometric normalization computations that yield a loss of 3.9% performance efficiency, which means that at the particular installed angle the solar power system will be performing at 96.9% efficiency.

Solar Power System Construction Costing

Figure 8.5 is a spreadsheet that computes construction cost of a hypothetical 1000-watt or 1-kW DC solar power system in Southern California. The exercise is intended to demonstrate basic cost components of a 1-megawatt solar power system. It should be noted that the standard norm used for estimating solar power system integration is based on par amount or value of the DC power capacity. This is because solar power manufacturers quote the sale value of PV modules, which is based on the STC power performance capacity of their products. As a result, cost estimating of solar power systems is commonly unitized on a cost per DC watt or kilowatt basis. In the following exercise, we would be using 1 watt as our basis for costing all cost components that are summed up to yield the final installed cost of our solar power system.

SOLAR POWER SYSTEM FIRST YEAR POWER PERFORMANCE PER PVWATTS V.2 - FIXED AXIS

ENERGY COST PER KILOWATT $ 0.180

COORDINATES : LATTITUDE 34.4 N,
LONGITUDE 118.2 N

YEAR ONE ECONOMETRIC ANALYSIS

MONTH	SOLAR RADIATION	DAYS PER MONTH	AC POWER KWatts/hr	AC POWER Kw/MONTH	ENERGY CONTRIB.
JANUARY	4.63	31	**910**	130625	$ 23,512.54
FEBRUARY	5.23	29	910	138033	$ 24,846.00
MARCH	5.8	31	910	163634	$ 29,454.15
APRIL	6.24	30	910	170369	$ 30,666.39
MAY	6.64	31	910	187333	$ 33,719.92
JUNE	7.05	30	910	192484	$ 34,647.12
JULY	7.02	31	910	198054	$ 35,649.68
AUGUST	7.12	30	910	194395	$ 34,991.14
SEPTEMBER	6.57	31	910	185358	$ 33,364.44
OCTOBER	5.86	30	910	159994	$ 28,798.88
NOVEMBER	5.24	31	910	147835	$ 26,610.30
DECEMBER	4.44	30	910	121224	$ 21,820.32

AVERAGE DAILY INSOLATION   **5.99**
TOTAL YEAR ONE   1989338 $ 358,080.88

OVERHEAD & MAINTENANCE   1.00%   PERCENT OF INSTALLED COST
COSPER YEAR
O & M PER YEAR $ 48,994
AVERAGE ESTIMATED YEARLY INFLATION   2%

Figure 8.6. First-year power output performance and economic contribution for a fixed axis solar power system.

The lower portion of the spreadsheet shown in Figure 8.5 consists of five columns. From left to right, the first column outlines various material, manpower, and miscellaneous expenses. The second column represents unit cost for each of the cost components. The third column represents the solar power system as 1000 watts, which forms the unit base that can be applied to a typical solar power system. The last column represents the calculated value of each of the cost items, and finally, the bottom left-most column establishes the installed cost of a 1-kW commercial system, which is calculated to be $4366.50.

It should be noted that construction costing of solar power systems in real settings includes considerably more costing elements and computational methodologies that are unique to each contractor, which are also subject to specific project requirements. In other words, there are no rule of thumb costing procedures.

First-Year Power Output Performance and Economic Contribution

Figure 8.6 is a spreadsheet representation of a 500-kW DC hypothetical solar power system. As mentioned earlier in this chapter, the exercise can also be scaled up to compute desired solar power system configuration.

The assumption used in calculating power performance and the economic contribution of our hypothetical solar power system are based on an electric tariff of $0.18/kWh. The geographic location of the solar power system installation is

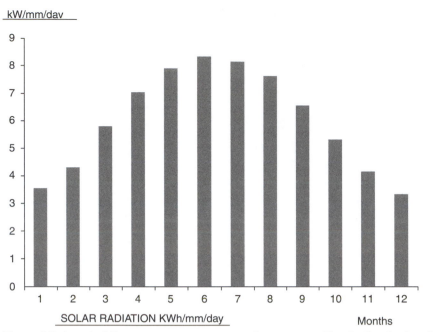

Figure 8.7. A typical first-year power output performance profile of a fixed axis solar power system.

assumed to be Huntington Beach, California, with a latitude of 34.4 degrees north and a longitude of 118.2.

The spreadsheet shown in Figure 8.6 consists of six columns. Commencing from the left-most column, the first column represents months of a calendar year. The second column shows solar insolation, or the average exposure time of the PV module in each day of the month. The third column represents days in each of the months. The fourth column is the AC power production capacity of the solar power system. The fifth column represents the monthly solar power production and the sixth column the energy contribution value of the solar power system based on the first-year grid electrical utility tariff. The bottom of the second column is the computed mean value of daily solar insolation in hours. The fifth column is the accumulated annual energy production, and the sixth column the annual economic contribution. The lower portion of the spreadsheet is the annual overhead and maintenance cost for the upkeep of the solar power system, which in this case is assumed to represent 1% of the installed system cost. Figure 8.7 depicts the typical yearly power output performance profile of a solar power system.

Solar Power System First and Second 25-Year Life Cycle Economic Contribution

Prior to discussing the economic contribution of solar power system, it is essential to note specific attributes of mono-silicon-based solar photovoltaic systems and more specifically understand their unique long-term power performance capabilities.

First, it is imperative to highlight the following features, which are uniquely associated with solar photovoltaic power generation systems:

- Mono-silicon- and poly-silicon-based solar power cells are constructed from glass-based elements that are inert and do not deteriorate or become oxidized.
- The etching processes used to construct the solar cells are identical to those used in the manufacture of transistors and solid state electronic devices.
- Solar cells, upon interconnection, form a PV module that is laminated with impermeable material that protects the cells from exposure to environmental conditions and prevents penetration of radical oxygen that corrodes soldered components.
- Solar power modules, as a result of their intrinsic power output performance and longevity, are guaranteed by manufacturers up to 25 years.
- Even though solar PV modules exhibit an annual performance degradation of about 1–1.5%, design methodologies used account for such losses over the guaranteed life span of the solar power projects.
- Solar power modules are manufactured to withstand environmental wear and tear as well as the impact of hail.
- Solar power modules require a minimum amount of maintenance and because of the absence of moving parts do not require repair and replacement. DC to AC conversion equipment used in solar power systems, which has warranty periods of 5 to 10 years, can be configured in a manner such that the failure of units can have a minimal effect on the overall power performance of the entire solar power system.
- Solar power PV modules that have been in operation for several decades are still functioning properly, supplying power to satellites and communication repeater stations.
- Solar power systems with worst case 1.5% annual power output degradation at the end of their system warranty period of 25 years still are capable of performing above 60% of their original power production capacity. This is a significant factor that is not taken into account through econometric analysis. This feature alone represents a significant salvage value that is generally not taken into consideration. At present it is not clear as to what extent the life cycle of solar power systems could be extended.
- Among all renewable energy technology systems, the solar power system is the only one that does not have a scheduled system retrofit or even a retirement schedule.

In view of the preceding specific mono-silicon PV technology features, we would be exploring economic analysis of solar power systems in light of two life cycles. The first 25-year life cycle computations presented in the spreadsheet (Figure 8.8) account for the economic contribution during the conventional warranty period of the project, and the second 25-year life cycle (Figure 8.9) accounts for the economic contribution of the solar power system in its extended life cycle.

The spreadsheet shown in Figure 8.8 represents computational procedures for establishing future values of the energy contribution during the first 25-year life cycle of a hypothetical solar power system. Assumptions made and parameters used in calculating the solar power system are as follows:

- The system is assumed to have 910 kW AC power production capacity (approximately 1 MW).
- Annual energy escalation rate is set at 6% and electrical tariff at $0.18/kWh.

SOLAR POWER SYSTEM BASE OUTPUT – KW AC	910	INSOLATION	5.99	HOURS
ENERGY ESCALLATION RATE	6%	UTILITY RATE $	0.18	9645039
ANNUAL SYSTEM DERATE	1.5%	ANNUAL KWH	1988668	

	n		EFFICIENCY	ADJUSTED	ANNUAL ENERGY
	YEAR	UTIL RATE	DERATE	ANN. Kw – AC	VALUE – FV
UTILITY RATE / KW AC – YEAR 1	1	$ 0.18	100%	1988668	$ 357960
UTILITY RATE / KW AC – YEAR 2	2	$ 0.19	99%	1958838	$ 373,746
UTILITY RATE / KW AC – YEAR 3	3	$ 0.20	97%	1929008	$ 390,138
UTILITY RATE / KW AC – YEAR 4	4	$ 0.20	96%	1899178	$ 384,105
UTILITY RATE / KW AC – YEAR 5	5	$ 0.21	94%	1869348	$ 400,756
UTILITY RATE / KW AC – YEAR 6	6	$ 0.23	93%	1839518	$ 418,023
UTILITY RATE / KW AC – YEAR 7	7	$ 0.24	91%	1809688	$ 435,919
UTILITY RATE / KW AC – YEAR 8	8	$ 0.26	90%	1779858	$ 454,457
UTILITY RATE / KW AC – YEAR 9	9	$ 0.27	88%	1750028	$ 473,651
UTILITY RATE / KW AC – YEAR 10	10	$ 0.29	87%	1720198	$ 493,512
UTILITY RATE / KW AC – YEAR 11	11	$ 0.30	85%	1690368	$ 514,051
UTILITY RATE / KW AC – YEAR 12	12	$ 0.32	84%	1660538	$ 535,279
UTILITY RATE / KW AC – YEAR 13	13	$ 0.34	82%	1630708	$ 557203
UTILITY RATE / KW AC – YEAR 14	14	$ 0.36	81%	1600878	$ 579,830
UTILITY RATE / KW AC – YEAR 15	15	$ 0.38	79%	1571048	$ 603,168
UTILITY RATE / KW AC – YEAR 16	16	$ 0.41	78%	1541218	$ 627218
UTILITY RATE / KW AC – YEAR 17	17	$ 0.43	76%	1511388	$ 651,983
UTILITY RATE / KW AC – YEAR 18	18	$ 0.43	75%	1481558	$ 639,115
UTILITY RATE / KW AC – YEAR 19	19	$ 0.46	73%	1451727	$ 663,822
UTILITY RATE / KW AC – YEAR 20	20	$ 0.48	72%	1421897	$ 689,192
UTILITY RATE / KW AC – YEAR 21	21	$ 0.51	70%	1392067	$ 715,218
UTILITY RATE / KW AC – YEAR 22	22	$ 0.54	69%	1362237	$ 741,885
UTILITY RATE / KW AC – YEAR 23	23	$ 0.58	67%	1332407	$ 769,178
UTILITY RATE / KW AC – YEAR 24	24	$ 0.61	66%	1302577	$ 797075
UTILITY RATE / KW AC – YEAR 25	25	$ 0.65	64%	1272747	$ 825,551
AVERAGE UTILITY RATE		$ 0.36			
CUMULATED VALUE				40767690	$ 14,092,035

Figure 8.8. Economic contribution of a fixed axis solar power system during its first 25-year life cycle.

- The solar power system is assumed to have maximum annual power output degradation of 1.5%.
- Daily insolation or solar power exposure is set at 5.5 hours per day.
- Future value is computed by the following equation: $PV = FV/(1 + i)^n$ where PV represents present value, FV represents future value, i represents the interest rate, and n represents the number of years in service.

Starting from the left-most columns, the first column of the spreadsheet represents the first 25-year life cycle service. The second column represents escalated cost of the grid energy cost; The third column represents cost of electrical energy at compound escalation rate of 6%; the fourth column shows computed percent efficiency of production. The fifth column shows computed annual electrical energy production capacity of the solar power system, and the last column represents computed future value of its economic contribution.

Figure 8.9 represents the computational algorithm for the second 25-year life cycle, which is identical in procedure to the first life cycle.

What should be noted is that at the start of the second life cycle, the solar power system commences its electrical power output capacity at 63% of the original potential and continues to produce a considerable amount of energy that hypothetically may amount to 11 gigawatts of power, a hefty economic contribution.

Even though some of the assumptions regarding an additional 25 years of extended service may be subject to dispute, however, the fact remains that solar power systems, if properly maintained, do not discharge as batteries do or become exhausted to an extent that their salvage value can be considered to be worthless.

SOLAR POWER SYSTEM YEAR 1 OUTPUT – KW AC	910	INSOLATION	5.99	HOURS
ENERGY ESCALLATION RATE	6%	UTILITY RATE $	0.180	
ANNUAL SYSTEM DERATE	1.5%			

	n YEAR	UTIL RATE	DERATE	ADJUSTED ANN.Kw – AC	ANNUAL ENERGY VALUE – FV
UTILITY RATE / KW AC – YEAR 26	26	$ 0.69	63%	1242917	$ 854,574
UTILITY RATE / KW AC – YEAR 27	27	$ 0.73	61%	1213087	$ 884,108
UTILITY RATE / KW AC – YEAR 28	28	$ 0.77	60%	1183257	$ 914,110
UTILITY RATE / KW AC – YEAR 29	29	$ 0.77	58%	1153427	$ 891,065
UTILITY RATE / KW AC – YEAR 30	30	$ 0.82	57%	1123597	$ 920,101
UTILITY RATE / KW AC – YEAR 31	31	$ 0.87	55%	1093767	$ 949,414
UTILITY RATE / KW AC – YEAR 32	32	$ 0.92	54%	1063937	$ 978,933
UTILITY RATE / KW AC – YEAR 33	33	$ 0.98	52%	1034107	$ 1,008,575
UTILITY RATE / KW AC – YEAR 34	34	$ 1.03	51%	1004277	$ 1,038,250
UTILITY RATE / KW AC – YEAR 35	35	$ 1.10	49%	974447	$ 1,067,856
UTILITY RATE / KW AC – YEAR 36	36	$ 1.16	48%	944617	$ 1,097,276
UTILITY RATE / KW AC – YEAR 37	37	$ 1.23	46%	914787	$ 1,126,383
UTILITY RATE / KW AC – YEAR 38	38	$ 1.31	45%	884957	$ 1,155,032
UTILITY RATE / KW AC – YEAR 39	39	$ 1.38	43%	855127	$ 1,183,065
UTILITY RATE / KW AC – YEAR 40	40	$ 1.47	41%	825297	$ 1,210,303
UTILITY RATE / KW AC – YEAR 41	41	$ 1.55	40%	795467	$ 1,236,550
UTILITY RATE / KW AC – YEAR 42	42	$ 1.65	38%	765637	$ 1,261,590
UTILITY RATE / KW AC – YEAR 43	43	$ 1.65	37%	735807	$ 1,212,437
UTILITY RATE / KW AC – YEAR 44	44	$ 1.75	35%	705977	$ 1,233,082
UTILITY RATE / KW AC – YEAR 45	45	$ 1.85	34%	676147	$ 1,251,838
UTILITY RATE / KW AC – YEAR 46	46	$ 1.96	32%	646317	$ 1,268,407
UTILITY RATE / KW AC – YEAR 47	47	$ 2.08	31%	616487	$ 1,282,457
UTILITY RATE / KW AC – YEAR 48	48	$ 2.21	29%	586657	$ 1,293,627
UTILITY RATE / KW AC – YEAR 49	49	$ 2.34	28%	556827	$ 1,301,520
UTILITY RATE / KW AC – YEAR 50	50	$ 2.48	26%	526997	$ 1,305,703
AVERAGE UTILITY RATE		**$ 0.69**			
CUMU8LATED VALUE				**22123929**	**$ 27,926,258**

Figure 8.9. Economic contribution of a fixed axis solar power system during its second 25-year life cycle.

Solar Power System First and Second Life Cycle Net PV Economic Contribution

The following exercise, like the previous one, is intended to demonstrate the significance of the economic value of solar power system technology and its long-term financial contribution. In order to proceed with the exercise we will use the preceding exercise and compute its net present value. To proceed we will make the following assumptions:

- Solar power energy production capacity remains (Figure 8.4) unchanged at 1122 kW.
- Construction unit cost per kilowatt remains at $4366.50.00/kW.
- Annual inflation rate will be set at 2%.
- Annual interest rate will be set at 6%.
- Annual overhead and maintenance cost will remain unchanged at 1%.

The spreadsheets shown in Figures 8.10 and 8.11 represent the net future value (NFV) economic computation. Both spreadsheets consist of eight columns. The first four columns from the left list parameters used for computing net PV, which is expressed by the compound interest given by $PV = FV/((1 + (i - r))^n$ where

FV represents future value
PV represents value
i represents annual interest rate
r represents average inflation rate
n represents years

The first column of the spreadsheet represents $(1 + (i - r))$ more clearly stated $(100\% + 5\% - 2\%)$, the second column (n) represents years of service, the third column represents the result of the $(1 + (i - r))^n$ factor, and the fourth column computes future value. The sixth column represents the annual overhead and maintenance (O&M) cost with an escalation rate of 1% per year. The sixth column represents the net annual economic contribution (net economic contribution less O&M). The seventh column represents the computed value of grid electrical energy, starting with a tariff rate of \$0.18/kWh and annual escalation rate of 6%. This column effectively represents the energy cost expenditure in the absence of the same solar power system, and the last column shows the respective annual saving.

Solar Power System Second Life Cycle

Unlike conventional capital equipment expenditures, which have a single life expectancy cycle, mono-silicon and poly-silicon PV modules have operational longevity that extends beyond the guaranteed 25-year warranty period. Such long life is inherent in PV technology, which uses materials that are hermetically sealed and ruggedly constructed and readily withstand harsh environmental operational conditions. Except for minimal degradation of the laminate material such as Tedlar (which sandwiches silicon cells and tempers glass covers) solar photovoltaic modules can continue their operational life cycle with a minimal annual degradation of about 1–1.5%. In order to meet power performance warranties of 25 years, solar power systems are designed to be oversized accordingly.

What also needs to be kept in mind is that the weakest links in any solar power system installations are the inverters, which have a warranty period of 5 years. This is because inverter designs incorporate filtering and timing capacitors, which intrinsically have shorter life spans, causing the failure of inverters. In addition to capacitor failure, large inverters deploy forced cooling fans that are subject to bearing wear and tear; as such, they also have a limited life span when compared to PV modules. In essence, with proper inverter maintenance and parts replacement procedures, at the end of the 25-year warranty life cycle, solar power output production capacity, at 1.5% annual depreciation, is reduced by 37% of its original power output performance capacity. This is a significant factor, which is commonly ignored or not taken into consideration when assessing economic viability of solar power systems. In essence at the end of a 25-year life cycle, a solar power system if maintained properly has an economic value or a salvage value equivalent to 63% of its original value.

The preceding long-term operational power output potential can be referred to as the second or extended life cycle. Figures 8.9 and 8.11 represent the second life cycle of solar power systems. To appreciate the merits of the solar power systems we must consider their extended life span, which may total an additional 25 years of service. Such extended service life will undoubtedly render solar power systems financially viable investments that yield superior returns on investment.

Long-Term Solar Power System Energy Savings

Another significant aspect of solar power system economic analysis often ignored is the energy saving that results not only from direct electrical energy cost reduction, but also from savings that result from not expending additional funds for purchase of grid energy. The energy cost saving computations are computed and listed in the

| SOLAR POWER | 910 | Kw | INTEREST RATE | 5% | ANNUAL MAINT | 1.0% |
| COST/KW | $ 4,899 | INST.COST | $ 4,899,423 | | r-INFLATION RATE | 2.0% |

| | | | PV = FV/(1+(i-r))>n | ANN. MAINT. | NET PRESENT | ENERGY | FUT. VALUE NET COST |
| | Years | | | | ENERGY | | |
1+(i-r)	<n	((1+(i-r))<n	PRESENT VALUE	CUM O&M	VALUE	COST	GRID ENERGY
103.0%	0	1.03	$ 357,960	$ (48,994)	$ 308,966	$ 0.180	$ 357,960
103.0%	1	1.030	$ 362,860	$ (49,484)	$ 313,376	$ 0.185	$ 363,169
103.0%	2	1.093	$ 357,032	$ (49,979)	$ 307,052	$ 0.197	$ 379,418
103.0%	3	1.093	$ 351,510	$ (50,479)	$ 301,032	$ 0.197	$ 373,551
103.0%	4	1.159	$ 345,696	$ (50,984)	$ 294,712	$ 0.209	$ 390,076
103.0%	5	1.159	$ 360,590	$ (51,493)	$ 309,097	$ 0.209	$ 383,851
103.0%	6	1.230	$ 354,442	$ (52,008)	$ 302,433	$ 0.221	$ 400,624
103.0%	7	1.230	$ 369,515	$ (52,528)	$ 316,987	$ 0.221	$ 394,020
103.0%	8	1.305	$ 363,014	$ (53,054)	$ 309,960	$ 0.235	$ 411,010
103.0%	9	1.305	$ 378,236	$ (53,584)	$ 324,652	$ 0.235	$ 404,004
103.0%	10	1.384	$ 371,362	$ (54,120)	$ 317,241	$ 0.249	$ 421,176
103.0%	11	1.384	$ 386,697	$ (54,661)	$ 332,035	$ 0.249	$ 413,743
103.0%	12	1.469	$ 379,428	$ (55,208)	$ 324,220	$ 0.264	$ 431,055
103.0%	13	1.469	$ 394,836	$ (55,760)	$ 339,076	$ 0.264	$ 423,170
103.0%	14	1.558	$ 387,150	$ (56,318)	$ 330,833	$ 0.280	$ 440,575
103.0%	15	1.558	$ 402,587	$ (56,881)	$ 345,707	$ 0.280	$ 432,210
103.0%	16	1.653	$ 394,460	$ (57,450)	$ 337,011	$ 0.298	$ 449,657
103.0%	17	1.653	$ 386,675	$ (58,024)	$ 328,651	$ 0.298	$ 440,782
103.0%	18	1.754	$ 378,568	$ (58,604)	$ 319,964	$ 0.316	$ 458,210
103.0%	19	1.754	$ 393,037	$ (59,190)	$ 333,846	$ 0.316	$ 448,795
103.0%	20	1.860	$ 384,465	$ (59,782)	$ 324,683	$ 0.335	$ 466,138
103.0%	21	1.860	$ 398,800	$ (60,380)	$ 338,420	$ 0.335	$ 456,149
103.0%	22	1.974	$ 389,736	$ (60,984)	$ 328,752	$ 0.355	$ 473,332
103.0%	23	1.974	$ 403,871	$ (61,594)	$ 342,278	$ 0.355	$ 462,735
103.0%	24	2.094	$ 394,288	$ (62,210)	$ 332,078	$ 0.377	$ 479,673
103.0%	25	2.094	$ 394,288	$ (63,454)	$ 330,834	$ 0.377	$ 479,673
			$ 9,446,816	$ (1,383,754)	$ 8,063,062		$ 10,555,082

Figure 8.10. Future value of a fixed axis solar power system during the first 25-year life cycle.

last four left-hand columns of Figures 8.9 and 8.11. The fourth column from the left represents the net future value of the solar power contribution. The third column from the left represents the grid energy inflation multiplier. The second column from the left represents the annual cost expenditure that would result from use of the same electrical energy amount in the second column from the left of Figures 8.8 and 8.10 (net annual solar power production). The future values of grid energy are simple multiples of the grid energy escalation factor and the amount of solar power energy production, as discussed previously. Net energy savings result from the difference of values in columns 4 and 2.

The preceding econometric exercise indicates that during the first 25-year life cycle the 1-MW hypothetical solar power system at an annual energy escalation rate of 6%, an interest rate of 5%, and inflation of 2% yields a remarkable $2,492,062 of savings, and in its second life span (however hypothetical), it results in an additional $4,307,364.

What is of significant value is that the extended life span of solar power systems represents a significant contribution to the overall economics of the solar power system, as such; if considered as a long-term investment, solar power can be considered as an extremely favorable financial investment, which should be taken into serious consideration by banking and lending institutions.

Figure 8.12 and 8.13 are graphic representation of economic contribution of the above solar power system's first and second life cycle energy cost profiles.

| SOLAR POWER | 701280 | Kw | r-INFLATION RATE | 5% | ANNUAL MAINT | 1.0% |
| ADJ. CST/KW | $ (985) | SYSTEM . COST | $ (1,224,855.65) | | r-INFLATION RATE | 2.0% |

	years		PV = FV/(1+i)<n	ANN. MAINT.	NET PRESENT	ENERGY	FUT.VALUE NET COST
1+(i-r)	<n	((1+(i-r))<n	PRESENT VALUE	CUM O&M	ENERGY VALUE	COST	GRID ENERGY
103.0%	25	2.094	$ 408,149	$ (62,832)	$ 345,318	$ 0.377	$ 468,431
103.0%	26	2.221	$ 398,016	$ (63,460)	$ 334,556	$ 0.400	$ 485,031
103.0%	27	2.221	$ 411,522	$ (64,095)	$ 347,428	$ 0.400	$ 473,104
103.0%	28	2.357	$ 378,120	$ (64,736)	$ 313,385	$ 0.424	$ 489,263
103.0%	29	2.357	$ 390,442	$ (65,383)	$ 325,059	$ 0.424	$ 476,610
103.0%	30	2.500	$ 379,754	$ (66,037)	$ 313,717	$ 0.450	$ 492,211
103.0%	31	2.500	$ 391,560	$ (66,697)	$ 324,863	$ 0.450	$ 478,787
103.0%	32	2.652	$ 380,259	$ (67,364)	$ 312,895	$ 0.477	$ 493,704
103.0%	33	2.652	$ 391,448	$ (68,038)	$ 323,410	$ 0.477	$ 479,462
103.0%	34	2.814	$ 379,498	$ (68,718)	$ 310,780	$ 0.506	$ 493,553
103.0%	35	2.814	$ 389,954	$ (69,405)	$ 320,548	$ 0.506	$ 478,444
103.0%	36	2.985	$ 377,319	$ (70,099)	$ 307,220	$ 0.537	$ 491,552
103.0%	37	2.985	$ 386,916	$ (70,800)	$ 316,116	$ 0.537	$ 475,524
103.0%	38	3.167	$ 373,557	$ (71,508)	$ 302,048	$ 0.570	$ 487,478
103.0%	39	3.167	$ 382,157	$ (72,223)	$ 309,934	$ 0.570	$ 470,473
103.0%	40	3.360	$ 368,032	$ (72,946)	$ 295,086	$ 0.605	$ 481,084
103.0%	41	3.360	$ 375,485	$ (73,675)	$ 301,809	$ 0.605	$ 463,043
103.0%	42	3.565	$ 340,141	$ (74,412)	$ 265,729	$ 0.642	$ 472,103
103.0%	43	3.565	$ 345,932	$ (75,156)	$ 270,776	$ 0.642	$ 452,964
103.0%	44	3.782	$ 331,034	$ (75,908)	$ 255,127	$ 0.681	$ 460,245
103.0%	45	3.782	$ 335,416	$ (76,667)	$ 258,749	$ 0.681	$ 439,940
103.0%	46	4.012	$ 319,664	$ (77,433)	$ 242,230	$ 0.722	$ 445,191
103.0%	47	4.012	$ 322,448	$ (78,208)	$ 244,240	$ 0.722	$ 423,649
103.0%	48	4.256	$ 305,793	$ (78,990)	$ 226,803	$ 0.766	$ 426,596
103.0%	49	4.256	$ 306,775	$ (79,780)	$ 226,996	$ 0.766	$ 403,743
103.0%	25	2.09	$ 13,337,736	$ (1,774,570)	$ 7,394,821		$ 11,702,185

Figure 8.11. Future value of a fixed axis solar power system during the second 25-year life cycle.

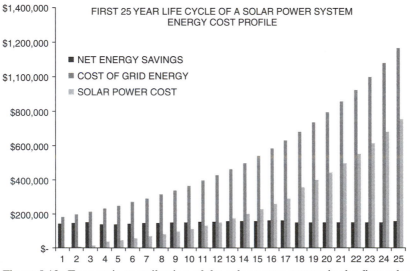

Figure 8.12. Economic contribution of the solar power system in the first solar life cycle.

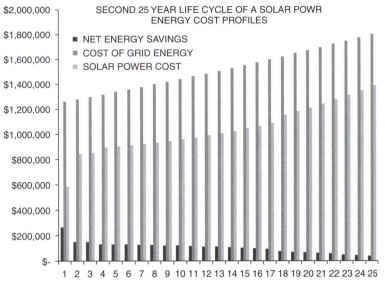

Figure 8.13. Economic contribution of the solar power over the second life cycle.

The bottom of the columns shows totalized values. Once again it must be noted that the previous exercise is intended to demonstrate the viability of solar power systems as long-term investment instruments.

The construction cost estimate per kilowatt of solar power used in the preceding exercise is extremely conservative. Nowadays installation cost for multimegawatt solar power installations in the United States has approached $4000.00/kW or less. Considering the fact that the cost of PV manufacturing over the last 5 years has decreased by nearly 50%, it is inevitable that within a short time, improvement in silicon purification techniques and introduction of less expensive solar power technologies will significantly enhance the financial perspective of solar power investment.

Solar Power System Economics Summary Overview

The spreadsheet shown in Figure 8.14 is a summary representation of the previous hypothetical solar power system economic evaluation exercise.

The spreadsheet in the figure consists of four separate segments. The upper segment represents the computed installation cost of construction, as discussed previously in this chapter. The second segment from the top represents computed summary value of the CSI PBI rebate at $0.20/ kWh for a period of 5 years. The totalized kilowatt hours are a summation of the first 5 years of energy generation capacity shown under the fifth column of Figure 8.8.

The left bottom segment represents projected O&M expense for the first and second life cycles. The lower segment is an account summary of the first and second life cycle operations of the econometric exercise.

What is of significant interest is that the net economic value of the solar power system in the first life cycle shows that when considering incentives, the economic value of the solar power system (sixth line), the accelerated depreciation and the net present value of grid energy savings, validates the viability of solar power investment.

ECONMIC COST SUMMARY
DUAL AXIS TRACKING SYSTEM

ESTIMATED INSTALLAED SOLARCOST /KW AC	$	4,366.50
INSTALLED COAS PER WATT	$	4.37
SYSTEM SIZE KW DC		1,122,048

PROJECTED TOTAL INSTALLED COST	$	4,899,423

PBI - 5 YEAR ELECTRICAL ENERGY - KW AC		9645039	kw/hr
PERFORMANCE BASED IREBATE / KW			
CSI REBATE/kWh (OVER FIVE YEARS)	$	0.20	PER kWhr

REBATE ANF FEDERAL INCENTIVE SUMMARY

TOTAL CSI - PBI REBATE		$	1,929,008	
FEDREAL TAX CREDIT	30%	$	1,469,827	PERCENT OF INSTALLED COST
TOTAL INCENTIVE VALUE		$	3,398,835	
OUT OF POCKET COST IN 5 YEARS		$	1,500,588	INSTALLED COST LESS INCENTIVES OVER 5 TEARS

INCOME PERFORMANCE SUMMARY
FIRST LIFECYCLE

INATALLED COST	$	(4,899,423)			
TOTAL VALUE OF INCENTIVES			$	3,398,835	
DEPRECIATION VALUE - 5 YEARS			$	1,224,856	%25 OF INSTALLED COST
ENERGY SAVINGS OVER FIRST 25 YEAR LIFE CYCLE			$	8,063,062	
FIRST 25 YEAR LIFECYCLE ENERGY SAVINGS			$	2,492,020	
SYSTEM VALUE PV 1ST LIFECYCLE			$	15,178,772	
NET PRESENT VALUE - IN FIRST 25 YEARS	$	10,279,350			
SECOND LIFECYCLE					
ENERGY VALUE OVER SECOND 25 YEAR LIFE CYCLE	$	7,394,821			
SECOND 25 YEAR LIFECYCLE ENERGY SAVINGS	$	4,307,364			
NET PRESENT VALUE - IN FIRST 25 YEARS	$	11,702,185			
SYSTEM COMBINED VALUE 1ST & 2ND LIFECYCLE			$	21,981,534	

Figure 8.14. Solar power system economic contribution summary overview.

US 5 YEAR ACCELERATED DEPRECIATION CALCULATION

CONSTRUCTION COST	$	4,899,423

YEAR	PERCENT DEPRECIATION	DEPRECIATION AMOUNT
1	20%	$ 979,885
2	32%	$ 1,567,815
3	19%	$ 940,689
4	12%	$ 564,413
5	12%	$ 564,413
6	6%	$ 282,207
TOTAL		$ 4,899,423

ESTIMATED % DEPRECIATION IN 25 YEARS		25%
SYSTEM VALUE IN 25 YEARS	$	1,224,856

Figure 8.15(a). U.S. accelerated depreciation calculation for the solar power system.

RETURN ON INVESTMENT

YEARS	ACCUM. PRES. VALUE	NET CONSTR. COST	PAYOFF IN YEARS
1	$ 308,966	$ 1,500,588	$ 1,191,622
2	$ 721,129		$ 779,459
3	$ 1,022,160		$ 478,428
4	$ 1,316,873		$ 183,716
5	$ 1,625,969		$ (125,381)
6	$ 1,928,403		$ (427,815)
7	$ 2,245,389		$ (744,801)
8	$ 2,608,404		$ (1,107,815)
9	$ 2,986,639		$ (1,486,051)
10	$ 3,358,001		$ (1,857,413)
11	$ 3,744,698		$ (2,244,110)
12	$ 4,124,125		$ (2,623,537)
13	$ 4,518,962		$ (3,018,374)

Figure 8.15(b). Return on investment for the fixed axis solar power system.

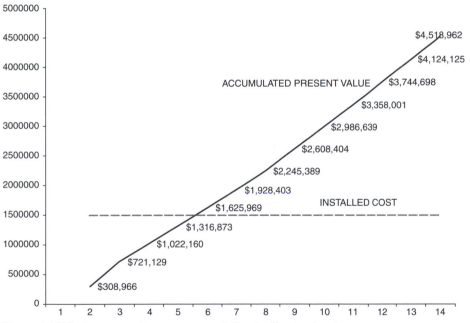

Figure 8.15(c). Return on investment graph for the fixed axis solar power system.

Furthermore, even discounting the combined incentive value indicates that the solar power system can still become a viable investment. Figure 8.15(a) represents accelerated depreciation of solar power capital investment that is allowed in the United States.

Figure 8.15(b) shows return on investment in years. Figure 8.15(c) is a graphic representation of the return on investment inflection point. The lower part of the figure is a projection of the amount of depreciated value of the solar power system in

ENVIRONMENTAL IMPACT	kWhr	CO2 MULTIPLIER	TONS CO2	POUNDS CO2
ELECTRICITY / KILOWATT HOUR PER YEAR	939183	0.009	8452.65	16905294
ELECTRICITY - KWH FIRST 25 YEARS	40767690	0.009	366909.21	733818415
ELECTRICITY - KWH SECOND 25 YEARS	22123929	0.009	199115.36	398230725
TOTAL EXTENDED LIFE CYCLE	**62891619**		**566025**	**1132049140**

Figure 8.16. Carbon dioxide abatement of the solar power system over the first and second life cycles of operation.

the first 25-year life cycle as having a salvage value equivalent to 25% of the original construction cost. In view of the speculative nature of the assumption, the salvage value has not been taken into consideration in the preceding econometric exercise.

It should be noted that this indicates that the solar power "out of pocket" installed cost amounts to less than 50% of the system cost. Furthermore, when considering income summary performance, the net present value in the first life cycle exceeds twice the projected installed cost of the solar power system.

Therefore, when considering extended life expectancy of the solar power system, even though somewhat hypothetical, the economic analysis shown in Figure 8.14 suggests that extended financial contributions cannot be dismissed, invalidated or discounted as a mere theoretical exercise.

Solar Power System as CO_2 Free Energy Source

In view of the fact that solar power systems do not use fossil fuel feedstock for generating electrical energy, they do not contribute to atmospheric pollution. Even though during the fabrication and silicon purification process insignificant amounts of atmospheric pollution (CO_2) are created, however, when considering the overall system longevity of the solar power systems, they are indeed considered as the best energy production technology, which creates the lowest atmospheric pollution footprint. Figure 8.16 represents a table of CO_2 abatement that can result from the deployment of a solar power system in their first and second 25-year life cycles.

In general electrical power plants (with the exception of hydropower), generate CO_2 air pollution, which results from burning of various types of fossil fuel feedstocks such as coal and natural gas. The carbon dioxide multiplier index listed in the third column of the table represents an average of mixed use of coal and natural gas in the generation of 1 kilowatt hour (kWh) of grid electrical power.

As indicated in the table, significant amounts of atmospheric pollution can be abated by the use of a 1-megawatt solar power system. To quantify the amount of CO_2 abated, one should consider the fact that a single pound of the gas in frozen state, commonly referred to as dry ice, occupies roughly a volume equivalent to a gallon milk container. It is therefore not difficult to visualize the thousands of tons of atmospheric pollution that can be prevented by maximizing the use of solar power technology.

Single and Dual Axis Solar Power Tracking Systems

As discussed earlier in Chapter 3, single and dual axis solar power systems have the potential to provide 25–50% more electrical power output production capability

MATERIALS

	COUNT	DESCRPTION	RATING
INVERTER	2	SATCON	500 KW
RECOMBINER BOXES - 4 POINT	4		
COMBINER BOXES - 12 POINT	48	SANYO N210	210 W
NUMBER OF STRINGS/COMBINER		12	
PV STRING LENGH		12	
PV MODULES	**6912**		

SOLAR POWER INTEGRATION/INSTALLATION COST BREAKDOWN – SINGLE AXIS COMMERCIAL

	MULIPLIER	UNIT COST PER WATT	WATTS	TOTAL 1 KW
PV MODULE		$ **1.40**	1000	$ 1,400.00
DUAL AXIS TRACKER SYSTEM		$ 0.50	1000	$ 500.00
ROOF MOUNT SUPPORT		$ 0.35	1000	$ 350.00
INVERTER + COMBINER BOX		$ 0.35	1000	$ 350.00
ELECTRICAL MATERIALS		$ 0.35	1000	$ 350.00
LABOR		$ 0.30	1000	$ 300.00
PERMITS AND TAX		$ 0.15	1000	$ 150.00
TRANSPORTATION AND STORAGE		$ 0.20	1000	$ 200.00
MISCELL. FIELD WORKS		$ 0.20	1000	$ 200.00
CONSTRUCTION LOAN		$ 0.15	1000	$ 150.00
CONSTRUCTION BOND		$ 0.10	1000	$ 100.00
SUB TOTAL		$ 4.05	1000	$ 4,050.00
OVERHEAD AND PROFIT	15%	$ 0.61	1000	$ 607.50
INSTALLED COST	8.0%	$ 0.32	1000	$ 324.00
ENGINEERING SUPPORT				
ELECT-STRUCTURAL				
INSTALLED COST		$ 7.39		$ **4,981.50**

Figure 8.17. Single axis solar power system costing ledger.

compared to fixed axis solar power systems. Even though somewhat more costly to install, such systems do add a considerable amount to the return on investment and have a more accelerated return as well. Therefore, when considering a solar power investment, lending institutions as well as owners must undertake thorough evaluation through feasibility studies. Figures 8.17, 8.18, and 8.19 represent single axis solar power system costing, first-year power output contribution, financial contribution performance, and the summary of the economics. Figures 8.23, 8.24, 8.25 and 8.26 represent a dual axis solar power system costing, first-year power output contribution, financial contribution performance, and the summary of the economics. Figure 8.20 shows the graph of return on investment for a single axis solar power system. Figure 8.21 is a photograph of a single axis solar power system, and Figure 8.22 is a photograph of a dual axis solar power system.

NREL – Solar Advisor Model 2010

In order to substantiate long-term economic investment benefits of large-scale solar power systems, we will use the National Renewable Energy Laboratories (NREL)

SOLAR POWER SYSTEM FIRST YEAR POWER PERFORMANCE PER PVWATTS V.2 – SINGLE AXIS

ENERGY COST PER KILOWATT $ 0.180

COORDINATES : LATTITUDE 34.4 N,
LONGITUDE 118.2 N

YEAR ONE ECONOMETRIC ANALYSIS

MONTH	SOLAR RADIATION	DAYS PER MONTH	AC POWER kWatts/hr	AC POWER Kw/MONTH	ENERGY CONTRIB.
JANUARY	4.63	31	1213	174108	$ 31,339.37
FEBRUARY	5.23	29	1213	183982	$ 33,116.72
MARCH	5.80	31	1213	218105	$ 39,258.82
APRIL	6.24	30	1213	227081	$ 40,874.59
MAY	6.64	31	1213	249692	$ 44,944.58
JUNE	7.05	30	1213	256558	$ 46,180.43
JULY	7.02	31	1213	263982	$ 47,516.71
AUGUST	7.12	30	1213	259105	$ 46,638.96
SEPTEMBER	6.57	31	1213	247060	$ 44,470.77
OCTOBER	5.86	30	1213	213252	$ 38,385.43
NOVEMBER	5.24	31	1213	197046	$ 35,468.32
DECEMBER	4.44	30	1213	161577	$ 29,083.84

AVERAGE DAILY INSOLATION TOTAL YEAR ONE	5.99			2651548	$ 477,278.55

OVERHEAD & MAINTENANCE COS PER YEAR		1.00%	PERCENT OF INSTALLED COST	
O & M PER YEAR	$	55,895		
AVERAGE ESTIMATED YEARLY INFLATION		2%		

Figure 8.18. Single axis solar power system first-year energy production and economic contribution.

solar power system economic computations program called the Solar Advisor Model (SAM) to validate the results yielded by the preceding economic spreadsheet program.

Since SAM is a Web-based software program (http://www.sam.com), we will not discuss details of the program formulation. Instead, we will explore results of economic computations and associated graphs. Results of SAM calculations are based upon parameters identical to those used in the previous econometric program.

Figures 8.17 through 8.32 represent SAM solar photovoltaic computational results. As mentioned, for comparison of economic evaluation, main entry parameters such as system PV and equipment count, utility energy escalation rate, interest rate, and so forth, are identical to the previous exercise. For clarity purposes, the life cycle economic computations in addition to 50-year life cycle are shown in Figure 8.32(a) through 8.32(g). The first and second 25-year life cycles are analyzed separately.

In order to demonstrate the economic viability of the solar power systems of the second 25-year life cycle shown, expenses taken into account include overhead and maintenance and inverter replacement. All other parameters remain unchanged. The following are descriptions of the SAM analytical software results:

• Figure 8.27 represents SAM program data entry subsections.

ESTIMATED INSTALLAED SOLAR COST/KW AC	$	4,981.50		
INSTALLED COST PER WATT	$	4.98		
SYSTEM SIZE KW DC		1,122,048		
PROJECTED TOTAL INSTALLED COST	$	**5,589,482**		
PBI - 5 YEAR ELECTRICAL ENERGY - KW AC		**12855671**	kw/hr	ENERGY PRODUCTION OVER 5 YEARS
PERFORMANCE BASED REBATE / KW				
CSI REBATE/kWh (OVER FIVE YEARS)	$	0.20	PER kWhr	

REBATE AND FEDERAL INCENTIVE SUMMARY

TOTAL CSI - PBI REBATE			$	2,571,134	
FEDERAL TAX CREDIT	30%	$	1,676,845	PERCENT OF INSTALLED COST
TOTAL INCENTIVE VALUE		$	**4,247,979**	
OUT OF POCKET COST IN 5 YEARS		$	**1,341,503**	**INSTALLED COST LESS INCENTIVES OVER 5 YEARS**

INCOME PERFORMANCE SUMMARY
FIRST LIFECYCLE

INSTALLED COST	$	(5,589,482)		
TOTAL VALUE OF INCENTIVES			$	4,247,979
DEPRECIATION VALUE - 5 YEARS			$	**1,397,371 25%**
				OF INSTALLED COST
ENERGY SAVINGS OVER FIRST 25 YEAR LIFE CYCLE			$	11,012,816
FIRST 25 YEAR LIFECYCLE ENERGY SAVINGS			$	3,055,833
SYSTEM VALUE (PV 1ST LIFECYCLE)			$	**19,713,999**
NET PRESENT VALUE – IN FIRST 25 YEARS	$	**14,124,517**		
SECOND LIFECYCLE				
ENERGY VALUE OVER SECOND 25 YEAR LIFE CYCLE	$	10,197,181		
SECOND 25 YEAR LIFECYCLE ENERGY SAVINGS	$	5,400,418		
NET PRESENT VALUE – FIRST 25 YEARS	$	**15,597,598**		
SYSTEM COMBINED VALUE 1ST & 2ND LIFECYCLE		$	**29,722,115**	

Figure 8.19. Summary of economic contribution from the single axis solar power system.

- Figure 8.28 represents field entry location of the project site. SAM has limited geographic data for representative locations within the United States, which are selectable through a pull-down menu. For instance, the project location selected for this exercise is Huntington Beach, California; however, the representative location closest to geographic vicinity available is Long Beach, California. As shown in the figure, in addition to geographic coordinates, the report displays weather information.
- Figures 8.29 and 8.30 represent SAM selected PV module and inverter equipment specifications and output performance curves. The program includes pull-down menus for most California Energy Commission (CEC) listed solar power components and equipment. As in the previous exercise, the PV module selected from the pull-don menus is SANYO-HIT series 210-watt module. The inverter selected is a quantity of four Satcon PVS 250-kW units. Both report sheets outline all significant parameters that are used in computation of the solar power system's projects energy production potential, used in econometric analysis.

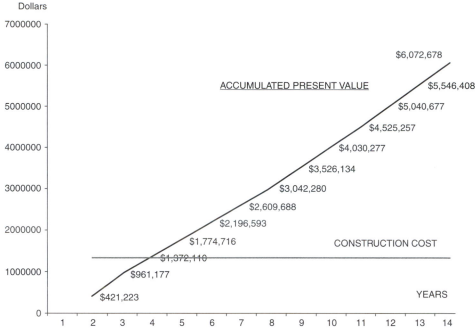

Figure 8.20. Return on investment graph for the single axis solar power system.

Figure 8.21. A single axis solar power system.
Source: Photo courtesy of Vector Delta Design Group, Inc.

- Figure 8.31 displays five entry fields: system layout information, system derate information, system PV module support platform, radiation model (for beam or diffused), and PV platform tilt angle ratio calculation model (not discussed in this book). The layout or system information field entries shown are identical in PV count and inverter capacity to those used in the prior exercise. The inverter selected is shown to have 97% efficiency.
- Figure 8.32 displays system installation and costing information entry fields. In order to maintain equivalency, cost parameters have been adjusted to represent identical installation costs to the one used in our previous exercise. Operation

MATERIALS

	COUNT	DESCRPTION	RATING
INVERTER	2	SATCON	500 KW
RECOMBINER BOXES - 4 POINT	4		
COMBINER BOXES - 12 POINT	48	SANYO N210	210 W
NUMBER OF STRINGS/COMBINER		12	
PV STRING LENGTH		12	
PV MODULES	6912		

SOLAR POWER INTEGRATION/INSTALLATION COST BREAKDOWN -DUAL AXIS COMMERCIAL

	MULIPLIER	UNIT COST PER WATT	WATTS	TOTAL 1 KW
PV MODULE		$ 1.40	1000	$ 1,400.00
DUAL AXIS TRACKER SYSTEM		$ 1.00	1000	$ 1,000.00
ROOF MOUNT SUPPORT		$ 0.35	1000	$ 350.00
INVERTER + COMBINER BOX		$ 0.35	1000	$ 350.00
ELECTRICAL MATERIALS		$ 0.35	1000	$ 350.00
LABOR		$ 0.30	1000	$ 300.00
PERMITS AND TAX		$ 0.15	1000	$ 150.00
TRANSPORTATION AND STORAGE		$ 0.20	1000	$ 200.00
MISCELL. FIELD WORKS		$ 0.20	1000	$ 200.00
CONSTRUCTION LOAN		$ 0.15	1000	$ 150.00
CONSTRUCTION BOND		$ 0.10	1000	$ 100.00
SUB TOTAL		$ 4.55	1000	$ 4,550.00
OVERHEAD AND PROFIT	15%	$ 0.68	1000	$ 682.50
INSTALLED COST	8.0%	$ 0.36	1000	$ 364.00
ENGINEERING SUPPORT ELECT-STRUCTURAL				
INSTALLED COST		$ 7.39		$ 5,596.50

Figure 8.22. Dual axis solar power system costing ledger.

SOLAR POWER SYSTEM FIRST YEAR POWER PERFORMANCE PER PVWATTS V.2

ENERGY COST PER KILOWATT $ 0.180

COORDINATES : LATTITUDE 34.4 N, LONGITUDE 118.2 N

YEAR ONE ECONOMETRIC ANALYSIS

MONTH	SOLAR RADIATION	DAYS PER MONTH	AC POWER kWatts/hr	AC POWER Kw/MONTH	ENERGY CONTRIB.
JANUARY	4.63	31	1638	235125	$ 42,322.57
FEBRUARY	5.23	29	1638	248460	$ 44,722.80
MARCH	5.80	31	1638	294541	$ 53,017.47
APRIL	6.24	30	1638	306664	$ 55,199.50
MAY	6.64	31	1638	337199	$ 60,695.86
JUNE	7.05	30	1638	346471	$ 62,364.82
JULY	7.02	31	1638	356497	$ 64,169.42
AUGUST	7.12	30	1638	349911	$ 62,984.05
SEPTEMBER	6.57	31	1638	333644	$ 60,056.00
OCTOBER	5.86	30	1638	287989	$ 51,837.99
NOVEMBER	5.24	31	1638	266103	$ 47,898.54
DECEMBER	4.44	30	1638	218203	$ 39,276.57
AVERAGE DAILY INSOLATION	5.99				
TOTAL YEAR ONE				3580809	$ 644,545.58

OVERHEAD & MAINTENANCE COSPER YEAR		1.00%	PERCENT OF INSTALLED COST	
O & M PER YEAR	$ 75,354			
AVERAGE ESTIMATED YEARLY INFLATION	2%			

Figure 8.23. Dual axis solar power system first-year power production and economic contribution.

ECONMIC COST SUMMARY
DUAL AXIS TRACKING SYSTEM

ESTIMATED INSTALLAED SOLARCOST /KW AC	$	5,596.50
INSTALLED COST PER WATT	$	5.60
SYSTEM SIZE KW DC		1,346,458
PROJECTED TOTAL INSTALLED COST	$	7,535,450

PBI - 5 YEAR ELECTRICAL ENERGY - KW AC		17361070	kW/hr	ENERGY PRODUCTION OVER 5 YEARS
PERFORMANCE BASED REBATE / KW			
CSI REBATE/kWh (OVER FIVE YEARS)	$	0.20	PER kWhr

REBATE AND FEDERAL INCENTIVE SUMMARY

TOTAL CSI - PBI REBATE			$	3,472,214
FEDERAL TAX CREDIT	30%		$	2,260,635 PERCENT OF INSTALLED COST
TOTAL INCENTIVE VALUE			$	5,732,849
OUT OF POCKET COST IN 5 YEARS			$	1,802,601 INSTALLED COST LESS INCENTIVES OVER 5 YEARS

INCOME PERFORMANCE SUMMARY
FIRST LIFECYCLE

INATALLED COST	$	7,535,450		
TOTAL VALUE OF INCENTIVES			$	5,732,849
DEPRECIATION VALUE - 5 YEARS			$	1,883,862 25% OF INSTALLED COST
ENERGY SAVINGS OVER FIRST 25 YEAR LIFE CYCLE			$	14,876,017
SYSTEM VALUE PV (1ST LIFECYCLE)	$	22,492,728		
NET PRESENT VALUE - IN FIRST 25 YEARS	$	14,957,278		

SECOND LIFECYCLE

ENERGY VALUE OVER SECOND 25 YEAR LIFE CYCLE	$	17,812,074
SYSTEM COMBINED VALUE 1ST & 2ND LIFECYCLE	$	32,769,352

Figure 8.24. Dual axis solar power system economic summary.

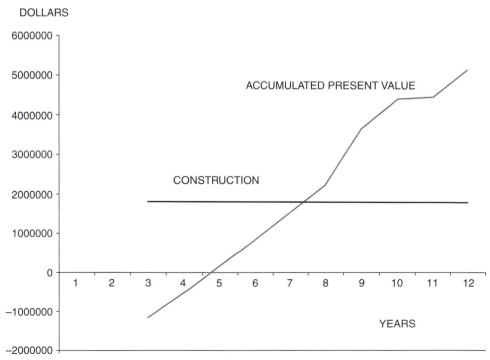

Figure 8.25. Dual axis solar power system return on investment graph.

Figure 8.26. A dual axis solar power system.
Source: Photo courtesy of AMONIX

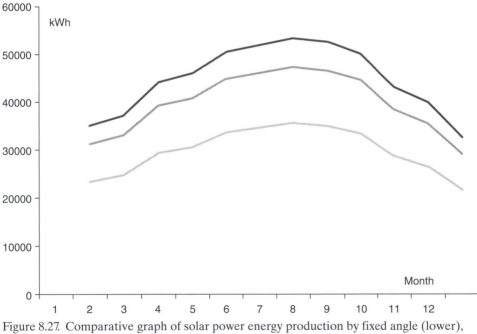

Figure 8.27. Comparative graph of solar power energy production by fixed angle (lower),
single axis (middle), and dual axis (upper) tracking platforms.

and maintenance cost shown in the lower part of the figure is automatically
computed by the program.

• Figure 8.33 shows five entry fields: the analysis period (in this case a 50-year
life cycle), inflation rate, tax and insurance (which may or may not apply),
accelerated depreciation, and additional data, which are used to compute Power
Purchase Agreement rates as well as the internal rate of return (IRR).

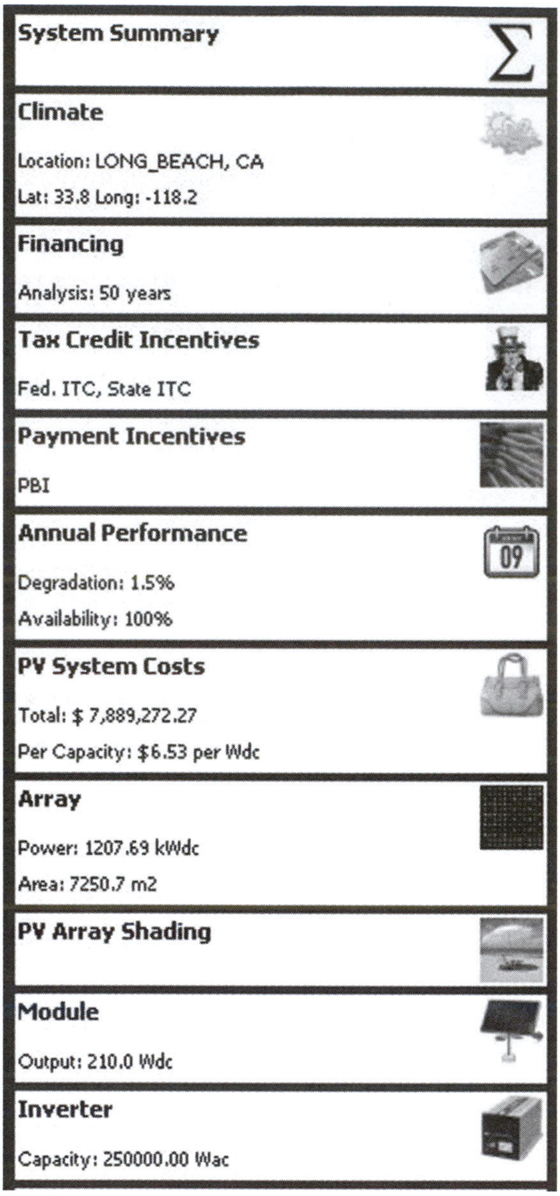

System Summary	Σ
Climate Location: LONG_BEACH, CA Lat: 33.8 Long: -118.2	
Financing Analysis: 50 years	
Tax Credit Incentives Fed. ITC, State ITC	
Payment Incentives PBI	
Annual Performance Degradation: 1.5% Availability: 100%	
PV System Costs Total: $ 7,889,272.27 Per Capacity: $6.53 per Wdc	
Array Power: 1207.69 kWdc Area: 7250.7 m2	
PV Array Shading	
Module Output: 210.0 Wdc	
Inverter Capacity: 250000.00 Wac	

Figure 8.28. NREL-SAM solar power system economic computation program overview page.

- Figures 8.34, 8.35, 8.36 represent the entry field for Investment Tax Credit (ITC), anticipated annual system performance figures, and PBI data entry.
- Figure 8.37 represents the average monthly insolation hours, presented as 0 and 1. In the shading matrix map, "0" represents no solar exposure and "1" represents full solar exposure. In view of an average of 6 hours of insolation, for simplicity, months of the year are marked to represent 6 hours of solar exposure.
- Figure 8.38 depicts the graph of the first-year solar power output profile.
- Figure 8.39 depicts the graph of a 50-year life cycle.
- Figure 8.40 depicts the financial contribution of a 50-year life cycle.

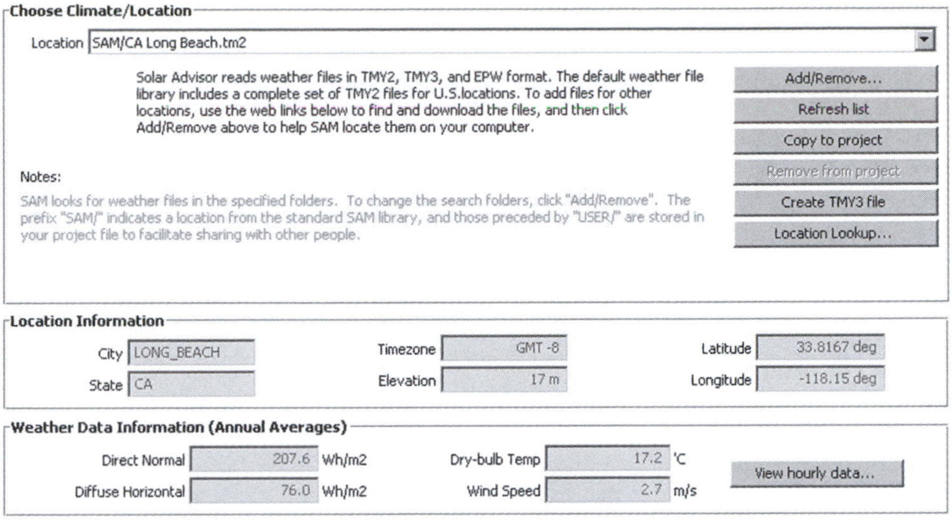

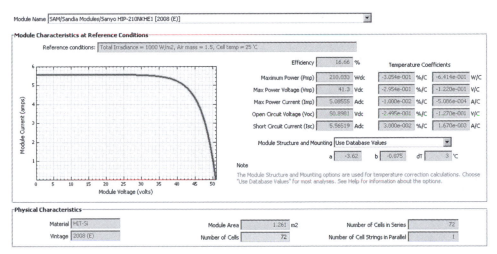

Figure 8.29. SAM solar power project location entry fields.

Figure 8.30. SAM PV module specification.

- Figure 8.41 shows revised system entry information for the second 25-year life cycle of the solar power system. As shown in the layout field, PV module count has been reduced from 575 units to 362 units, which lowers the capacity of the solar power system to 63% of its original power output production potential, which matches our previous exercise.
- Figure 8.42 shows a revised cost entry for the second 25-year lifecycle of the solar power system. Costs incurred in this period include replacement of major components for all four inverters and maintenance and operation expenses.
- Figure 8.43 depicts the graphic plot of the economic contribution of the solar power system in its second 25-year life cycle. It should be noted that the graph, because of reduced system upkeep cost and zero investment dollars, shows a significant return on investment. As stipulated previously, even though assumptions used in calculation of the extended or second life cycle may be

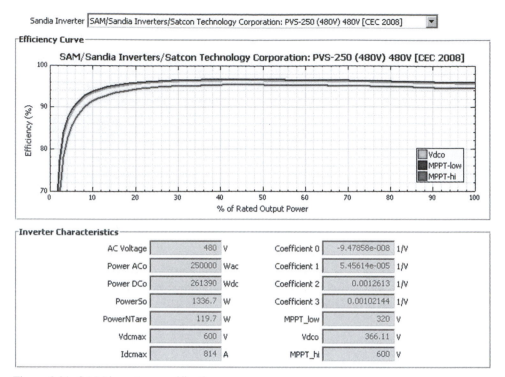

Figure 8.31. SAM inverter specification.

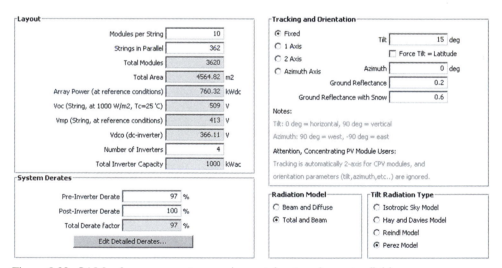

Figure 8.32. SAM solar power system equipment derate value entry fields.

somewhat hypothetical, nevertheless, economic contributions as shown in both exercises should be considered to be of significant importance when evaluating the economic contribution of a solar power system investment.

- Figures 8.44(a) through 8.44 (g) are SAM report results printed in spreadsheet format.

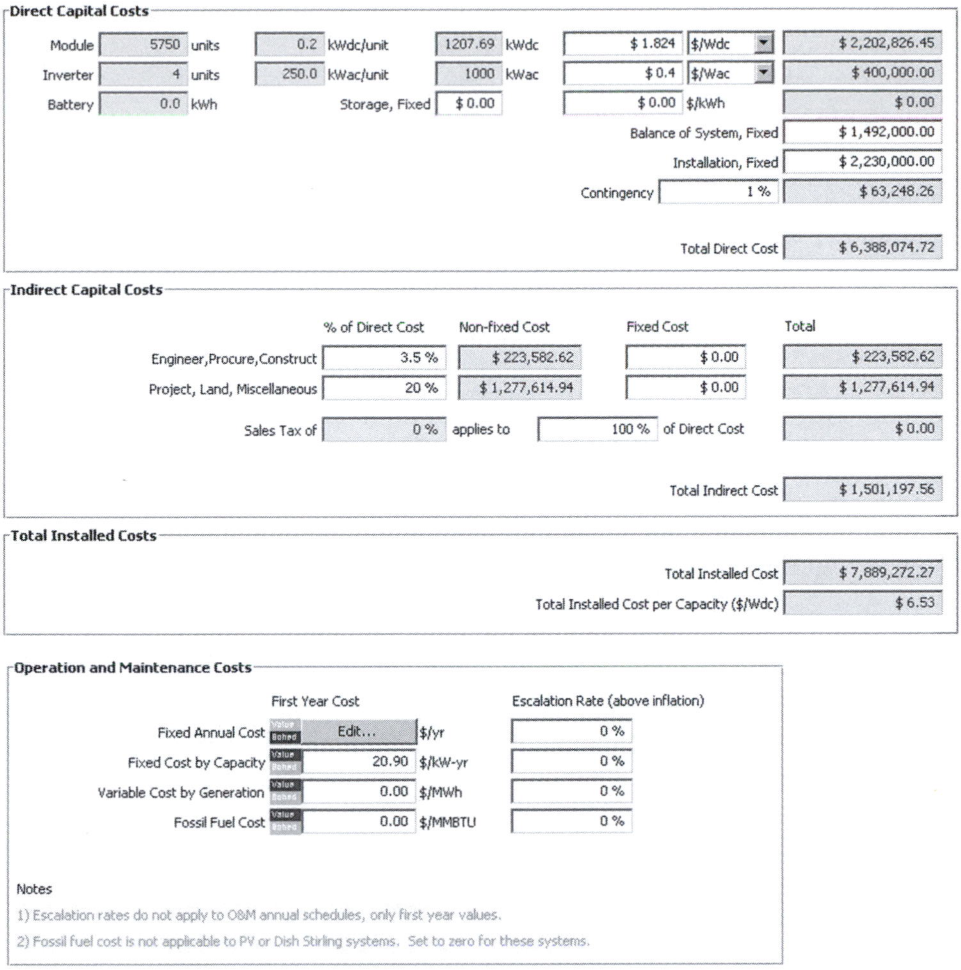

Direct Capital Costs

Module	5750	units	0.2	kWdc/unit	1207.69	kWdc	$1.824	$/Wdc	$2,202,826.45
Inverter	4	units	250.0	kWac/unit	1000	kWac	$0.4	$/Wac	$400,000.00
Battery	0.0	kWh		Storage, Fixed	$0.00		$0.00	$/kWh	$0.00

Balance of System, Fixed	$1,492,000.00
Installation, Fixed	$2,230,000.00
Contingency 1 %	$63,248.26
Total Direct Cost	**$6,388,074.72**

Indirect Capital Costs

	% of Direct Cost	Non-fixed Cost	Fixed Cost	Total
Engineer,Procure,Construct	3.5 %	$223,582.62	$0.00	$223,582.62
Project, Land, Miscellaneous	20 %	$1,277,614.94	$0.00	$1,277,614.94
Sales Tax of 0 % applies to	100 % of Direct Cost			$0.00

Total Indirect Cost	**$1,501,197.56**

Total Installed Costs

Total Installed Cost	$7,889,272.27
Total Installed Cost per Capacity ($/Wdc)	$6.53

Operation and Maintenance Costs

	First Year Cost	Escalation Rate (above inflation)
Fixed Annual Cost	Value Bahed Edit... $/yr	0 %
Fixed Cost by Capacity	Value Bahed 20.90 $/kW-yr	0 %
Variable Cost by Generation	Value Bahed 0.00 $/MWh	0 %
Fossil Fuel Cost	Value Bahed 0.00 $/MMBTU	0 %

Notes

1) Escalation rates do not apply to O&M annual schedules, only first year values.

2) Fossil fuel cost is not applicable to PV or Dish Stirling systems. Set to zero for these systems.

Figure 8.33. SAM solar power system costing entry fields.

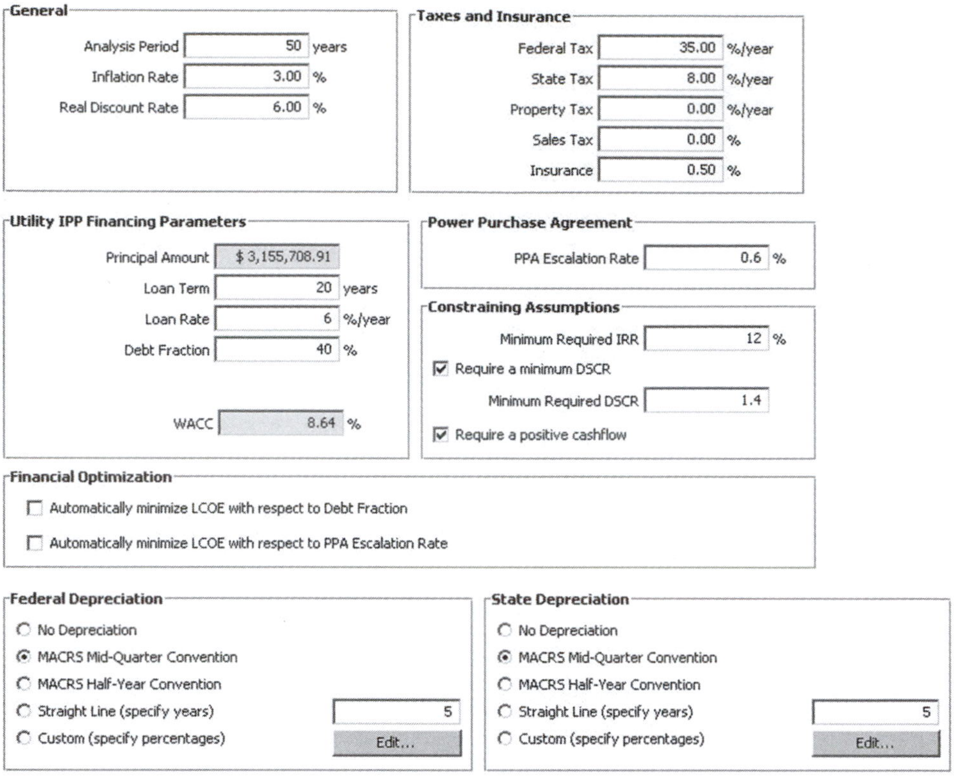

General

Analysis Period	50	years
Inflation Rate	3.00	%
Real Discount Rate	6.00	%

Taxes and Insurance

Federal Tax	35.00	%/year
State Tax	8.00	%/year
Property Tax	0.00	%/year
Sales Tax	0.00	%
Insurance	0.50	%

Utility IPP Financing Parameters

Principal Amount	$ 3,155,708.91	
Loan Term	20	years
Loan Rate	6	%/year
Debt Fraction	40	%
WACC	8.64	%

Power Purchase Agreement

PPA Escalation Rate	0.6	%

Constraining Assumptions

Minimum Required IRR	12	%
☑ Require a minimum DSCR		
Minimum Required DSCR	1.4	
☑ Require a positive cashflow		

Financial Optimization

☐ Automatically minimize LCOE with respect to Debt Fraction

☐ Automatically minimize LCOE with respect to PPA Escalation Rate

Federal Depreciation

○ No Depreciation
◉ MACRS Mid-Quarter Convention
○ MACRS Half-Year Convention
○ Straight Line (specify years) — 5
○ Custom (specify percentages) — Edit...

State Depreciation

○ No Depreciation
◉ MACRS Mid-Quarter Convention
○ MACRS Half-Year Convention
○ Straight Line (specify years) — 5
○ Custom (specify percentages) — Edit...

Figure 8.34. Example of an SAM solar power system, general tax, and purchase for a 20-year life cycle.

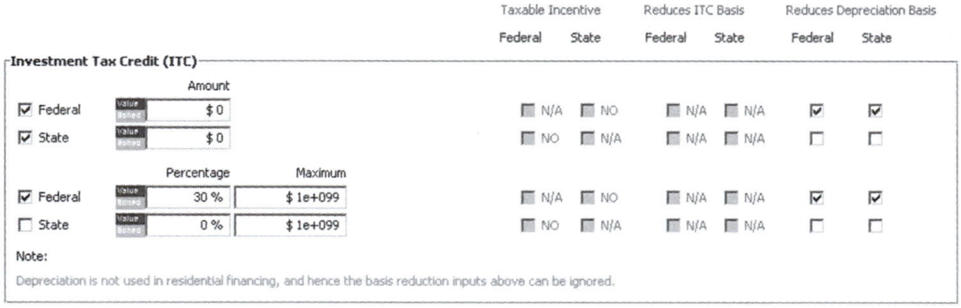

Investment Tax Credit (ITC)

			Taxable Incentive		Reduces ITC Basis		Reduces Depreciation Basis	
			Federal	State	Federal	State	Federal	State
		Amount						
☑ Federal	Value	$ 0	☐ N/A	☐ NO	☐ N/A	☐ N/A	☑	☑
☑ State	Value	$ 0	☐ NO	☐ N/A	☐ N/A	☐ N/A	☐	☐
		Percentage / Maximum						
☑ Federal	Value	30 % / $ 1e+099	☐ N/A	☐ NO	☐ N/A	☐ N/A	☑	☑
☐ State	Value	0 % / $ 1e+099	☐ NO	☐ N/A	☐ N/A	☐ N/A	☐	☐

Note:

Depreciation is not used in residential financing, and hence the basis reduction inputs above can be ignored.

Figure 8.35. SAM solar power system investment tax credit entry fields.

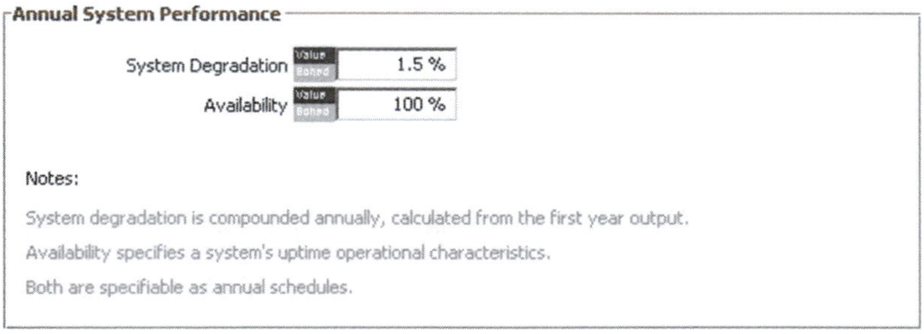

Annual System Performance

System Degradation	Value	1.5 %
Availability	Value	100 %

Notes:

System degradation is compounded annually, calculated from the first year output.

Availability specifies a system's uptime operational characteristics.

Both are specifiable as annual schedules.

Figure 8.36. SAM solar power system annual performance entry fields.

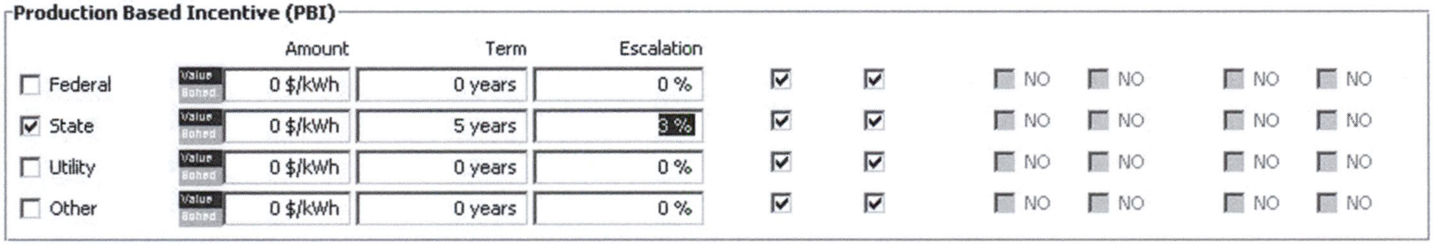

Figure 8.37. SAM solar power system PBI entry fields.

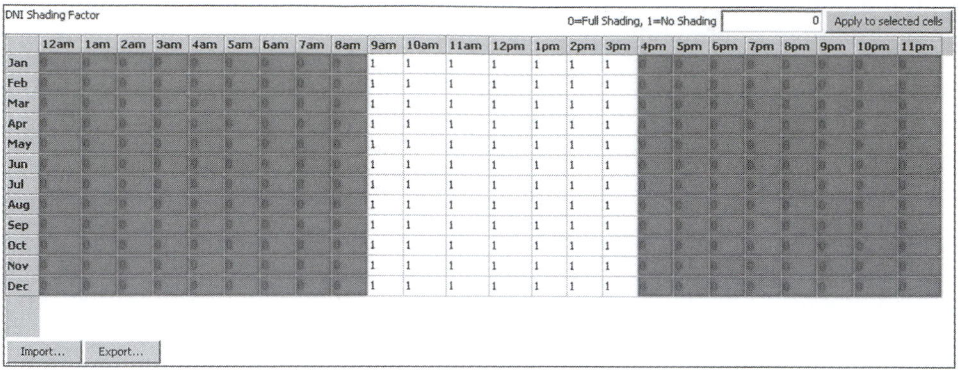

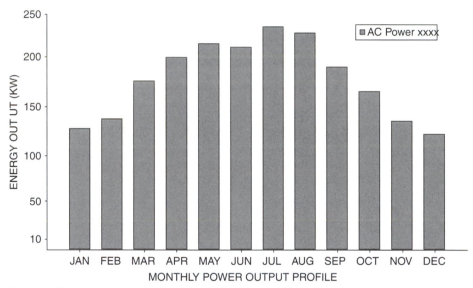

Figure 8.38. SAM solar power system shading map.

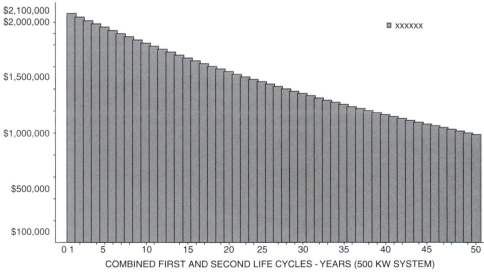

Figure 8.39. SAM solar power system first-year energy production graph.

Figure 8.40. SAM solar power system 50-year life cycle energy production graph.

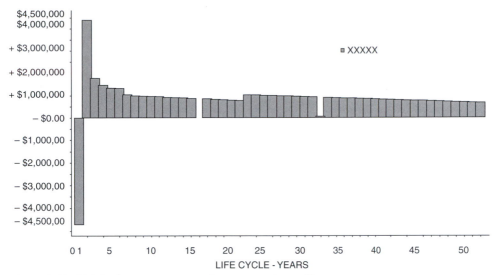

Figure 8.41. SAM solar power system first economic profile.

Layout

Modules per String	10
Strings in Parallel	575
Total Modules	5750
Total Area	7250.75 m2
Array Power (at reference conditions)	1207.69 kWdc
Voc (String, at 1000 W/m2, Tc=25 'C)	509 V
Vmp (String, at reference conditions)	413 V
Vdco (dc-inverter)	366.11 V
Number of Inverters	4
Total Inverter Capacity	1000 kWac

System Derates

Pre-Inverter Derate	97 %
Post-Inverter Derate	100 %
Total Derate factor	97 %
Edit Detailed Derates...	

Tracking and Orientation

- (•) Fixed
- ( ) 1 Axis
- ( ) 2 Axis
- ( ) Azimuth Axis

Tilt 15 deg
☐ Force Tilt = Latitude
Azimuth 0 deg
Ground Reflectance 0.2
Ground Reflectance with Snow 0.6

Notes:

Tilt: 0 deg = horizontal, 90 deg = vertical

Azimuth: 90 deg = west, -90 deg = east

Attention, Concentrating PV Module Users:

Tracking is automatically 2-axis for CPV modules, and orientation parameters (tilt, azimuth, etc..) are ignored.

Radiation Model

- ( ) Beam and Diffuse
- (•) Total and Beam

Tilt Radiation Type

- ( ) Isotropic Sky Model
- ( ) Hay and Davies Model
- ( ) Reindl Model
- (•) Perez Model

Figure 8.42. SAM solar power system second life cycle system data entry fields.

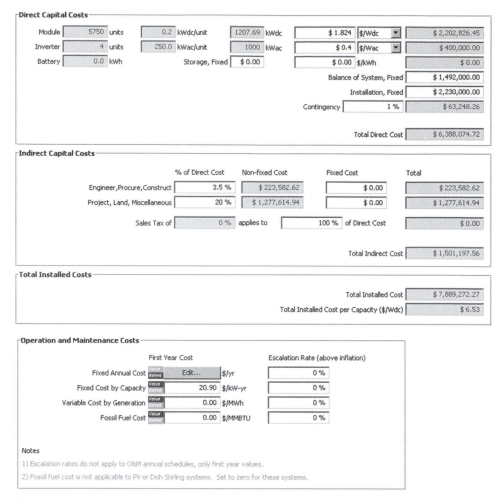

Figure 8.43. SAM solar power system second life cycle operating cost showing inverter replacement and miscellaneous maintenance expenses.

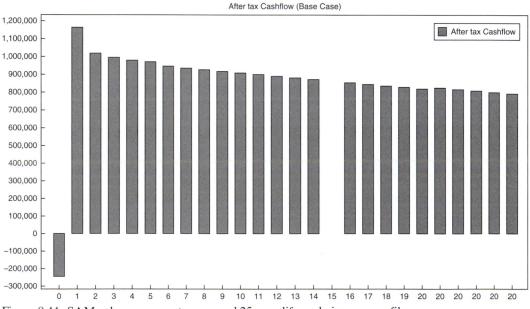

Figure 8.44. SAM solar power system second 25-year life cycle income profile.

9 Economics of Carbon Dioxide Sequestration and Carbon Trading

Introduction

Carbon trading is an economic trading instrument similar to securities or commodities in a marketplace. In other words, carbon dioxide generation is given an economic value allowing companies or nations to trade it. Entities or nations that buy carbon credit are effectively buying the rights to burn fossil fuels that emit carbon dioxide and, conversely, an entity selling carbon is giving up its rights to burn fossil fuel. The value of the carbon therefore is based on the ability of an entity owning the carbon to store it or to prevent it from being released into the atmosphere.

A carbon trading market thus created facilitates the buying and selling of the rights to emit greenhouse gases. The industrialized nations for which reducing emissions is difficult could, under the guideline of carbon trade, buy the emission rights from another nation whose industries do not produce as much of these gases.

The carbon trade came about in response to the Kyoto Protocol, signed in Kyoto, Japan, by some 180 countries in December 1997. The Kyoto Protocol members consist of 38 industrialized countries that jointly established guidelines to reduce their greenhouse gas emissions between the years 2008 and 2012 to levels 5.2% lower than those of 1990.

Properties of Carbon Dioxide

Carbon dioxide is composed of two oxygen atoms and a single carbon atom. In ambient temperature and atmospheric pressure, CO_2 is a gas. At present, the average concentration of CO_2 in the Earth's atmosphere is 387 parts per million (ppm). The atmospheric concentration of carbon dioxide varies by location and time. Carbon dioxide is an important element because it not only transmits but also strongly absorbs in part of the electromagnetic spectrum.

The main sources of carbon dioxide production are from all animals, plants, fungi, and microorganisms, including mineral water reservoirs, volcanoes, and the anthropogenic causes discussed in Chapter 1. In plants, the photosynthetic process, which involves the crucial role of the pigment chlorophyll, uses light energy and CO_2 to make sugar, which is consumed as energy in respiration or is used as the raw material for plant growth. This process is referred to as the *carbon cycle*. Carbon dioxide is also generated as a by-product of the combustion of vegetable matter, as

Figure 9.1. Carbon dioxide emission from coal fired electrical power generation plants.

a result of the chemical oxidization processes. Figure 9.1 is a photograph of carbon dioxide emission from a coal fired plant.

A unique physical property of carbon dioxide is that it is in a gaseous form. However, it is in solid form at temperatures below –78 °C. In its solid state, CO_2 is referred to as *ice*. Dry ice is commonly used as a cooling agent, and it is relatively inexpensive and often used in grocery stores, laboratories, and the shipping industry. Some of the chemical properties of carbon dioxide are as follows:

- CO_2 is an acid that if diluted in a water-based solution turns from blue to pink.
- CO_2 is toxic in concentrations higher than 1% (10,000 ppm), the inhalation of which can cause a feeling of drowsiness. At concentrations of 5%, carbon dioxide is directly toxic and can become deadly.
- Carbon dioxide is a colorless, odorless gas. When inhaled at concentrations much higher than regular atmospheric levels, it can produce a sour taste in the mouth and a stinging sensation in the nose and throat.
- Amounts above 5,000 ppm are considered very unhealthy, and those above about 50,000 ppm (5% by volume) are considered dangerous to animal life.
- At sea level the density of carbon dioxide is around 1.98 kg/m³, about 1.5 times that of air.
- The carbon dioxide molecule (O=C=O) contains two oxygen and one carbon, has a linear shape.
- Carbon dioxide is moderately nonflammable and supports the combustion of metals such as sodium and magnesium.
- At –78.51° (-109.3°), through a process of sublimation, carbon dioxide changes from a solid to a gaseous state.
- Liquid carbon dioxide forms only at above 5.1 atmospheres (atm).
- Carbon dioxide under extreme pressure and supercool temperatures (40–48 or about 400,000 atmospheres) takes on a solid, glasslike form. This synthetic form is referred to as Carbonia. The phenomenon, which was discovered in 2006, implies that carbon dioxide could exist in a glass state similar to that of other members of its elemental family, such as silica. However, unlike silicon and germanium glasses, Carbonia glass under normal atmospheric temperature and pressure reverts to a gaseous state.

Carbon Dioxide Production and Use

Carbon dioxide may be obtained from air, however, in minimal amounts. Thus yield is very small. In general, the mass production of carbon dioxide for industrial use is achieved by a large variety of chemical processes, which mainly involve the reaction of acids and metal carbonates. In one such example sulfuric acid reacts with calcium carbonate (commonly referred to as limestone or chalk), as shown in the following chemical equation:

$$H_2SO_4 + CaCO_3 \rightarrow CaSO_4 + H_2CO_3$$

In the reaction, H_2CO_3 decomposes to water and CO_2. Such reactions usually result in foaming and bubbling.

The production of CaO, a chemical that has widespread use in industry, is formed by heating limestone ($CaCO_3$) at 850 °C, which yields carbon dioxide as well:

$$CaCO_3 \rightarrow CaO + CO_2$$

In general, all carbon-containing fossil fuels, such as petroleum distillates such as coal and wood, yield carbon dioxide and water. As an example, the chemical reaction between methane and oxygen is given in the following equation:

$$CH_4 + 2\,O_2 \rightarrow CO_2 + 2\,H_2O$$

The following is an example of the chemical reaction when iron oxide (Fe_2O_3) is reduced to iron during oxidization with carbon (3C):

$$2\,Fe_2O_3 + 3\,C \rightarrow 4\,Fe + 3\,CO_2$$

In nature, in the production of wines, beers, and other spirits, CO_2 is metabolized to produce carbon dioxide and

$$\rightarrow 2\,CO_2 + 2\,C_2H_5OH$$

Organisms that require oxygen for survival produce CO_2 when they oxidize and form proteins in the mitochondria of cells. In nature, plants absorb CO_2 in the air and water, forming carbohydrates when they react:

$$nCO_2 + nH_2O \rightarrow (CH_2O)_n + nO_2$$

Industrial Manufacture of Carbon Dioxide

Carbon dioxide is manufactured mainly from the following processes. It is also a by-product in ammonia and hydrogen plants, where methane is converted to CO_2:

- From the thermal decomposition of limestone ($CaCO_3$) in the manufacture of quicklime
- Directly from natural carbon dioxide, where it is produced by the action of acidified water on ore

Carbon Sequestration

CO_2 sequestration is the storage of carbon dioxide in a solid material through biological or physical processes. CO_2 can also be captured as a pure by-product in processes related to petroleum refining (upgrading) and power generation. CO_2 sequestration can then be seen as being synonymous with the "storage", a term that refers to large-scale, permanent artificial capture and storage. It is also referred to as sequestration. The sequestration of industrial CO_2 is achieved by storage of it in liquid form in subsurface saline aquifers, reservoirs, ocean water, or other sinks.

Carbon Dioxide Use in Commercial Applications

Carbon dioxide is used extensively by the food, oil, and chemical industries. It is used in many consumer products that require pressurized, nonflammable gas.

Carbon dioxide is one of the main leavening agents that cause dough to rise. The familiar yeast produces carbon dioxide by the fermentation of sugars within the dough, while other products, such as soda, act as chemical leavens to release carbon dioxide when heated or exposed to heat.

Carbon dioxide is commonly used in life jackets, often in the form of embedded canisters of pressured carbon dioxide that allow for rapid inflation. Carbon dioxide is also sold as compressed gas in aluminum capsules used for markers, tire inflation., The rapid vaporization of liquid carbon dioxide is also useful for blasting in coal mines. High-concentration carbon dioxide, which is toxic, is frequently used to kill pests, such as the cloth moth.

Another interesting use of liquid carbon dioxide is as a catalyst for many fat absorbing, lipophilic organic compounds, such as coffee. In such cases, liquid CO_2 can be used to remove caffeine from coffee. In this decaffeination process, the green coffee beans are soaked in water and then placed in the top of a long column. Carbon dioxide fluid is then introduced at the bottom of the column, which diffuses caffeine out of the beans into the carbon dioxide liquid.

Carbon dioxide also has extensive use in chemical processing industries and is frequently used in dry cleaning.

Carbon Dioxide Sequestration Proposals in Ocean Waters

In addition to plants, aquatic species such as phytoplankton use photosynthesis to absorb large amounts of dissolved CO_2 in the upper layers of the world's oceans and are responsible for providing much of the oxygen in the Earth's atmosphere.

There is about 50 times as much CO_2 dissolved in the oceans as hydration products that exist in the atmosphere. The oceans act as an enormous reservoir, which absorb about one-third of all human-generated CO_2 emissions. It should be noted that since gas solubility decreases as water temperature increases, absorption decreases as well.

Most of the CO_2 absorbed by the oceans is transformed into various forms of carbonic acid. Some of this carbonic acid is used up by the photosynthesis of various aquatic organisms, and a relatively small proportion is absorbed by algae, forming a carbon life cycle.

One of the major concerns of environmental scientists is that the considerable increase of CO_2 in the atmosphere results in the acidity of seawater, which adversely affects aquatic organisms. As water acidity increases, the number of shellfish with carbonated exoskeletons decreases in alarming numbers.

Pesticides and Plastics

As is well known, all plants require carbon dioxide for respiration and metabolism. Moderate amounts of greenhouse gases enrich the atmosphere with CO$_2$, boosting plant growth. However, high carbon dioxide concentrations in the atmosphere create a "suffocation" level for green plants. Under such conditions photosynthesis is drastically reduced, eventually killing green plants.

As mentioned earlier, at high concentrations, carbon dioxide is toxic to animal life, so raising the concentration to 10,000 ppm (1%) for several hours can eliminate pests such as silverfish.

Another interesting use of carbon dioxide is in the production of polymers and plastics. In this process, orange peels are combined with oxide to create these polymers and plastics.

Carbon Dioxide Use in the Oil and Chemical Industries

Carbon dioxide also finds another interesting use in recovery in the crude oil industry. In this process, carbon dioxide under high pressure and supercritical temperatures is injected into oil wells. Under such conditions, carbon dioxide acts as a pressurizing agent as well as a solvent that significantly reduces crude oil viscosity, enabling rapid flow through the earth.

Carbon Dioxide Use in Production of Fertilizer

The ammonia-soda process, also referred to as the Solvay process, is the worldwide industrial process that is used for the production of soda ash. Ingredients used in this process include salt brine, which is essentially seawater, and limestone.

The process produces soda ash, or Na$_2$CO$_3$, by a mixture of, as a source of NaCl, and, CaCO$_3$. The chemical equation for the process is as follows:

$$2\,NaCl + CaCO_3 \rightarrow Na_2CO_3 + CaCl_2$$

In this process, CO$_2$ is passed through a concentrated aqueous solution of sodium chloride NaCl and NH$_3$.

The ammonia NH$_3$ shown in the formula acts as a catalyst that maintains the solution's balance. Without the ammonia, a by-product would render the solution inert and arrest precipitation.

When properly designed and operated, a Solvay plant can reclaim almost all of its ammonia and consumes only small amounts of additional ammonia to make up for losses.

Soda ash is a significant product used in numerous industrial processes. The production and use of soda ash are sometimes viewed as a barometer of the economic health of a country. The following are some of the most important uses of soda ash.

Carbon Dioxide Use in Glass Making

More than 50% of the worldwide production of soda ash is used to make glass, which is created by melting a mixture of sodium carbonate, calcium carbonate, and silica sand SiO$_2$.

Water Treatment

Sodium carbonate is used extensively as a water softener. In this process, sodium carbonate precipitates manganese Mg^{2+} and calcium Ca^{2+}. It is used both industrially and domestically in the form of washing powders.

Manufacture of Soaps and Detergents

Sodium carbonate is used as an alternative NaOH in shampoo and soap products.

Papermaking

Sodium carbonate is used in the paper manufacturing process. Soda ash is used to make $NaHSO_3$, which is the principal agent used for separating lignin from carbon cellulose.

$NaHCO_3$ is also used in baking soda and fire extinguishers.

Carbon Economics

Introduction

As discussed in previous chapters, fossil fuels used in the electrical power generation, cement, steel, textile, and fertilizer industries represent a major source of industrial emissions. Greenhouse gases emitted by these industries generally consist of ozone and hydrogen sulfate, which contribute to atmospheric pollution, climate change, and environmental degradation.

The concept of carbon credit trading was developed as a result of the global need for reducing CO_2 and controlling harmful emissions. International trade policies are intended to provide tangible monetary value for carbon emissions, which could create incentives for producers and consumers alike. International carbon emission trading policies include sophisticated economic instruments, government funding, and regulations.

The preliminary fundamentals of carbon trading were established in Kyoto, Japan, on December 11, 1997, and are referred to as the Kyoto Protocol. It established an international agreement between more than 170 countries, which resulted in mechanisms to reduce industrial pollutants.

Carbon Credit Trading

Carbon credits are national and international financial instruments devised for trading carbon dioxide emissions. Carbon trading schemes essentially provide the means by which to encourage the reduction of emissions on an industrial scale. The scheme involves the establishment of caps on the amount of total annual emissions, allowing international markets to assign special monetary values for excess emissions through trading. Emission credits established by the cap are allowed to be exchanged between businesses or transacted in international markets at the prevailing market price. Emission credits, like monetary instruments, are used as financial instruments around the world.

Nowadays, large financial institutions, such as banks and hedge fund management companies, sell carbon credits to commercial and individual customers who wish to reduce and offset their emissions on a voluntary basis.

Emission Allowances

In principle, the Kyoto Protocol established caps on the maximum amount of gas emissions for industrialized and developing countries. The agreement stipulates that each participating country set quotas on the emissions by local businesses and other organizations, referred to as "operators." Under the protocol, each is allowed to manage emission assessment and validation through its own national "registries," which are in turn subject to verification, monitoring, and compliance by the United Nations (the United Nations Framework Convention on Climate Change). Under the Kyoto Protocol, each operator is allotted a certain amount of emission credits, whereby each unit provides the owner the right to emit 1 metric ton of carbon dioxide or an equivalent amount of emission.

Under the agreement, operators that have not used up their quotas can trade or sell their unused allowances as *carbon credits*. Likewise, businesses that exceed their quotas of greenhouse gas emissions can buy the extra allowances as credits. The trade usually can take place privately between the buyer and seller or can be in the open market.

The key features of emission caps or allowances are the promotion of global economic transactions and the introduction of an effective means of reducing greenhouse gas (GHG) emissions.

Since its enactment in 2005, the Kyoto Protocol has prompted numerous countries in the European Union to adopt CO$_2$ trading regulations. At present, the EU has the protocol and the trade of GHG in the stock market. In 2008, Australia, which did not ratify the Kyoto Protocol in 2005, became an international carbon trading partner. The United States, arguably the largest polluter in the world, has refused to adopt the Kyoto Protocol.

Kyoto Protocol Cap and Trade Mechanisms

Essentially, the Kyoto Protocol provides three trading mechanisms that enable developed countries to acquire GHG credits. Allocated carbon credit emission allowances are auctioned by a number of designated national and international administrators under the cap and trade program. Approved credits, which are traded in the international market, are referred to as certified emission reduction, or CER, units.

Types of CER Credit Transactions

Under CER a developed country that generates excessive amounts of GHGs can set up a credit transaction agreement with another developed country. Under the agreement, a developed country is allowed to sponsor a GHG reduction project in another developing country, which can offer a certain amount of economic advantage. Under such an agreement, the sponsor country receives the required credits for meeting its emission reduction targets, while the developing country receiving the economic assistance can use the capital investment to build clean technology or enhance beneficial ecological projects.

Under the international agreement, developing countries are allowed to trade their carbon credits in the international market. Countries with an accumulated surplus of carbon credits can enter into direct financial transactions with countries that are subject to capped emissions.

Carbon Dioxide Emission Economics

As mentioned earlier, a unit of carbon trade, a CER unit, is considered equivalent to 1 metric ton of CO_2 emissions. The CER units, as with commodity markets, are traded in the international market at the prevailing market price. International trading and carbon credit transfer are subject to validation by the European Union. The transfer of ownership of carbon credits within the European Union is validated by the European Commission.

Currently, there are several CER exchanges and financial trading organizations that actively transact carbon credits in the United States and Europe. Carbon trading is one of the fastest-growing segments in the financial services industry in Europe. In 2011, transactions of CER in London were worth about 30 billion Euros. It is estimated that within the next decade, carbon trading could exceed 1 trillion Euros

Economics of Global Warming

As discussed, carbon credits create a financial market that is intended to reduce global greenhouse gas emissions. Gas emission trading is treated as a tangible expense, and gas emission is traded, alongside capital expenses and production materials, as an asset or a liability.

A company with a surplus of carbon credit may offer it to a client that generates emissions above its set quota. In such a situation, the buyer would pay the seller the equivalent amount of carbon credit value with regard to annual metric tons of carbon dioxide emissions.

For example, consider a business that owns a factory putting out 100,000 metric tons of greenhouse gas emissions each year. Its government is an Annex I country that enacts a law to limit the emissions that the business can produce. Let us assume the factory is given a quota of 80,000 metric tons per year. The factory either reduces its emissions to 80,000 metric tons or is required to purchase carbon credits to offset the excess.

After costing up alternatives, the business may decide that it is uneconomical or infeasible to invest in new machinery with fewer emissions for that year. Instead it may choose to buy carbon credits on the open market from organizations that have been approved as being able to sell legitimate carbon credits.

One seller might be a company in the developing world, that recovers methane from a swine farm to feed a power station that previously would have used fossil fuel. So although the factory continues to emit gases, it can pay another group to reduce the equivalent of 20,000 metric tons of carbon dioxide emissions from the atmosphere for that year. Figure 9.2 is a bar chart of carbon footprint emissions from various energy production sources.

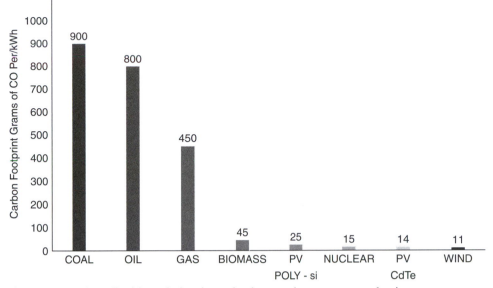

Figure 9.2. Carbon dioxide emission footprint from various energy production sources.

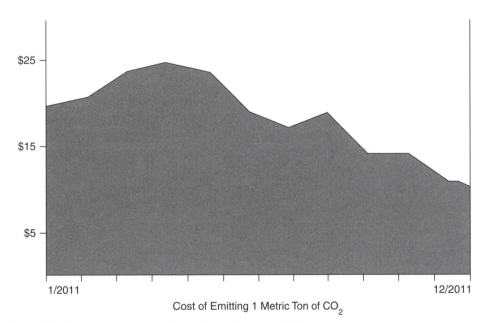

Figure 9.3. Cost of carbon dioxide emission for year 2011.
Source: The Bloomberg Economic Report 2011

In another scenario, another seller may have already invested in new low-emissions machinery and have a surplus of allowances as a result. The high-emissions factory could make up for its emissions by buying 20,000 metric tons of allowances from the lower-emissions seller. The cost of the seller's new machinery would be subsidized by the sale of allowances. Both the buyer and the seller would submit accounts for their emissions to prove that their allowances were met correctly. Figure 9.3 illustrates the cost of carbon dioxide emission for years 2011.

Carbon Dioxide as Tax Credit

Participants and signatories of the Kyoto Protocol, in evaluating the merits of credit versus taxation, chose carbon credit as a superior alternative to a tax. It was reasoned that a taxation scheme by governments would not be as efficient or beneficial in promoting the protection of the global environment.

By treating carbon emissions as a marketable, carbon credit lends itself as an easier business transaction and management tool, also allowing traders to predict future pricing fluctuations and adjustments. Moreover, the pricing mechanism established by the Kyoto Protocol provides added insurance that economic transactions are verifiable, thus promoting carbon dioxide reduction. The main advantages of tradable carbon credit, therefore, are the following:

- The price is more likely to be perceived as fair by those paying it, as the cost of carbon is set by the market and not by politicians. Investors in credits have more control over their own costs.
- The flexible mechanisms of the Kyoto Protocol ensure that all investment goes into genuine, sustainable carbon reduction schemes through its internationally agreed upon validation process.

The principle of living within the Kyoto Protocol means that the internal abatement of emissions should take precedence before a country buys carbon credits. Emissions trading and other such actions should be seen as a supplement to domestic preventative measures in emissions reduction. However, it also established the Clean Development Mechanism as a means by which capped entities could develop measureable, permanent emissions reductions voluntarily in sectors outside the cap. Many criticisms of carbon credits stem from the question of whether or not CO_2-equivalent greenhouse gas has truly been reduced, which involves a complex process of verification. This process has evolved as the concept of verification has been refined over the past 10 years.

The first step in determining whether or not this has legitimately led to the reduction of real, measurable, permanent emissions is understanding CDM (Clean Development Mechanism) methodology. The project's sponsors submit, through a Designated Operational Entity (DOE), their concepts for emissions reduction. The CDM Executive Board, with the CDM Methodology Panel and their expert advisers, review each project and decide whether and how the project results in reductions under the concept of "additionality," described in the following.

It is also important for each carbon credit to be proven under a concept called additionality. Additionality is a term used by Kyoto's Clean Development Mechanism, meaning that a carbon dioxide reduction project would not have occurred had it not been for concern about the mitigation of climate change. Succinctly, a project has to be proven additionality beyond-business-as-usual project.

It is generally agreed that voluntary carbon offset projects must also prove additionality to ensure the legitimacy of the environmental stewardship claims resulting from the retirement of the carbon credit (offset). According to the World Resources Institute/World Business Council for Sustainable Development (WRI/WBCSD), "GHG emission trading programs operate by capping the emissions of a fixed number of individual facilities or sources. Under these programs, tradable 'offset credits' are issued for project-based GHG reductions that occur at sources not covered by the program. Each offset credit allows facilities whose emissions are capped to emit more, in direct proportion to the GHG reductions represented by the credit. The idea is to achieve a zero net increase in GHG emissions, because each metric ton of increased emissions is 'offset' by project-based GHG reductions. The difficulty is that many projects that reduce GHG emissions (relative to historical levels) would happen regardless of the existence of a GHG program and without any concern for climate change mitigation. If a project 'would have happened anyway,' then issuing offset credits for its GHG reductions will actually allow a positive net increase in GHG emissions, undermining the emissions target of the GHG program. Additionality is thus critical to the success and integrity of GHG programs that recognize project-based GHG reductions."

The Kyoto Protocol and Its Shortcomings

In the past, all environmental greenhouse gas emissions restrictions have been voluntarily imposed on businesses through regulation. Even though the concept of carbon cap and trade has found acceptance in a vast number of countries, market-based carbon trading is still being scrutinized and rejected by some.

Currently, the Kyoto carbon trading mechanism is the only mechanism accepted for regulating carbon credit activities. Its supporting organization, the United Nations Framework Convention on Climate Change (UNFCCC), is the only organization with a global mandate on the overall effectiveness of emission control systems, although the enforcement of decisions relies on international and national cooperation. It should be noted that the Kyoto trading period only applies for 5 years, and will only be effective from 2008 to 2012. Since international business investment cycles operate over several decades, the limitations of the 5-year period add a certain risk and uncertainty for investors. The first phase of the EU ETS system started earlier and is expected to continue in a third phase afterward. It may coordinate with whatever is internationally agreed upon, but there is general uncertainty as to exactly what will be decided. Since large proportions of global emissions are produced by the United States, China, and India, many consider mandatory carbon caps a competitive disadvantage when compared to these uncapped countries. Thus, they have refused to be signatories of the Kyoto Protocol, also weakening international reinforcement of the cap and trade.

Another shortcoming of the cap and trade process concerns the accurate assessment and monitoring of the Clean Development Mechanism (CMD), which could be subject to manipulation. Establishing a meaningful offset project is complex, and voluntary offsetting activities outside the CDM mechanism are effectively unregulated. This particularly applies to some voluntary corporate schemes in uncapped countries and to some personal carbon offsetting schemes.

In addition, the governments of capped countries may seek to weaken their commitments..

A question has also been raised over the restriction of allowances. EU ETS nations have granted their respective businesses most or all of their allowances at no cost. This can be seen as a preventative and protectionist obstacle to new entrants into their markets.

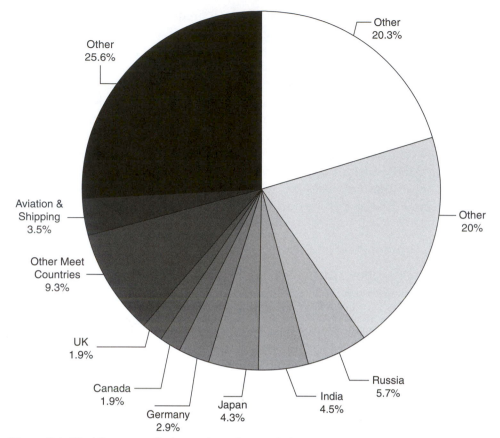

Figure 9.4. World's most polluting nations. *Source:* DOE

One concern is related to the accurate assessment of additionality, and others relate to the effort and time taken to get a project approved. Questions may also be raised about the validation of the effectiveness of some projects. It appears that many projects do not achieve the expected benefit after they have been audited, and the CDM board can only approve a lower amount of CER credits. For example, it may take longer to roll out a project than originally planned, or a forestation project may be reduced by disease or fire. For these reasons, some countries place additional restrictions on their local implementations and will not allow credits for some types of activity, such as forestry or land use projects. Figure 9.4 is a chart of the world's most polluting nations.

Carbon Trading Highlights of the Kyoto Protocol

The Kyoto Protocol originated in Kyoto, Japan, in 1997. It specifies emission obligations for various countries and defines the three so-called Kyoto mechanisms:

- Carbon Credit Accreditation By Independent Entity (AIE) – According to Kyoto Protocol carbon accreditation is entrusted to a joint independent counsel, which is responsible for the determination of whether a project and the ensuing reductions of anthropogenic emissions by sources or enhancements of anthropogenic removals by sinks meet the relevant requirements of the Kyoto Protocol.

- A 2% levy imposed on each project is established to assist least developed countries (LDCs) through an adaptation fund, which is imposed on all projects except those implemented in least developed countries.
- Under the Kyoto Protocol, certificates are to be awarded only to project-based activities where emissions reductions are "additional to those that otherwise would occur."
- Afforestation of nonforested areas is a project classification that refers to establishment of forests on lands that have not been forested for a period of at least 50 years. Reforestation refers to projects that use planting or seeding and/or promote natural seeding sources.
- The protocol establishes an Alliance of Small Island States (AOSIS), which comprises a coalition of some 43 low-lying and small island countries that are particularly vulnerable to sea-level rise. AOSIS countries are the first to have proposed a draft text during the Kyoto Protocol that calls for cuts in carbon dioxide emissions of 20% from 1990 levels by 2005.
- The protocol establishes allocation of emissions permits or allowances among greenhouse gas emitters to establish an emission trading market. An allocation plan established by the European Union member states has instituted rules that issue allowances for the installations under the EU-Emissions Trading Schemes for Afforestation and Reforestation.
- An international nontreaty agreement among the countries of Australia, India, Japan, the People's Republic of China, South Korea, and the United States was announced July 28, 2005, at an Association of South East Asian Nations (ASEAN) Regional Forum meeting. The partnership's objective was to focus on investment and trade in cleaner energy technologies, goods, and services in key market sectors.
- Assigned Amount (AA) and Assigned Amount Units (AAUs) are amounts of total greenhouse gases that each country is allowed to emit during the first commitment period of the Kyoto Protocol. An AAU is a tradable unit of 1 ton of carbon dioxide.
- The protocol has established rules for auctioning allocation of greenhouse gas emissions among emitters within a domestic emissions trading scheme. This is a voluntary measure that requires willingness of polluters to pay for permits.
- Parties to the Kyoto Protocol may bank some emissions allowances or credits not exceeding a maximum limit of 2.5% of the country's target, which can be used in subsequent commitment periods.
- The protocol establishes a baseline and baseline scenario for forecasting emissions under a business-as-usual (BAU) scenario, which is referred to as the "baseline scenario."
- A term referred to as a "bubble" is a regulatory concept whereby two or more emission sources are treated as if they were a single emission source. Under a bubble combination of several small-scale project activities, a single decrease transaction cost per unit of emission reductions is formed.
- "Debundling" is defined as the fragmentation of a large project activity into smaller parts.
- "Cap and Trade" is an emissions trading system whereby total emissions are limited or "capped." The Kyoto Protocol is a cap and trade system in the sense that emissions from various countries are capped and excess permits can be

traded. However, normally cap and trade systems are not supposed to include mechanisms that will allow for more permits to enter the system beyond the established cap.

- Carbon Dioxide Capture and Storage (CDCS) is a process consisting of the separation of CO_2 from industrial and energy-related sources, its transport to a storage location, and long-term isolation from the atmosphere.

- Carbon Dioxide Equivalent (CO_2e) is a measurement unit used to indicate the global warming potential (GWP) of greenhouse gases. Carbon dioxide is the reference gas against which other greenhouse gases are measured.

- Carbon neutral or zero CO_2 emissions from sources are referred to currently but not addressed or only inadequately addressed by climate policies related to private households, public administrations, most small and medium businesses, and air travel. Carbon neutrality is a voluntary market mechanism to encourage the reduction of emissions.

- The Kyoto Protocol certification process is the phase of a project when permits are issued on the basis of calculated emissions reductions and verification, possibly by a third party.

- Under the mechanism established for project-based emission reduction activities in developing countries, referred to as Clean Development Mechanism (CDM) Certificates, certificates are generated through the projects that lead to certifiable emissions reductions that would otherwise not occur. The CDM Executive Board (EB), which is accountable to the Conference of the Parties to the Kyoto Protocol, registers and validates project activities and issues certified emission reductions to relevant projects.

- The Clear Skies Act Initiative, created under the protocol established in the United States, delineates federally enforceable emissions limits or "caps" for three pollutants, namely, sulfur dioxide SO_2, nitrogen oxide NO_x, and mercury Hg for the period of 2008–18. Clear Skies' NO_x and SO_2 requirements affect all fossil fuel–fired electric generators greater than 25 MW that sell electricity.

- Under the Kyoto Protocol, the crediting period is defined as the duration when a project generates carbon credits. The crediting period cannot extend beyond the operational lifetime of the project. For CDM projects the crediting period continues for either a 7-year period, which can be renewed twice to make a total of 21 years, or a one-time 10-year period.

- Under the protocol, in order to participate in CDM, a party needs to appoint a Designated National Authority (DNA). The DNA issues the Letter of Approval (LoA) needed for registration of a project. A project is required to have both a host country approval as well as an investor country approval.

- A domestic legal entity or an international organization accredited and designated by the CDM EB, the DOE, validates and requests registration of a proposed Achievement by a Party as to its quantified emission limitation and reduction commitments under the Kyoto Protocol.

- The directive on Landfill of Waste Council is a subcommittee that has the responsibility to prevent or reduce negative effects on the environment from the landfilling of waste by introducing stringent technical requirements for waste and landfills.

- Early credits can be given for projects implemented between 2000 and 2008 to achieve compliance in the first commitment period.
- Emissions-to-cap are calculated by subtracting the seasonally adjusted cap from emissions (actual or forecasted). This metric gives an indication of whether the market (for a specific period) is producing more or less than the seasonally adjusted cap for that same period.
- Emissions trading allows for transfer of carbon credits across international borders or emission allowances between companies covered by a cap and trade scheme.
- An Emission Reduction Purchase Agreement (ERPA) is a binding purchase agreement signed by a buyer and seller.
- Global Warming Potential (GWP) refers to the impact of a greenhouse gas (GHG) on global warming. By definition, CO_2 is used as the reference case; hence it always has the GWP of 1. As GWP changes with time, the Intergovernmental Panel on Climate Change (IPCC) has suggested using a 100-year GWP scale for comparison purposes. The following is a list of 100-year GWPs:

Carbon dioxide (CO_2) GWP: 1
Methane (CH_4) GWP: 21
Nitrous oxide (N_2O) GWP: 310
Hydrofluorocarbons (HFCs) GWP: 150–11,700
Perfluorocarbons (PFCs) GWP: 6500–9,200
Sulfur hexafluoride (SF_6) GWP: 23,900

Note: HFC-23 (trifluoromethane) emissions are created as a by-product in the production of HCFC-22, which is used mostly as the refrigerant for stationary refrigeration and air-conditioning.

- Greenhouse gases (GHGs) are trace gases that control energy flows in the Earth's atmosphere by absorbing infrared radiation. Some GHGs occur naturally in the atmosphere, while others result from human activities. There are six GHGs covered under the Kyoto Protocol – carbon dioxide (CO_2), methane (CH_4), nitrous oxide (N_2O), hydrofluorocarbons (HFCs), perfluorocarbons (PFCs), and sulfur hexafluoride (SF_6). CO_2 is the most important GHG released by human activities.
- A Green Investment Scheme (GIS) is a proposed green investment schemes to promote the environmental effectiveness of carbon credit transfers, by earmarking revenues from these transfers for environmentally related purposes in the seller countries.
- The International Transaction Log (ITL) is a planned centralized database of all tradable credits under the Kyoto Protocol and the application that verifies all international transactions and their compliance with Kyoto rules and policies.
- The IPCC was established by the World Meteorological Organization (WMO) and the United Nations Environmental Program (UNEP) in 1988 to assess scientific, technical, and socioeconomic information relevant for the understanding of climate change, its potential impacts, and options for adaptation and mitigation. It is open to all Members of the UN and of WMO (www.ipcc.ch).
- A host country is the country where a project is physically located. A project has to be approved by the host country to receive CERs or ERUs.

- An Integrated Pollution Prevention and Control (IPCC) Directive is a measure to minimize pollution from various industrial sources throughout the European Union.

  - The JUSSCANNZ Group is an active Kyoto negotiations team that represents Japan, the United States, Switzerland, Canada, Australia, Norway, and New Zealand.
  - Marginal Abatement Cost (MAC) is an abatement cost in the context of the carbon market as the cost of reducing emissions with one additional unit. Aggregated marginal costs over a number of projects or activities define the marginal abatement.
  - The Regional Greenhouse Gas Initiative (RGGI) is a cooperative effort by Northeastern and Mid-Atlantic states of the United States of America to reduce carbon dioxide emissions by establishing a regional cap-and-trade program.
  - Public stakeholders include individuals, groups, or communities affected, or likely to be affected, by the project. Comments of stakeholders have to be included in a Project Design Document according to specific established rules.
  - Terrestrial sequestration refers to removal of carbon dioxide from the atmosphere or the prevention of carbon dioxide emissions from leaving terrestrial ecosystems. Sequestration can be enhanced in such ways as reducing the decomposition of organic matter, increasing photosynthetic carbon fixation of different types of vegetation, and creating energy offsets by using biomass for fuels.
  - The Umbrella group is an informal group of industrialized countries. The group was formed after the adoption of the Kyoto Protocol; it consists of Japan, the United States, Canada, Australia, Norway, New Zealand, Iceland, Russia, and Ukraine.

10 The Smart Grid Systems Deployment and Economics

Introduction

The following is a compendium of numerous treaties and discussions. In the past several decades, they have been rigorously explored worldwide. As a result of significant advances in computers and communication technologies, smart grid systems construction worldwide has gone through numerous developmental changes.

In view of accelerated proliferation of large-scale solar and wind power installations worldwide, existing electrical power transmission lines and grids can no longer sustain the extended burden of additional power transmission capacity. In addition, existing grid networks lack intelligence to regulate and manage dynamic supply and demand loads essential for solar and wind energy power generation systems interconnection.

The principal objective of smart grid systems is to deliver electricity from various sources of supplies such as electrical power generating stations and geothermal, wind, and solar power farms to consumers. These supplies use two-way digital technologies to control end user loads such as appliances at consumers' homes to save energy, reduce cost, and increase reliability and transparency. In essence, smart grid systems overlay electrical distribution grids with an information and net metering system. Currently, such grid modernizations are being promoted worldwide as a means for addressing energy independence, global warming, and national security.

Smart grid systems also include intelligent monitoring systems that keep track of all electricity flowing in the system. Additionally, smart grid systems will incorporate use of innovative superconductive transmission lines that can conduct significantly larger currents with minimal power losses. In essence smart grid systems are essential for integrating a wide network of future renewable electricity systems such as solar and wind. Smart grid end user power consumption management is achieved through selective control of home appliances such as washing machines or factory processes that can run at low peak energy demand hours. This results in a reduction in energy consumption at peak energy hours. Figure 10.1 is a diagram of an electrical power generation and distribution grid system.

In principle, the smart grid is an upgrade of twentieth-century power grids. These grids broadcast power from central power generating stations to a large number of

POWER GENERATION AND DISTRIBUTION

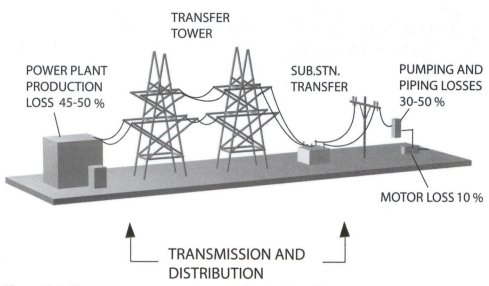

Figure 10.1. Electrical power generation and distribution grid system.

users. However, the new system will be capable of routing power in more optimal ways to respond to a very wide range of conditions and will also be capable of regulating the grid peak power demand through imposition of high premium utility rates on customers that use energy at peak hours.

Transitioning to a Smart Grid System

At present, a significant portion of our electrical energy is produced by burning fossil fuels such as coal or natural gas to generate electrical energy via mechanically driven steam turbines. Electrical power generated from the power plants is termed dispatchable energy; it is highly predictable, stable, and controllable. However, in order to connect less predictable renewable energy sources such as wind and solar power to the grid, a power transmission system must have a considerable amount of smart computerized technology to cope with so-called less-dispatchable energy.

In view of the significance of climate control and carbon emission control, it is inevitably true that renewable energy power generation will gradually constitute an important component of the electrical grid energy. In order to mitigate dispatchability of wind and solar power energy, the future electrical grid will incorporate considerable computer and communication system technologies that will permit delivery of stable and reliable combined conventional and renewable electrical energy from suppliers to consumers.

In order to accomplish that objective, smart grid systems must have intelligent communications capability to control a steady flow of electrical energy delivery during all times of the day, and through all seasonal climatic variations that affect consistency of wind and solar power generation. In addition, in order to ensure security of electrical power supply, variation in electricity generation patterns must be matched with energy demand from consumers. For instance, if wind farms become a significant source of electrical energy, the smart grid must be capable of responding rapidly to changes in wind speed or extended periods of low wind during wintertime.

At present, consumers in most distribution networks throughout the industrialized world have unrestricted access to electrical energy and can use electrical power whenever required or needed. In order to meet the electrical energy demand power generation plants automatically increase or reduce energy supply by closing or opening steam input to the generator turbines; thus output power is adjusted to meet the power demand frequency.

At present intermittent renewable power generation from wind and photovoltaic solar power represents a small percentage of the total grid power; as such, its effect on power stability is quite negligible in balancing grid power fluctuations. However, in the near future, with greater demand for renewable energy the smart grid systems must develop a mechanism to balance the supply and demand of electrical energy without resorting to fossil fuel power generation. Hence smart grid systems in the near future would enable total integration of renewables and reduce total dependence on fossil fuels, while maintaining a balance between supply and demand.

Smart Grid Technology Components

Energy Storage Technology

In order to eliminate intermittency of energy generated by renewable energy resources, the smart grid technologies must have a mechanism to store the excess power generated that could be used later on as needed. The excess stored will be used to feed the energy back to the grid at times of peak demand. Electrical energy storage facilities can be centralized within specific grid zones or can form distributed storage throughout.

Electrical Energy Demand and Supply Control

In addition to an energy storage mechanism, in order for smart grid systems to be able to ensure electrical energy distribution stability, the system must be responsive to energy demand and available supply. For instance, in the event renewable generation production remains low for extended periods, stored energy in conventional grid systems could be used up completely, resulting in electricity supply problems. In order to mitigate such a scenario, consumers of electrical energy on a smart grid system will be provided with Smart Home energy management systems that can receive real time energy pricing information, thereby allowing control of smart appliances that can operate only at preprogrammed local low energy costs. To balance the supply, demand, and storage, smart electricity grids must therefore incorporate an intelligent communication system. As such, a smart grid system would provide real time electricity pricing to smart meters installed in consumers' premises that could provide an integrated means of managing electrical power consumption.

For instance, consumers could program their machinery to operate within a period of 24 hours when the energy costs are low, thereby reducing the amount of peak electrical power demand by automatically shifting machinery use to off-peak periods, which could result in significant energy cost savings.

In order to balance the supply, demand, and energy storage, smart grid systems would be required to incorporate significant intelligent communication systems. Intelligent communication systems will be essential in providing real time electricity pricing information that must be supplied to customers' smart meters.

Smart Grid Technology Challenges

As discussed, smart grids will, in the near future, become the infrastructure of the future electrical energy transmission and distribution system. In order for smart grids to accommodate intermittent electricity generation by wind and solar renewable energy power generation, the technology must resolve the several technical challenges that will be required to maintain secure and reliable delivery of electrical energy.

Some of the challenges that must be mitigated, as mentioned, include intermittent and variable supply of energy generated by renewable energy supply sources. In addition to meeting variances of electrical energy demand, the smart grid systems will need to be able to adjust to increases in peak demand due to increases in electric vehicle charging. In order to meet the added burden, the smart grid systems will be required to deploy advanced energy storage technologies as well as communication technologies to manage added complexity of the mixed energy use. As such, one of the important challenges that must be overcome will be development of sophisticated communication systems that can reliably process significant volumes of data that will be needed to transfer information to any location requiring it.

Furthermore, smart grids must also accommodate increased distributed generation and voltage and current fault current surges, which are common on transmission lines.

Advantages and Challenges of the Smart Grid System Economics

One of the most significant challenges of the smart grid system lies in managing energy demand and supply response during peak power demand variation, an issue that must addressed with considerable care to assure acceptability. Variations in energy demand and supply patterns must be controlled and balanced in such a manner as to allow and facilitate user-friendly technology deployment. Furthermore, smart grid technologies must be implemented in a manner to provide energy price incentives that will equally be suitable for customers at varying economic levels.

At present electrical power producers and distributors of electricity sell their electrical power at prices that fluctuate depending on the time of use and continuity of supply. As a result, the price of electrical energy fluctuates during low and peak demand hours or seasons. In the near future, deployment of smart grid systems will inevitably be adjusted according to real time variations of energy demand and supply and will also depend on weather conditions, as well as the manner in which local customers use the power. Within the domain of the smart grid system, consumers in order to minimize their energy use cost will be required to change their energy use pattern, which will result in energy distribution and balancing of demand and supply of the electrical grid energy distribution.

Residential customers within smart grid systems would be required to have a domestic energy management computer that would automatically maximize the energy demand when the electricity price was low, which could result in considerable reduction of their electrical bill.

Low-Voltage Ride-Through (LVRT)

Another challenge facing the smart grid system is a transmission phenomenon referred to as low-voltage ride-through (LVRT). All electric grids once in a while experience sudden drops in voltage that are caused by short-circuits that occur within the transmission systems. In general such faults are usually isolated by automatic protection systems that take a few milliseconds, until such time as the voltage is restored. The effect of such voltage drops becomes less severe in locations farther away from the source of the problem; however, the effects can be significant over several hundred miles.

Currently conventional electrical energy generating systems use special types of technology referred to as synchronous generators, which respond quite rapidly to voltage dips or brownouts and are able to adjust the amount of reactive power required to maintain the grid voltage balance. Reactive power that is produced by conventional generating plants involves significant complexity; however, it is commonly used to control voltage levels.

However, many wind turbines currently in use employ induction type generators, which are not capable of reacting to line voltage drops; as a result, they consume large amounts of reactive power, which can, in fact, create additional voltage dip. Newer wing turbines deploy power electronics that, in the event of a voltage dip, instantaneously shut down power production. In situations where power production by windmills represents a small percentage of grid connected power, transmission line operators have minimal concern about renewable energy generation; however, it can become a significant problem if the amount of energy produced exceeds 10% of transmission line capacity.

To resolve the issue, wind turbine manufacturers have developed power generation technologies that inject reactive power into the transmission line during occurrence of voltage dip.

It is expected that, in the near future, renewable energy intermittency will be resolved by development of advanced energy storage technologies.

The United States Unified Smart Grid

The Unified National Smart Grid is a proposal for a United States–wide area grid that is a nationally interconnected network relying on a high-capacity backbone of electric power transmission lines. These lines could link all the nation's local electrical networks that have been upgraded to smart grids. Europe's analogous project is sometimes referred to as the Super Smart Grid, a term that also appears in the literature describing the Unified Smart Grid. President Barack Obama asked the U.S. Congress "to act without delay" to pass the Unified National Smart Grid as a proposal for a wide area grid that would be a national interconnected network relying on a high-capacity backbone of electric power transmission lines. These lines would link all the nation's local electrical networks that have been upgraded to smart grids. President Obama asked the Congress "to act without delay" to pass legislation that included doubling renewable energy production in the next three years and building a new electricity smart grid.

The International Smart Grid Systems

The concept of intelligent grid systems was explored in early 1950s in Europe and was intended to unify and synchronize the European Continental Grid system serving 24 countries and spanning 13 time zones, which would unify the European grid with that of the Russia, Ukraine, and other countries of the former Soviet Union. In view of the immense scale of the project and its associated problems such as network complexity, transmission congestion, and the need for rapid diagnostic, coordination, and control systems, the program has yet to materialize. However, advocates of the smart grid schemes are convinced that such a major technological upgrade is essential for establishing a transcontinental megagrid system.

Prospect of Smart Grid Systems and Special Features

Proposed future smart transmission systems in the United States will most probably be configured as high-voltage capacity transmission that will be capable of transmitting direct current at 800-kV potential. Such transmission networks would link all local electric utilities and power generation facilities throughout the country.

Existing long-distance interconnections in the United States consist of 1400 kilometers of transmission lines that intertie Los Angeles and the Pacific Northwest. The Pacific Intertie carries up to 3.1 gigawatts on two 500-kV overhead lines. Another high-voltage DC transmission system spanning 1200 kilometers interconnects Quebec, Canada, with New England; it has 2 gigawatts of capacity.

The proposed United States Unified Smart Grid is not intended to be a collection of point to point interconnections between regional systems with some communications intelligence. Rather, the present conceptual topology incorporates many grid node access points that would allow formation of virtual power generation clusters, which would consist of local electric utility providers, solar and wind farms, or grid energy storage facilities.

Super Smart Grid

At present the supergrid system is a conceptual plan that is intended to be a grid system that would be capable of delivering an inexpensive, high-capacity, low-loss electrical transmission system that would interconnect producers and consumers of electricity across vast distances. Smart grid capabilities use the local grid's transmission and distribution network to coordinate distributed generation, grid storage, and consumption in a cluster that appears to the supergrid as a virtual power plant.

The wide area supersmart grid involves two concepts, a centralized smart grid control system and a small-scale, local, and decentralized smart grid system. These two approaches are often perceived as being mutually exclusive alternatives. The Super Smart Grid system, in essence, is intended to reconcile the two approaches and considers them complementary and necessary to realize a transition toward a fully decarbonized electricity system. It should be noted that a SuperSmart Grid system only refers to a network superimposed on top of local grid networks and should not be confused with a Super Grid system, which refers to evolving technology in an electricity distribution system.

Grid Power Control and Surplus Power Management

The amount of data required to monitor and switch appliances off automatically is very small compared to residential or commercial voice, security, Internet, and TV system services. Most smart grid bandwidth upgrades are already supported by consumer communication services. However, since government power and communications companies are generally separate commercial enterprises in North America and Europe, smart grid system implementation will require considerable government aid and large vendors to facilitate and encourage various enterprises to cooperate in specific communication methodologies. For instance, a large communication enterprise such as Cisco will be in a position to provide specific smart grid communication and control devices to consumers such as those currently offered by Silver Spring Networks or Google (enterprises that in principle are data integrators rather than vendors of equipment).

Scope of Smart Grid System

As referenced earlier, smart grid systems would be the intelligent interconnecting backbone of the electrical power distribution system that provide layers of coordination above the local grids. Regardless of development of specific terminologies, smart grid project objectives are to allow continental and national interconnection that prevent a local or regional grid failure to cause local smart grid shutdown. Therefore, all power distribution networks within the smart grid would have the capability to function independently and ration whatever power was available to critical needs.

Types of Electrical Power Grids

Municipal Grid

Municipalities, whether generating or purchasing electrical power, have the primary responsibility and a legal mandate to control power distribution during emergencies. They must also frequently ration power to ensure the distribution of power to critical clients such as hospitals, fire stations, and emergency shelters during power outages.

In fact, most large municipalities in the United States have actively taken a lead in enforcing integration standards for smart grid metering. In recent years a number of municipalities generating electricity have installed fiber optic communication networks and power control transit exchange mechanisms that will provide smooth integration with future smart grid systems.

Residential Networking System

A residential or home grid network consists of electrical and electronic hardware communication and control devices that use an electrical power distribution power line to establish communication with equipment and appliances within a house. Currently, smart grid network communication is established through the use of radio frequency (RF) standards by a number of organizations such as Zigbee, INSTEON, Zwave, Wi-Fi, and others. The most prevalent smart grid communication standard developed by National Institute of standards and Technology (NIST) is promoting interoperability between the different standards. Another communication standard developed by OSHAN http://en.wikipedia.org/wiki/Smart_grid – cite_note-7 enables device interoperability at home. In general, communication standards developed for smart power grids and home area networks support more bandwidth than is required for power control; as such they may impose a residual utility cost increase.

The smart grid communication will be using the existing 802.11 home networking system, which has a wide bandwidth and multimegabits that accommodate a wide variety of communication services used by burglar, fire, medical, and environmental alarm systems, as well as closed circuit television (CCTV), local area networks (LANs), and cable TV networks.

At present, consumer electronics devices consume more than half the power in a typical U.S. home. Consequently, the ability to shut down or to enforce stand-by power mode devices during peak power hours could result in substantial curtailment of energy use. However, the main issue of concern is that electric service providers could at their discretion decide to turn power off without prior notice. Figure 10.2 is a graphic representation of residential appliance use time shift. Figure 10.3 is a bar chart of yearly electrical energy consumption by residential housing.

Home devices that could aid in the utilities' efforts to shed load during times of peak demand include air-conditioning units, electric water heaters, pool pumps, and other high-wattage devices. In the near future, smart grid companies will represent one of the biggest and fastest growing sectors in the clean technology markets. In the past, smart grid technologies have received a substantial portion of venture capital investment. Figure 10.4 is a pie chart representation of U.S. household energy use.

Principal Function and Architecture of Smart Grid Systems

There are currently multiple networks, power generation companies, and power distribution operational centers employing varying levels of manual communication and control protocols. Smart grids, on the other hand, increase power transmission and distribution connectivity by means of automation. This promotes coordination among electrical power providers, consumers, and networks that perform local or long-distance power transmission tasks.

In general, power transmission networks transport bulk electricity over long distances, which operate from 345 kV to 800 kV over AC and DC lines. On the other hand, local networks move power in one direction, distributing the bulk power to consumers and businesses via transmission lines that operate under 132 kV.

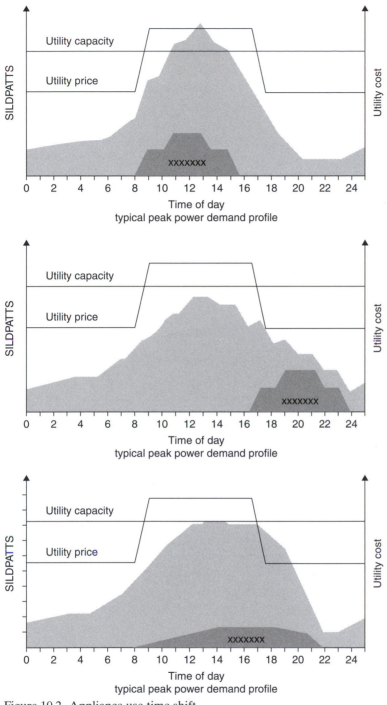

Figure 10.2. Appliance use time shift.

In the past, residential and commercial solar power systems, as well as wind turbines, generated an energy surplus, which was sold back to utilities. To modernize the existing power generation and distribution system, it is necessary to incorporate real time power flow management that would enable the bidirectional metering needed to compensate local producers of power. Even though there currently

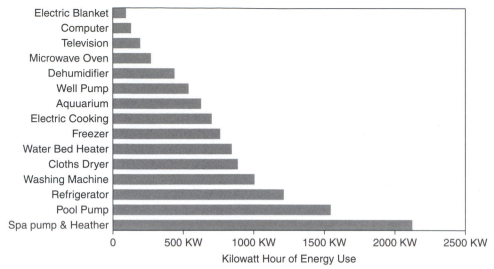

Figure 10.3. Yearly electrical energy consumption by typical U.S. residential housing.

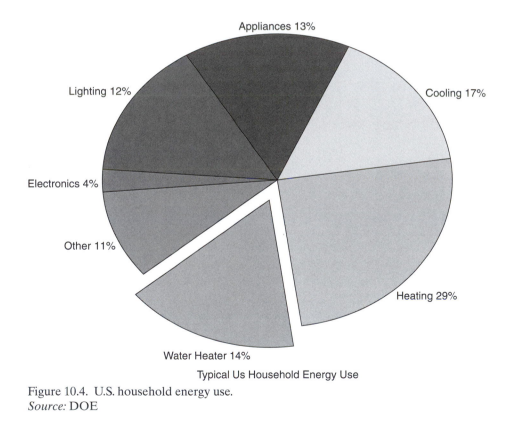

Figure 10.4. U.S. household energy use.
Source: DOE

exist transmission networks that are already controlled in real time, many U.S. and European countries continue to operate under antiquated standards, which would be incompatible to be integrated with smart grid networking systems.

Smart grid systems, as discussed, are modernizations of both the transmission and distribution grids. The main goal of grid modernization is to facilitate greater competition among providers, enabling greater use of variable energy sources by establishing automation and monitoring capabilities. These would allow for bulk electrical power transmission across continental distances, which would force markets to enact energy conservation.

Smart grids will enable energy suppliers to charge variable electric rates and tariffs that will reflect differences in the cost of generating electricity during peak or off-peak periods. Such measures will allow load control switches to make large energy consuming devices, such as air-conditioning systems and hot water heaters, operate at low peak energy cycles.

Peak Power Consumption Control

In order to reduce electrical power demand during the high-cost peak usage periods, which mostly occur between 12 noon and 5 pm in the summertime, smart grid communications and metering technologies inform smart residential or industrial devices as to when energy demand is high and track how much electricity is used by the device. In order to motivate clients to reduce or cut back energy use during peak hours, also referred to as curtailment of peak power, prices of electricity will be considerably increased during high-demand periods and decreased during low-demand periods. As a result, consumers and businesses will be motivated to consume less during high-demand periods.

Impact of Smart Grid on Renewable Energy Production

As discussed in earlier chapters, renewable energy sources such as solar or wind power generating systems (because of natural environmental phenomena) produce intermittent electrical energy that is not fully compatible with present electrical grid systems. Consequently, clients who plan to use intermittent renewable energy power produced by solar power systems must have power consumption schemes that can automatically control their electrical loads. They would do this by arming and disarming synchronously with power outputs of solar photovoltaic systems whose output power is constantly affected by phenomena. By setting lower electrical utility tariffs for peak solar power output periods and higher tariffs for conventional electrical power, consumers will be encouraged to schedule their power consumption. The main drawback of such tariff variations is that they may create unpredictable energy cost control that will be subject to climatic and environmental conditions.

Synchronized Grid Interconnections

Forthcoming plans to synchronize and interconnect the North American grid, also referred to as the wide area synchronous grid, will be integrated on a regional scale with networks of electrical power transmission systems that will operate at a synchronized frequency. At present, such synchronized zones interconnect Continental Europe (ENTSO-E) with 603 GW of electrical power generation. The widest segment of the synchronized grid, referred to as the IPS/UPS system, serves most countries of the former Soviet Union. In 2008 the ENTSO-E grid transmitted more than 350,000 megawatt hours of energy per day on the European Energy Exchange (EEX). Figure 10.2 depicts the appliance use time shift system that is planned to interconnect Europe and African renewable electrical energy production centers

Some of the interconnections in North America are synchronized at an average of 60 Hz, while those in Europe run at 50 Hz. Interconnections can be tied to each other either via high-voltage direct current power transmission lines referred to as DC ties or with variable frequency transformers (VFTs), which permit a controlled flow of energy while also functionally isolating independent AC frequencies of each side (60 or 50 Hz).

Significant benefits of synchronous zones include the collectivization or pooling of electrical power generation, which results in lower energy costs; the lowering of transmission line load burden; power distribution equalization; common provisioning of reserves; avoidance of load disturbances; and development of new energy markets.

Western American Interconnection

The Western American Interconnection is currently one of the two major alternating current (AC) power grids in North America. The other wide area synchronous grid is the Eastern Interconnection. Currently, there are also three minor interconnections, the Hydro Québec (Canada) Interconnection, the Texas Interconnection, and the Alaska Interconnection.

The Western Interconnection presently extends from western Canada through the Rockies, the Great Plains of the western United States to Baja California in Mexico. All of the electric grid interconnections operate at a synchronized frequency of 60 Hz. The interconnections are tied to each either by high-voltage DC current, or with VFTs, which permit functional isolation of independent AC frequencies for each zone. The Western Interconnection is also coupled to the Eastern Interconnection with six DC ties.

All of the electric utilities in the Eastern Interconnection are electrically tied together during normal system conditions and operate at a synchronized frequency operating at an average of 60 Hz. The Eastern Interconnection reaches from central Canada eastward to the Atlantic coast, south to Florida, and back west to the foot of the Rockies (the intertie excludes both Quebec and most of Texas). Figure 10.5 is a map of the United States smart grid network system. Figure 10.6 is a US and Canada regional transmission system organization map.

Smart Grid Advanced Services and Devices

In order to allow customer load grid connection and control, industries involved in smart grid technology will in the near future be required to develop a variety of communications network protocols, wireless communication actuators, advanced energy consumption and generation sensors, and distributed computing technology. This technology would have to provide efficient, reliable, and safe power delivery and use. In the near future, smart grid technologies will open up great prospects for new services and instrumentation such as fire alarm monitoring and power control management systems that will allow power shutdown and load shedding. Figure 10.6 is the map of U.S. and Canadian regional transmission organizations.

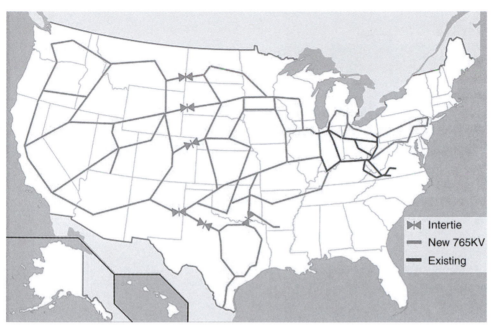

Figure 10.5. The U.S. smart grid network system.
Source: DOE

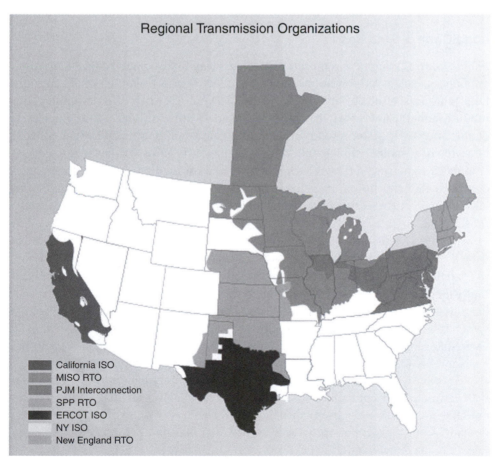

Figure 10.6. U.S. and Canadian regional transmission system organizations.
Source: DOE

Figure 10.7. A smart residential meter system.
Source: courtesy of METERUS

Load Control Switches

A load control switch is a remotely controlled relay that is placed adjacent to home appliances such as air-conditioner units and electric water heaters, which consume large amounts of electricity. A load control switch consists of a communication module and the relay switch, which are used to turn on and off appliances to enable smart grid systems to improve energy consumption efficiency. Such a switch operates in a manner similar to a telephone pager. It allows the system to receive command signals from power companies to turn off appliance power during peak electrical demand hours. In general, such devices also incorporate a timer that allows automatic resetting of the load control switch back to a normal state after a preset time.

Most load control switches are designed to have only one-way communication with a power company. However, some advanced load control switches also have the capability for two-way communication, which allows power companies to locate faulty load control switches. Most power companies in the United States offer special incentives and free load control switch installations to their customers. Load control switches are excellent devices that can prevent power brownouts and blackouts to enable electrical power service companies to respond to emergencies and will considerably minimize and avoid shutting off all power to their customers.

Another important feature of load control switch deployment is that the switch will permit power companies to reduce costs for electric tariffs by reducing the amount of expensive electricity that they must purchase during peak power energy demand from bulk energy providers. Figure 10.7 is a photograph of a residential energy management system.

High-Temperature Conductors (HTCs)

In addition to the smart grid control and communication technologies discussed previously, to increase power transportation and distribution efficiency modern transmission lines will be required to carry substantially larger currents. These currents will surpass the physical capacities of present-day conventional transmission cables. To carry larger currents, transmission line cables must be replaced by conductors capable of withstanding very high temperatures. They will also have to be strong enough to span long crossings.

At present there are three types of conductors designed to withstand high temperatures. There are aluminum conductor composite reinforced (ACCR), aluminum conductor composite core (ACCC™), and aluminum conductor steel supported (ACSS). The significant benefit of these types of conductors is that they can replace existing overhead power line cables without the need for replacing existing transmission line towers and structures. The temperature ratings of the conductors are quite high, so maintenance crews must take extreme care during repairs.

Aluminum Conductor Composite Reinforced (ACCR)

ACCR conductors are composed of heat-resistant aluminum-zirconium alloy outer strands and aluminum oxide matrix core strands. The core of the ACCR is composed of a stranded fiber reinforced metal matrix, constructed from an aluminum oxide fiber embedded in high-purity aluminum. The outer strands are either round or trapezoidal in shape. In addition to strength, the cables can withstand extremely high temperatures without softening or losing strength. Moreover, the cable has less thermal expansion than steel and retains its strength at high temperatures. ACCR conductors use similar stranding to conventional aluminum conductor steel reinforced ACSR cables. The main features of ACCR cables are their lightweight core and their resistance to heat, which permits higher electrical conductivity and lower thermal expansion. This results in less sag, higher operating temperatures, and conductor ampacity. ACCR conductors and hardware are rated for 210 °C continuous operating temperatures.

Aluminum Conductor, Composite Core (ACCC)

Aluminum conductor, composite core conductors are composed of trapezoidal aluminum stranded wires that are wrapped around the composite core. The core of the ACCC conductor is a solid with no voids and is constructed from a carbon glass fiber polymer matrix material core. This solid polymer matrix core in turn is composed of carbon fibers surrounded by an outer shell of boron free glass fibers that insulate the carbon from the aluminum conductor. The trapezoidal aluminum wires are fully annealed; as a result they are softer than the hardened aluminum wires used in conventional transmission conductors. The aluminum strands, even though soft, are tempered by the composite core, which allows the conductor to convey considerably more current at high temperatures with minimal expansion.

Similar to ACCR cables, ACCC conductors are rated for 180 °C continuous operating temperatures. The main disadvantage of ACCC cables, because of the softer temper of the aluminum wires, is that the outer wires become more susceptible to damage from improper installation and handling.

Aluminum Conductor, Steel Supported (ACSS)

Aluminum conductor, steel supported (ACSS) is another type of high-temperature conductor that is constructed with round or trapezoidal aluminum strands. ACSS conductors have similar operational characteristics to ACSRs; however, the aluminum strands in ACSS are fully annealed and strengthened by steel outer wires that minimize sag characteristics. ACSS conductors are rated for 250 °C continuous operating temperatures.

Similar to ACCC cables, due to softer temper of the aluminum wires, the outer wire shell is more susceptible to damage by improper installation and handling.

Smart Grid and Mesh Networking

The principal backbone of all smart grid systems are the interwoven communication networks referred to as mesh networks that allow instantaneous global communication among all power generation, distribution, and energy consumption centers. Without mesh networking it would be impossible to establish synchronized electrical power interconnections, as described previously.

Mesh networking is a type of networking in which each node in the network, such as a power generating center, may act as an independent communication router regardless of whether it is connected to another network or not. Such communication systems allow for continuous connections and reconfigurations around failed or congested paths by bypassing or hopping from node to node until information is reached at the destination. A mesh communication network where all nodes are connected to one another is referred to as a fully connected network.

Mesh networks differ from one another since the component parts forming the mesh can be connected at various stationary nodes to each other via multiple hops. Mesh networks can also be seen as ad hoc communication system networks. A category of mesh networks referred to as mobile ad hoc networks, or, for short, MANET, is closely interconnected by mobile communication nodes. Most often mesh networks are self-healing, a characteristic that allows them to operate when one or more nodes break down or become dysfunctional. In such a communication scheme, the network there is often more than one path between a source and a destination. MANET systems generally deploy wireless and hard-wired communication systems, which are entirely controlled by intricate software programs. Figure 10.8 is a depiction of a smart grid system mesh network layout and nodal layout.

Meshed network architecture using a wireless communication system was originally developed for military applications. Nowadays most nodal radio communication systems within each network are capable of supporting multiple functions such as client access, backhaul services, and scanning. Moreover, considerable miniaturization of transmitters and receivers has reduced mesh networking system costs and enabled mesh nodes to become more modular and

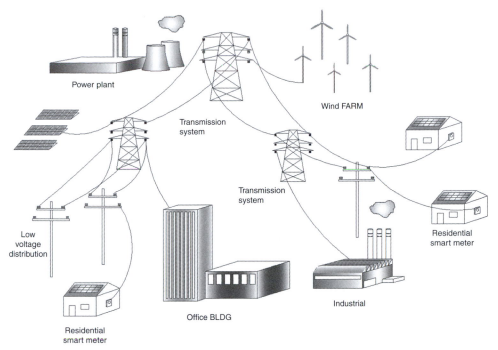

Figure 10.8. Smart grid system mesh network layout.

versatile. This allows one node or device to contain multifunction communication circuitry that allows nodes to be customized to perform specific set communication frequency band control.

Wireless Mesh Networks in the United States

In early 2007, an American communication company called Meraki undertook the installation of a small wireless mesh router in a rural location. The wireless mesh network operates at speeds of up to 50 megabits per second. The radio within the Meraki Mini mesh network, which uses IEEE 802.11 communication standards, was later optimized for long-distance communication, which covers a span of 250 meters. Such small communication systems using a simple single-radio mesh network will in the near future be used within communities that provide inexpensive multifunctional communication system infrastructures.

Recently, an MIT Media Lab project has developed an inexpensive mesh networking system on a laptop computer called the XO-1 laptop or OLPC, which is intended for underprivileged schools in developing nations. The OLPC mesh communication system also uses the IEEE 802.11s standard to create a robust and inexpensive communication infrastructure. The feature of OLPS is that it establishes instantaneous connection via laptops that reduce the need for an external communication infrastructure such as the Internet, since the node can readily establish connections with nodes nearby.

SMesh is another mesh communication system; developed by Distributed Systems and Network Labs at John Hopkins University, it uses the IEEE 802.11 communication standard and provides multihop wireless networking. A significant feature of the SMesh is that it provides a fast handoff scheme that allows mobile clients to roam within the network without interruption in connectivity, which is quite valuable for real-time applications.

Recent standards for wired communications incorporate innovative concepts that allow networking speeds of up to 1 gigabit per second. These can readily be deployed in a local area network using existing home wiring, power lines, phone lines, and coaxial cables.

Smart Grid and Climate Change

In addition to intelligent power transmission and demand load control, it is estimated that smart grids could also reduce greenhouse gas emissions by 5–9% from 2005 levels. They would also facilitate atmospheric pollution reduction to nearly one-quarter of the proposed Waxman-Markey GHG reduction targets set for 2030 through energy conservation and end-use efficiency improvements, by grid efficiency optimization, the integration of large-scale renewable energy production distribution, and provision of vehicle electrification that would be powered by renewable energies. and

Energy Conservation

The smart grid system will enhance end-user energy conservation by providing real time feedback on energy usage and communicating time-sensitive price information to customers. Studies have shown that reduction of 5–15% in electricity consumption can be achieved by implementing load management through feedback of energy usage. Just a 2% reduction in end-user energy consumption is expected to reduce about 100 million tons of CO_2 and GHG in the United States.

Grid Efficiency

At present, up to 10% of the electricity generated at power plants is lost during electrical power transmission from the source to destination. Smart grid infrastructure can improve grid efficiency to reduce these line losses by networking distribution automation devices that minimize reactive power flow through adaptive voltage controls. A 1% reduction in grid losses from a smart grid-network will translate into 30 million tons of CO_2 and GHG reductions in the United States.

Renewable Energy Integration

Renewable energy produced by solar and wind power and other generators will substantially benefit the smart grid. As discussed earlier, using the smart grid will inevitably enable demand response control that will transform static demand loads into active loads that can offset intermittency associated with renewable generation. Furthermore, smart grid networking is essential for utilities to identify and manage and therefore minimize safety risks to line workers that are associated with conventional decentralized power generation systems. In the long run both utility scale and distributed renewable energy power generation will benefit from smart grid–networked energy storage, which consequently will render renewable energy power by storing off-peak generation for on-peak sales.

Effects of Smart Grid on Electrical Vehicles

Electric vehicles (EVs) present a powerful opportunity for the electric grid to reduce U.S. GHG emissions significantly by displacing internal combustion with electric power. Pacific Northwest National Laboratory estimates that EVs could reduce *total* U.S. carbon emissions by as much as 27% by utilizing off-peak electrical energy, which could significantly reduce imported petroleum.

In fact, the smart grid networking will be necessary for EVs to take full advantage of off-peak renewable power generation since electrical power generated by renewable energy systems would effectively eliminate carbon emissions generated by transportation. Just a 50% increase in electrical vehicle use is estimated to reduce U.S. CO_2 and GHG by about 100 million tons.

Summary Overview of Smart Grid Systems

The smart grid systems are energy management and distribution systems that consist of collections of energy control and monitoring devices, software, and networking and communications infrastructures that are installed in homes, in businesses, and throughout the electricity distribution grid. Smart grid systems are intended to give customers the ability to monitor and control energy consumption in real time.

Essential Features and Benefits of Smart Grid Systems

The most important element of smart grid systems consists of two-way wireless communications networking equipment and software that provide linkage among numerous power generation and distribution nodes. The communications make use of well-established protocols that ensure compatibility and interoperability. The communication networks interconnect residential and commercial loads to the electric meters, which are connected to the electrical grid through a series of data communication equipment. The smart grid communication systems allow each utility company to interact with its clients' energy control devices from a remote distance.

Smart Grid Energy Control Devices

Residential Appliance Control Devices

Some of the smart grid residential appliance control devices include *smart thermostats* that can be adjusted remotely by the utility or by the consumer to save energy during peak load hours when the energy costs are the highest. The smart thermostats are designed to reduce household energy use by up to 40%. With appropriate setting of appliance controls and programming, customers could save up to 30% of their annual heating and cooling energy bill (see Figure 10.2, in-home energy usage and cost saving shown in real time). Such savings are estimated to exceed $25 billion per year in the United States.

Load control switches connected to various appliances and to pool pumps, electric water heaters, and other loads can result in considerable savings.

Grid Monitoring and Control Devices

Grid monitoring and control devices such as transformer monitors, voltage sensors, and other devices specifically designed to operate with smart grid systems provide fast response to power outages and allow enhanced grid response to renewable resources, such as solar and distributed power generation.

Electrical Vehicles

Electric vehicle integration within the residential or commercial electrical system in general draws significant power, which can also be controlled to act as backup power for the grid.

Benefits of the Smart Grid System

Smart grid systems in near future will enable customers as well as utility companies to achieve significant economic advantage as well as result in environmental benefits related to energy use and energy production.

Energy User and Consumer Benefits

Energy distribution and use control by means of real time communication will allow smart grid systems to distribute electrical energy more efficiently and result in reduction of energy losses that are experienced by present power transmission systems. Advanced electrical transmission line cable technologies, as discussed previously, will provide better power distribution from various power generation centers to any location within the grid with greater reliability and result in greater reduction in energy consumption, energy cost, as well as carbon footprint. Most significantly smart grid systems will enable consumers to manage their energy consumption and make significant progress on energy efficiency.

Utility Provider Benefits

The smart grid systems will enable utility companies to support increasing use of grid connected solar and wind systems as well as large-scale energy storage systems, as discussed in Chapter 12. Reliable communication and control systems will, in addition to integration of clean energy technologies and distributed energy generation systems, enable utilities to provide higher-quality and significantly more reliable power to their clients, which will result in greater economic growth. With greater integration of clean energy sources within the grid it would also be possible to minimize use of fossil fuel–fired generation power, which will result in significant reduction of atmospheric and environmental pollution.

Greater Use and Deployment of Clean Energy Resources

Among the most significant benefits of the smart grid system will be greater deployment and use of renewable energy systems. As such, smart grid systems will create greater initiative for funding and promotion of renewable energy resources.

11 Environmental Design Considerations

Introduction

The following segment is intended to introduce readers to the economics of environmental design, otherwise commonly referred to as Leadership in Energy and Environmental (efficiency) Design (LEED™). Essentially all energy utilization, whether from electrical grid or solar power systems, must be designed in a manner to minimize loss of energy in buildings by following guidelines recommended by the U.S. Green Building Council (USGBC). The following is a summary of guidelines that are currently adopted and practiced by architects and engineers worldwide. The subject matter, presented by the author in previous publications (titled *Solar Power in Building Design*), is reviewed to inform readers that LEED™ must be considered as an integral part of any solar power system design because of its significant impact on the economics of solar power energy generation. As such, when conducting a solar feasibility study and econometric analysis, energy and environmental efficiency must be incorporated in the design.

Ever since the creation of tools, the formation of settlements, and the advent of progressive development technologies, humankind has consistently harvested the abundance of energy that has been accessible in various forms. Up until the eighteenth-century Industrial Revolution, energy forms used by humans were limited to river or stream water currents, tides, and solar, wind, and to a very small degree geothermal energy, none of which had an adverse effect on the ecology.

Upon the discovery and harvesting of steam power and the development of steam-driven engines, humankind resorted to the use of fossil fuels. We commenced the unnatural creation of air, soil, water, and atmospheric pollutants with increasing acceleration to a degree that fears about sustaining life on our planet under the prevailing pollution and waste management control have come into focus.

Since global material production is made possible by the use of electric power generated from the conversion of fossil fuels, continued growth of the human population and the inevitable demand for materials within the next couple of centuries, if not mitigated, will tax the global resources and this planet's capacity to sustain life as we know it.

To appreciate the extent of energy use in human-made material production, we must simply observe that every object used in our lives from a simple nail to a supercomputer is made by using pollutant energy resources. The conversion of raw materials to finished products usually involves a large number of energy-consuming processes, but products made using recycled materials such as wood, plastics, water, paper, and metals require fewer process steps and therefore less pollutant energy.

In order to mitigate energy waste and promote energy conservation, the U.S. Department of Energy, Office of Building Technology, founded the U.S. Green Building Council. The council was authorized to develop design standards that provide for improved environmental and economic performance in commercial buildings by the use of established or advanced industry standards, principles, practices, and materials. It should be noted that the United States, with 5% of the world population, presently consumes 25% of the global energy resources.

The U.S. Green Building Council introduced the Leadership in Energy and Environmental Design (LEED™) rating system and checklist. This system establishes qualification and rating standards that categorize construction projects with certified designations, such as silver, gold, and platinum. Depending on adherence to the number of points specified in the project checklist, a project may be bestowed recognition and potentially a set amount of financial contribution by state and federal agencies.

Essentially the LEED™ guidelines discussed in this chapter, in addition to providing design guidelines for energy conservation, are intended to safeguard the ecology and reduce environmental pollution resulting from construction projects. There are many ways to analyze the benefits of LEED™ building projects. In summary, green building design is about productivity. A number of studies, most notably a study by Greg Kats of Capital-E, have validated the productivity value.

There are also a number of factors that make up this analysis. The basic concept is that if employees are happy in their workspace, such as having an outside view and daylight in their office environment, and a healthy environmental quality, they become more productive.

State of California Green Building

Action Plan

The following is adapted from the detailed direction that accompanies the California governor's executive order regarding the Green Building Action Plan, also referred to as Executive Order S-20–04. The original publication, which is a public domain document, can be found on the Californian Energy Commission's Web pages.

Public Buildings

State buildings: All employees and all state entities under the governor's jurisdiction must immediately and expeditiously take all practical and cost-effective measures to implement the following goals specific to facilities owned, funded, or leased by the state.

Green Buildings

The U.S. Green Building Council (USGBC) has developed green building rating systems that advance energy and material efficiency and sustainability, known as

Leadership in Energy and Environmental Design for New Construction and Major Renovations (LEED™-NC) and LEED™ Rating System for Existing Buildings (LEED™-EB).

All new state buildings and major renovations of 10,000 ft² and over and subject to Title 24 must be designed, constructed, and certified by LEED™-NC Silver or higher, as described later.

Life cycle cost assessments, defined later in this section, must be used in determining cost-effective criteria. Building projects less than 10,000 ft2 must use the same design standard, but certification is not required.

The California Sustainable Building Task Force (SBTF), in consultation with the Department of General Services (DGS), Department of Finance (DoF), and the California Energy Commission (CEC), is responsible for defining a life-cycle cost assessment methodology that must be used to evaluate the cost effectiveness of building design and construction decisions and their impact over a facility's life cycle.

Each new building or large renovation project initiated by the state is also subject to a clean on-site power generation.

All existing state buildings over 50,000 ft² must meet LEED™-EB standards by no later than 2015 to the maximum extent of cost effectiveness.

Energy Efficiency

All state-owned buildings must reduce the volume of energy purchased from the grid by at least 20 percent by 2015 as compared to a 2003 baseline. Alternatively, buildings that have already taken significant efficiency actions must achieve a minimum efficiency benchmark established by the CEC.

Consistent with the executive order, all state buildings are directed to investigate "demand response" programs administered by utilities, the California Power Authority, to take advantage of financial incentives in return for agreeing to reduce peak electrical loads when called upon, to the maximum extent cost effective for each facility.

All occupied state-owned buildings, beginning no later than July 2005, must use the energy-efficiency guidelines established by the CEC.

All state buildings over 50,000 ft² must be retro-commissioned, and then re-commissioned on a recurring 5-year cycle, or whenever major energy-consuming systems or controls are replaced. This is to ensure that energy and resource-consuming equipment are installed and operated at optimal efficiency. State facility leased spaces of 5000 ft² or more must also meet minimum U.S. EPA Energy Star standards guidelines.

Beginning in the year 2008, all electrical equipment, such as computers, printers, copiers, refrigerator units, and air-conditioning systems, that is purchased or operated by state buildings and state agencies must be Energy Star rated.

Financing and Execution

The Department of Energy, in consultation with the CEC, the State Treasurer's Office, the DGS, and financial institutions, will facilitate lending mechanisms for resource efficiency projects. These mechanisms will include the use of the life cycle cost methodology and will maximize the use of outside financing, loan programs, revenue bonds, municipal leases, and other financial instruments. Incentives for cost-effective projects will include cost sharing of at least 25 percent of the net savings with the operating department or agency.

Schools

New School Construction The Division of State Architect (DSA), in consultation with the Office of Public School Construction and the CEC in California, was mandated to develop technical resources to enable schools to be built with energy-sufficient resources. As a result of this effort, the state designated the Collaborative for High Performance Schools (CHPS) criteria as the recommended guideline. The CHPS is based on LEED™ and was developed specifically for kindergarten to grade 12 schools.

Commercial and Institutional Buildings

This section also includes private sector buildings, state buildings, and schools. The California Public Utilities Commission (CPUC) is mandated to determine the level of ratepayer-supported energy efficiency and clean energy generation so as to contribute toward the 20 percent efficiency goal.

Leadership

Mission of Green Action Team: The state of California has established an interagency team know as the Green Action Team, which is composed of the director of the Department of Finance and the secretaries of Business, Transportation, and Housing, with a mission to oversee and direct progress toward the goals of the Green Building Order.

LEED™

LEED™ project sustainable building credits and prerequisites are based on LEED™-NC2.1 New Construction. There are additional versions of LEED™ that have been adopted or are currently in development that address core or shell, commercial interiors, existing buildings, homes, and neighborhood development. In view of ongoing updates and changes to LEED™, over the past few years prerequisites and credit values have been updated and revised several times. To obtain updated values of credits, readers are advised to contact the US Green Building Council.

Sustainable Sites Construction activity pollution prevention – The intent of this prerequisite is to control and reduce top erosion and reduce the adverse impact on the surrounding water and air quality.

Mitigation measures involve the prevention of the loss of topsoil during construction by means of a storm water system runoff as well as the prevention of soil displacement by gust wind. It also imposes measures to prevent sedimentation of storm sewer systems by sand dust and particulate matter.

Some suggested design measures to meet these requirements include deployment of strategies, such as temporary or permanent seeding, silt fencing, sediment trapping, and sedimentation basins that could trap particulate material.

Credit No.1 – Site selection, credit – The intent of this credit is to prevent and avoid development of a site that could have an adverse environmental impact on the project location surroundings.

Sites considered unsuitable for construction include prime farmlands; lands that are lower than 5 ft above the elevation of established 100-year flood areas, as defined by the Federal Emergency Management Agency (FEMA); lands that are designated habitats for endangered species; lands within 100 ft of any wetland; or a designated public parkland.

To meet site selection requirements, it is recommended that the sustainable project buildings have a reasonably minimal footprint to avoid site disruption. Favorable design practices must involve underground parking and neighbor-shared facilities.

Credit No. 2 – Development density and community connectivity – The intent of this requirement is to preserve and protect green fields and animal habitats by means of increasing the urban density, which may also have a direct impact on reduction of urban traffic and pollution.

A specific measure suggested includes project site selection within the vicinity of an urban area with high development density.

Credit No. 3 – Brownfield redevelopment. The main intent of this credit is the use and development of projects on lands that have environmental contamination. To undertake development under this category, the Environmental Protection Agency (EPA) must provide a sustainable redevelopment remediation requirement permit.

Projects developed under Brownfield redevelopment are usually offered state, local, and federal tax incentives for site remediation and cleanup.

Credit No. 4 – Alternative transportation – The principal objective of this measure is to reduce traffic congestion and minimize air pollution.

Measures recommended include locating the project site within 1/2 mile of a commuter train, subway, or bus station; construction of a bicycle stand and shower facilities for 5 percent of building habitants; and installation of alternative liquid and gas fueling stations on the premises. An additional requisite calls for a preferred parking facility for car pools and vans that serve 5 percent of the building occupants, which encourages transportation sharing.

Credit No. 5 – Site development – The intent of this measure is to conserve habitats and promote biodiversity.

Under this prerequisite, one point is provided for limiting earthwork and the destruction of vegetation beyond the project or building perimeter, 5 ft beyond walkways and roadway curbs, 25 ft beyond previously developed sites, and restoration of 50 percent of open areas by planting of native trees and shrubs.

Another point under this section is awarded for 25 percent reduction of a building footprint by what is allowed by local zoning ordinances.

Design mitigations for meeting the preceding goals involve underground parking facilities, ride-sharing among habitants, and restoring open spaces by landscape architecture planning that uses local trees and vegetation.

Credit No. 6 – Storm water management – The objective of this measure involves preventing the disruption of natural water flows by reducing storm water runoffs and promoting on-site water filtration that reduces contamination.

Essentially these requirements are subdivided into two categories. The first one deals with the reduction of the net rate and quantity of storm water runoff that is caused by the imperviousness of the ground, and the second relates to measures undertaken to remove up to 80 percent of the average annual suspended solids associated with the runoff.

Design mitigation measures include maintenance of natural storm water flows that include filtration to reduce sedimentation. Another technique used is construction of roof gardens that minimize surface imperviousness and allow for storage and reuse of storm water for non-potable uses such as landscape irrigation and toilet and urinal flushing.

Credit No. 7 – Heat island effect – The intent of this requirement is to reduce the microclimatic thermal gradient difference between the project being developed and adjacent lands that have wildlife habitats.

Design measures to be undertaken include shading provisions on site surfaces such as parking lots, plazas, and walkways. It is also recommended that site or building colors a reflectance of at least 0.3 and that 50 percent of parking spaces be of the underground type.

Another design measure suggests use of Energy Star high-reflectance and high-emissivity roofing.

To meet these requirements the project site must feature extensive landscaping. In addition to minimizing building footprints, it is also suggested that building rooftops have vegetated surfaces and that gardens and paved surfaces be of light-colored materials to reduce heat absorption.

Credit No. 8 – Light pollution reduction – Essentially, this requirement is intended to eliminate light trespass from the project site, minimize the so-called night sky access, and reduce the impact on nocturnal environments. This requirement becomes mandatory for projects that are within the vicinity of observatories.

To comply with these requirements, site lighting design must adhere to Illumination Engineering Society of North America (IESNA) requirements. In California, indoor and outdoor lighting design should comply with California Energy Commission (CEC) Title 24, 2005 requirements.

Design measures to be undertaken involve the use of luminaries and lamp standards equipped with filtering baffles and low-angle spotlights that could prevent off-site horizontal and upward light spillage.

Water Efficiency Measures

Credit No. 1 – Water-efficient landscaping – Basically, this measure is intended to minimize the use of potable water for landscape irrigation purposes.

One credit is awarded for the use of high-efficiency irrigation management control technology. A second credit is awarded for the construction of special reservoirs for the storage and use of rainwater for irrigation purposes.

Innovative water technologies, credit no. 2: The main purpose of this measure is to reduce the potable water demand by a minimum of 50 percent. Mitigation involves the use of gray water by construction of on-site natural or mechanical wastewater treatment systems that could be used for irrigation and toilet or urinal flushing. Consideration is also given to the use of waterless urinals and storm water usage.

Water use reduction, credit no. 3: The intent of this measure is to reduce water usage within buildings and thereby minimize the burden of local municipal supply and water treatment. This measure provides one credit for design strategies that reduce building water usage by 20 percent and a second credit for a reducing water use by 30 percent.

Design measures to meet this requirement involve the use of waterless urinals, high-efficiency toilet and bathroom fixtures, and non-potable water for flushing toilets.

Energy and Atmosphere

Prerequisite No. 1 – Fundamental commissioning of building energy systems – This requirement is a prerequisite intended to verify intended project design goals and involves design review verification, commissioning, calibration, physical verification of installation, and functional performance tests, all of which are to be presented in a final commissioning report.

Prerequisite No. 2 Minimum energy performance – The intent of this prerequisite is to establish a minimum energy efficiency standard for a building. Essentially, the basic building energy efficiency is principally controlled by mechanical engineering heating and air-conditioning design performance principles, which are outlined by:

ASHRAE/IESNA and Local Municipal or State Codes

The engineering design procedure involves so-called building envelope calculations, which maximize energy performance. Building envelope computations are achieved by computer simulation models that quantify energy performance as compared to a baseline building.

Prerequisite No. 1 – Fundamental refrigerant management – The intent of this measure is the reduction of ozone-depleting refrigerants used in HVAC systems.

Mitigation involves replacement of old HVAC equipment with equipment that does not use CFC refrigerants.

Credit No. 2 – Optimize energy performance – The principal intent of this measure is to increase levels of energy performance above the prerequisite standard in order to reduce environmental impacts associated with excessive energy use. The various credit levels are intended to reduce the design energy budget for the regulated energy components described in the requirements of the ASHRAE/IESNA standard. The energy components include building envelope, hot water system, and other regulated systems defined by ASHRAE standards.

Similarly to previous design measures, computer simulation and energy performance modeling software is used to quantify the energy performance as compared to a baseline building system.

On-site renewable energy, credit no. 2: The intent of this measure is to encourage the use of sustainable or renewable energy technologies such as solar photovoltaic cogeneration, solar power heating and air-conditioning, fuel cells, wind energy, landfill gases, and geothermal and other technologies discussed in various chapters of this book. The credit award system under this measure is based on a percentage of the total energy demand of the building.

Credit No. 3 – Additional commissioning – This is an enforcement measure to verify whether the designed building is constructed and performs within the expected or intended parameters. The credit verification stages include preliminary design documentation review, construction documentation review when construction is completed, selective submittal document review, establishment of commissioning documentation, and finally the post-occupancy review.

It should be noted that all these reviews must be conducted by an independent commissioning agency.

Credit No. 4 – Enhanced refrigerant management – This measure involves installation of HVAC, refrigeration, and fire suppression equipment that do not use hydrochlorofluorocarbon (HCFC) agents.

Credit No. 5 – Measurement and verification – This requirement is intended to optimize building energy consumption and provide a measure of accountability. The design measures implemented include the following:

- Lighting system control, which may consist of occupancy sensors, photocells for control of daylight harvesting, and a wide variety of computerized systems that minimize the energy waste related to building illumination. A typical discussion of lighting control is covered under California Title 24 energy conservation measures, with which all building lighting designs must comply within the state.
- Compliance of constant and variable loads, which must comply with motor design efficiency regulations.
- Motor size regulation that enforces the use of variable-speed drives (VFD).
- Chiller efficiency regulation measures that meet variable-load situations.
- Cooling load regulations.
- Air and water economizer and heat recovery and recycling.
- Air circulation, volume distribution, and static pressure in HVAC applications.
- Boiler efficiency.
- Building energy efficiency management by means of centralized management and control equipment installation.
- Indoor and outdoor water consumption management.

Credit No. 6 – Green power – This measure is intended to encourage the use and purchase of grid-connected renewable energy cogenerated energy derived from sustainable energy such as solar, wind, geothermal, and other technologies described throughout this book. A purchase-and-use agreement of this so-called green power is usually limited to a minimum of a 2-year contract. The cost of green energy use is considerably higher than that of regular energy. Purchasers of green energy, who participate in the program, are awarded a Green-e products certification.

Material and Resources

Credit No. 1 – Storage and collection of recyclables – This prerequisite is a measure to promote construction material sorting and segregation for recycling and landfill deposition. Simply put, construction or demolition materials such as glass, iron, concrete, paper, aluminum, plastics, cardboard, and organic waste must be separated and stored in a dedicated location within the project for further recycling.

Credit No. 2 – Building reuse – The intent of this measure is to encourage the maximum use of the structural components of an existing building that will serve to preserve and conserve a cultural identity, minimize waste, and reduce environmental impact. It should be noted that another significant objective of this measure is to reduce the use of newly manufactured material and associated transportation which ultimately results into energy use and environmental pollution. Credit is given for implementation of the following measures:

- One credit for maintenance and reuse of 75 percent of the existing building.
- Two credits for maintenance of 100 percent of the existing building structure shell and the exterior skin (windows excluded).
- Three credits for 100 percent maintenance of the building shell and 50 percent of walls, floors, and the ceiling.

This simply means that the only replacements will be of electrical, mechanical, plumbing, and door and window systems, which essentially boils down to a remodeling project.

Credit No. 3 – Construction waste management – The principal purpose of this measure is to recycle a significant portion of the demolition and land-clearing materials, which calls for implementation of an on-site waste management plan. An interesting component of this measure is that the donation of materials to a charitable organization also constitutes waste management.

The two credits awarded under this measure include one point for recycling or salvaging a minimum of 50 percent by weight of the demolition material and two points for salvage of 75 percent by weight of the construction and demolition debris and materials.

Credit No. 3 – Material reuse – This measure is intended to promote the use of recycled materials, thus, reducing the adverse environmental impact caused by manufacturing and transporting new products. For using recycled materials in a construction, one credit is given to the first 5 percent and a second point for a 10 percent total use. Recycled materials used could range from wall paneling, cabinetry, bricks, construction wood, and even furniture.

Credit No. 4 – Recycled content – The intent of this measure is to encourage the use of products that have been constructed from recycled material. One credit is given if 25 percent of the building material contains some sort of recycled material or 40 percent of minimum by weight use of so-called postindustrial material content. A second point is awarded for an additional 25 percent recycled material use.

Credit No. 5 – Regional materials – The intent of this measure is to maximize the use of locally manufactured products, which minimizes transportation and thereby reduces environmental pollution.

One point is awarded if 20 percent of the material is manufactured within 500 miles of the project and another point is given if the total recycled material use reaches 50 percent. Materials used in addition to manufactured goods also include those that are harvested, such as rock and marble from quarries.

Credit No. 6 – Rapidly renewable materials – This is an interesting measure that encourages the use of rapidly renewable natural and manufactured building materials. Examples of natural materials include strawboards, woolen carpets, bamboo flooring, cotton-based insulation, and poplar wood. Manufactured products may consist of linoleum flooring, recycled glass, and concrete as an aggregate.

Credit No. 7 – Certified wood – The intent of this measure is to encourage the use of wood-based materials. One point is credited for the use of wood-based materials such as structural beams and framing and flooring materials that are certified by Forest Council Guidelines (FSC).

Indoor Environmental Quality

Prerequisite No. 1 – Minimum indoor air quality (IEQ) performance – This prerequisite is established to ensure indoor air quality performance to maintain the health and wellness of the occupants. One credit is awarded for adherence to ASHRAE building ventilation guidelines such as placement of HVAC intakes away from contaminated air pollutant sources such as chimneys, smoke stacks, and exhaust vents.

Prerequisite No. 2 – Environmental tobacco smoke (ETS) control – This is a prerequisite that mandates the provision of dedicated smoking areas within buildings, which can effectively capture and remove tobacco and cigarette smoke from the building. To comply with this requirement, designated smoking rooms must be enclosed and designed with impermeable walls and have a negative pressure (air being sucked in rather than being pushed out) compared to the surrounding quarters. Upon the completion of construction, designated smoking rooms are tested by the use of a tracer gas method defined by ASHRAE standards, which impose a maximum of 1 percent tracer gas escape from the ETS area.

This measure is readily achieved by installing a separate ventilation system that creates a slight negative room pressure.

Credit No. 5 – Outdoor air delivery monitoring – As the title implies, the intent of this measure is to provide an alarm monitoring and notification system for indoor and outdoor spaces. The maximum permitted carbon dioxide level is 530 parts per million.

To comply with the measure HVAC systems are required to be equipped with a carbon dioxide monitoring and annunciation system, which is usually a component of building automation systems.

Credit No. 2 – Increased ventilation – This measure is intended for HVAC designs to promote outdoor fresh air circulation for building occupants' health and comfort. A credit of one point is awarded for adherence to the ASHRAE guideline for naturally ventilated spaces where air distribution is achieved in a laminar flow pattern. Some HVAC design strategies used include displacement and low-velocity ventilation, plug flow or under-floor air delivery, and operable windows that allow natural air circulation.

Credit No. 3 – Construction (IAQ) air quality management plan – This measure applies to air quality management during renovation processes to ensure that occupants are prevented from exposure to moisture and air contaminants. One credit is awarded for installation of absorptive materials that prevent moisture damage and filtration media to prevent space contamination by particulates and airborne materials.

A second point is awarded for a minimum of flushing out of the entire space, by displacement, with outside air for a period of 2 weeks prior to occupancy. At the end of the filtration period a series of test are performed to measure the air contaminants.

Credit No. 4 – Low-emitting materials – This measure is intended to reduce indoor air contaminants resulting from airborne particulates such as paints and sealants. Four specific areas of concern include the following: (1) adhesives, fillers, and sealants; (2) primers and paints; (3) carpet; and (4) composite wood and agrifiber products that contain urea-formaldehyde resins.

Each of these product applications are controlled by various agencies such as the California Air Quality Management District, Green Seal Council, and Green Label Indoor Air Quality Test Program.

Credit No. 5 – Indoor chemical and pollutant source control – This is a measure to prevent air and water contamination by pollutants. Mitigation involves installation of air and water filtration systems that absorb chemical particulates entering a building. Rooms and areas such as document reproduction rooms, copy rooms, and blueprint quarters, which generate trace air pollutants, are equipped with dedicated air exhaust and ventilation systems that create negative pressure. Likewise, water circulation, plumbing, and liquid waste disposal are collected in an isolated container for special disposal. This measure is credited a single point.

Credit No. 6 – Controllability of systems – The essence of this measure is to provide localized distributed control for ventilation, air-conditioning, and lighting. One point is awarded for autonomous control of lighting and control for each zone covering 200 ft² of area with a dedicated operable window within 15 ft of the perimeter wall. A second point is given for providing air and temperature control for 50 percent of the non-perimeter occupied area. Both of these measures are accomplished by centralized or local area lighting control and HVAC building control systems. The above measures are intended to control lighting and air circulation. Each of the above two measures is awarded one point.

Credit No. 7 – Thermal comfort – The intent of this measure is to provide environmental comfort for building occupants. One credit is awarded for thermal and humidity control for specified climate zones, and another for the installation of a permanent central temperature and humidity monitoring and control system.

Credit No. 8 – Daylight and views – Simply stated, this measure promotes architectural space design that allows for maximum outdoor views and interior sunlight exposure. One credit is awarded for spaces that harvest indirect daylight for 75 percent of spaces occupied for critical tasks. A second point is awarded for direct sight of vision glazing from 90 percent of normally occupied work spaces. It should be noted that copy rooms, storage rooms, mechanical equipment rooms, and low-occupancy rooms do not fall into these categories. In other words 90 percent of the work space is required to have direct sight of a glazing window. Some architectural design measures taken to meet these requirements include building orientation, widening of building perimeter, the deployment of high-performance glazing windows, and the use of solar tubes.

Innovation and Design Process

Credit No. 1 – Innovation in design – This measure is in fact a merit award for an innovative design that is not covered by LEED™ measures and in fact exceeds the required energy efficiency and environmental pollution performance milestone guidelines. The four credits awarded for innovation in design are: (1) identification of the design intent, (2) meeting requirements for compliance, (3) proposed document submittals that demonstrate compliance, and (4) a description of the design approach used to meet the objective.

Credit No. 2 – LEED™-accredited professional One point is credited to the project for a design team that has a member who has successfully completed the LEED™ accreditation examination.

Credit summary	
Sustainable sites	10 points
Water efficiency	3 points
Energy and atmosphere	8 points
Material and resources	9 points
Indoor environmental quality	10 points
Innovation in design	2 points

Los Angeles Audubon Nature Center – A LEED™-Certified Platinum Project

The following project is one of the highest-ranked LEED™-certified buildings by the U.S. Green Building Council within the United States. The pilot project, known as Debs Park Audubon Center, which was commissioned in January, 2004, is a 282-acre urban wilderness that supports 138 species of birds. It is located in the center of the city of Los Angeles, California. Based on the Building Rating System™2.1, the project received a platinum rating, which is the highest possible.

The key to the success of the project lies in the design considerations given to all aspects of the LEED™ ranking criteria, which include sustainable building design parameters such as the use of renewable energy sources, water conservation, recycled building materials, and maintenance of native landscaping. The main office building of the project is entirely powered by an on-site solar power system that functions "off the grid." The building water purification system is designed such that it uses considerably less water for irrigation and bathrooms. To achieve the platinum rating the building design met the highest available LEED™ energy conservation points.

The entire building, from the concrete foundation and rebars to the roof materials, was manufactured from recycled materials. For example, concrete-reinforcement rebars were manufactured from melted scrap metal and confiscated handguns. All wood material used in the construction of the building and cabinetry were manufactured from wheat board, sunflower board, and Mexican agave plant fibers.

A 26-kW roof-mount photovoltaic system provides 100 percent of the center's electric power needs. A 10-ton solar thermal cooling system installed by SUN Utility Network Inc., provides a solar air-conditioning system believed to be the first of its kind in southern California. The HVAC system provides the total air-conditioning needs of the office building. The combination of the solar power and the solar thermal air-conditioning system renders the project completely self-sustainable requiring no power from the power grid. The cost of this pilot project upon completion was estimated to be about $15.5 million. At present, the project houses a natural bird habitat, exhibits, an amphitheater, and a hummingbird garden. The park also has a network of many hiking trails enjoyed by local residents.

The thermal solar air-conditioning system, which is used only in few countries, such as Germany, China, and Japan, utilizes an 800-ft^2 array of 408 vacuum tube solar collectors. Each tube measures 78 in long and has a 4-in diameter, and each encloses a copper heat pipe and aluminum nitride plates that absorb solar radiation. Energy trapped from the sun's rays heats the low-pressure water that circulates within and is converted into a vapor that flows to a condenser section. A heat exchanger compartment heats up an incoming circulating water pipe through the manifold which allows for the transfer of thermal energy from the solar collector to a 1200-gal insulated high-temperature hot-water storage tank. When the water temperature reaches 180°F, the water is pumped to a 10-ton Yamazaki single-effect absorption chiller. A lithium bromide salt solution in the chiller boils and produces water vapor that is used as a refrigerant, which is subsequently condensed; its evaporation at a low pressure produces the cooling effect in the chiller. This system also provides space heating in winter and hot water throughout the year. Small circulating pumps used in the chiller are completely energized by the solar photovoltaic system.

It is estimated that the solar thermal air-conditioning and heating system relieves the electric energy burden by as much as 15 kW. The cost of energy production at the Audubon Center is estimated to be $0.05 per kilowatt, which is substantially lower than the rates charged by the city of Los Angeles Department of Water and Power. It should be noted that the only expense in solar energy cost is the minimal maintenance and investment cost, which will be paid off within a few years. The following are architectural and LEED™ design measures applied in the Los Angeles Audubon Center.

Architectural green design measures include the following:

- Exterior walls are ground-faced concrete blocks, exposed on the inside, insulated and stuccoed on the outside.
- Steel rebars have 97 percent recycled content.
- 25 percent fly ash is used in cast-in-place concrete and 15 percent in grout for concrete blocks.
- More than 97 percent of construction debris is recycled.
- Aluminum-framed windows use 1-in-thick clear float glass with a low-emittance (low-E) coating.
- Plywood, redwood, and Douglas fir members for pergolas are certified by the Forestry Stewardship Council.
- Linoleum countertops are made from linseed oil and wood flour and feature natural jute tackable panels that are made of 100 percent recycled paper.
- Burlap-covered tackable panels are manufactured from 100 percent recycled paper.
- Batt insulation is formaldehyde-free mineral fiber with recycled content.
- Cabinets and wainscot are made of organic wheat boards and urea-formaldehyde-free medium-density fiberboard.
- Engineered structural members are urea-formaldehyde-free.
- Synthetic gypsum boards have 95 percent recycled content.
- Ceramic tiles have recycled content.
- Carpet is made of sisal fiber extracted from the leaves of the Mexican agave plant.
- Green energy operating system

- 100 percent of the electric power for lighting is provided by an off-grid polycrystalline photovoltaic solar power system. The system also includes a 3- to 5-day battery-backed power storage system.
- To balance the electric power provided by the sun, all lighting loads are connected or disconnected by a load-shedding control system.
- Heating and cooling is provided by a thermal absorption cooling and heating system.
- Windows open to allow for natural ventilation.
- Green water system
- All the wastewater is treated on-site without a connection to the public sewer system.
- Storm water is kept on-site and diverted to a water-quality treatment basin before being released to help recharge groundwater.
- Two-stage, low-flow toilets are installed throughout the center.
- The building only uses 35 percent of the city water typically consumed by comparable structures.

12 Energy Storage Systems

Introduction

Perhaps one of the most significant technical challenges facing renewable energy systems is development and deployment of large-scale energy storage. Presently all types of renewable energy sources generated by wind, solar, oceanic current, and tidal energy are harvested only during limited hours of each day. For instance, as discussed in the first chapter of this book, solar photovoltaic power is only generated during solar insolation periods, which average a few hours per day. Likewise, wind energy production, because of the unpredictable nature of wind streams and turbulence, is unpredictable, whereas use or demand of electrical power requires availability of 24 hours a day. Therefore, storage and availability of uninterruptible and consistent electrical power generation can be considered as one of the most important challenges and obstacles that must be overcome.

In general, large energy storage systems fall into the following main four categories:

- Mechanical technologies – Compressed air, flywheel, and pumped storage hydroelectric systems
- Electrochemical process – Batteries and charge capacitor systems
- Thermal process – Molten salt and solar ponds
- Chemical process – Hydrogen generation

Most renewable energy systems, such as wind, are capable of producing excess energy, which is typically available in off-peak energy demand hours such as late night and early morning. Likewise, solar photovoltaic energy can readily be overproduced during late morning or afternoon hours. In net metered grid connected residential and commercial installations, customers are allowed to export a limited amount of excess power to the grid and accumulate energy credit. However, large-scale grid connected renewable energy systems must comply with and deliver energy in a manner that meets grid capacity requirements, which depend upon minimum power generation conditions of other power generation sources. This means that electrical power supply to the grid is not continuous; rather it is used on demand. Therefore, in order to render renewable energy suitable for grid connection, it must be stored and dispensed on demand. As such, the only means to ensure uninterruptible energy is to have larger energy storage capability.

There are several ways that stored energy can be used. One of the basic methods, called load leveling, involves storage of energy during periods of low energy demand and delivery to the grid when demand is high. Another basic method is referred to as ramping; in it the stored energy is delivered to the grid to compensate for periodic voltage and frequency drops resulting from grid loading or excessive energy use. Ramping is used to compensate or relieve sudden periodic overloading of primary energy generators that power the grid. Technologies such as compressed air, pumped hydroelectric storage, and load leveling battery technologies have energy storage and discharge performance characteristics that are suitable to mitigate grid energy fluctuations.

Applications of large power storage fall into two classifications, namely, energy applications and power applications. Energy applications involve storage and discharge over periods of hours, typically last one cycle per day, and involve long charging periods. Power delivery to the grid involves short periods of discharge (a few seconds or minutes), which require short recharging periods, and generally operate many cycles per day.

Energy Applications include deployment of the storage system in peak shaving and load leveling as well as grid voltage and frequency regulation, power quality, renewable energy power output smoothing, and ramp rate control. Power applications, on the other hand, involve bursts of energy delivery to the grid.

At present, large energy storage system technologies are in various stages of development. In this chapter we will discuss several large battery system energy storage systems, namely, lead acid, lead carbon, lithium-ion, sodium sulfur (Na/S), vanadium redox (VRB), and flow type ion exchange zinc-bromine battery (ZBB) technologies, which hold promise for use in large utility scale deployment.

Lead Acid Storage Batteries

At present, most prevalent methods of grid energy storage or stabilization consist of large installations of lead acid batteries. The energy stored in the battery is used to inject electrical power into the grid in the event of power plant or transmission line failure. The excess power connection is used to provide temporary relief up until such time as main power sources are restored and brought on line. In such applications, the battery systems are charged from the grid during periods when the grid frequency and voltage are high and discharged when the voltage and frequency begin to sag. In such applications, even though power supply to the grid takes place for only a few minutes at a time, the technology involves intricate grid management controls that manage shifting peak loads, charge control, and grid power flow.

One of the most significant components of solar power systems consists of battery backup systems that are frequently used to store electric energy harvested from solar photovoltaic systems for use during the absence of sunlight (such as at night and during cloudy conditions). Because of the significance of battery storage, it is important for design engineers to have a full understanding of the technology since this system component represents a considerable portion of the overall installation cost. More importantly, the designer must be mindful of the hazards associated with handling, installation, and maintenance of lead acid batteries. To provide in-depth knowledge of lead acid battery technology, this section covers physical and chemical principles, manufacturing, design application, and maintenance procedures. In this section we will also attempt to analyze and discuss the advantages and disadvantages of different types of commercially available solar power batteries and their specific performance characteristics.

The battery is an electric energy storage device that in the terminology of physics can be described as a device or mechanism that can hold kinetic or static energy for future use. For example, a rotating flywheel can store dynamic rotational energy in its wheel that releases the energy when the primary mover such as a motor no longer engages the connecting rod. Similarly, a weight held at a high elevation stores static energy embodied in the object mass, which can release its static energy when dropped. Both of these batteries are examples of energy storage devices.

Energy storage devices can take a wide variety of forms, such as chemical reactors and kinetic and thermal energy storage devices. It should be noted that each energy storage device is referred to by its specific name; the word *battery,* however, is solely used for electrochemical devices that convert chemical energy into electricity by a process referred to as galvanic interaction. A galvanic cell is a device that consists of two electrodes, referred to as the anode and the cathode, and a solid or liquid electrolyte solution. Batteries consist of one or more galvanic cells.

It should be noted that a battery is an electrical storage reservoir and not an electricity-generating device. Electric charge generation in a battery is a result of chemical interaction, a process that promotes electric charge flow between the anode and the cathode in the presence of an electrolyte. A recharging process that can be repeated numerous times resurrects the electrogalvanic process that eventually results in the depletion of the anode and cathode plates. Batteries incur energy losses as heat when discharging or delivering stored energy or during chemical reactions when charging.

Major Types of Lead Acid Batteries

Solar power backup batteries are divided into two categories based on what they are used for and how they are constructed. The major applications where batteries are used as solar backup include automotive and marine systems, as well as deep-cycle discharge systems.

The major processes used to manufacture lead acid batteries include flooded or wet construction, gelled, and absorbed glass mat (AGM). Absorbed glass mat batteries are also referred to as "starved electrolyte" or "dry" type, because instead of containing wet sulfuric acid solution, the batteries contain a fiberglass mat saturated with sulfuric acid that has no excess liquid. Figure 12.1 depicts the storage battery operation principle of common flooded-type batteries, which are usually equipped with removable caps for maintenance-free operation. Gelled-type batteries are sealed and equipped with a small vent valve that maintains a minimal positive pressure. Absorbed glass mat batteries are also equipped with a sealed regulation-type valve that controls the chamber pressure.

Common automobile batteries are built with electrodes that are grids of metallic lead containing lead oxides that change in composition during charging and discharging. The electrolyte is dilute sulfuric acid. Lead acid batteries, even though invented nearly a century ago, are still the battery of choice for solar and backup

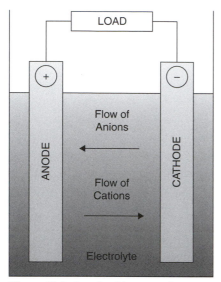

Figure 12.1. Lead acid storage battery operation principle.

power systems. With improvements in manufacturing, batteries could last as long as 20 years.

Deep-discharge batteries used in solar power backup applications in general have lower charging and discharging rate characteristics and are more efficient.

In general, all batteries used in PV systems are lead-acid type batteries. Alkaline-type batteries are used only in exceptionally low-temperature conditions of below –50°F.

Lead Acid Battery Life Span

The life span of a battery will vary considerably with how it is used, how it is maintained and charged, temperature, and other factors. In extreme cases, it can be damaged within 10 to 12 months of use when overcharged. On the other hand, if the battery is maintained properly, the life span could extend past 25 years. Another factor that can shorten the life expectancy by a significant amount is storage of uncharged batteries in a hot storage area. Even dry charged batteries when sitting on a shelf have a maximum life span of about 18 months; as a result, most are shipped from the factory with damp plates. As a rule, deep-cycle batteries can be used to start and run marine engines. When starting, engine combustion engines require a very large inrush of current for a very short time. Regular automotive starting batteries have a large number of thin plates for maximum surface area, which enables large current discharge. The plates are constructed from impregnated lead paste grids similar in appearance to very fine foam sponges. This gives a very large surface area, and when deep-cycled, the grid plates quickly become consumed and fall to the bottom of the cells in the form of sediment. Automotive batteries will generally fail after 30 to 150 deep cycles if deep-cycled, while they may last for thousands of cycles in normal starting use discharge conditions. Deep-cycle batteries are designed to be discharged fully time after time and are designed with thicker plates that hold larger charge capacity.

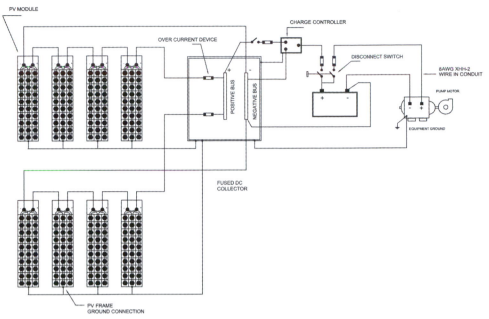

Figure 12.2. A typical solar power battery bank system.

Figure 12.3. Deep-cycle lead acid battery pack.
Source: Solar Integrated Technologies

The major difference between a true deep-cycle battery and a regular battery is that the plates in a deep-cycle battery are made from solid lead and are not impregnated with lead oxide paste. Figure 12.2 is a single line diagram of a battery backed solar power system. Figure 12.3 shows a typical solar battery bank system.

The stored energy in batteries in general is discharged rapidly. For example, short bursts of power are needed when starting an automobile on a cold morning, which result in transfer of high amounts of current from the battery to the starter. The standard unit for energy or work is the joule (J), which is defined as 1 watt-second of mechanical work performed by a force of 1 newton (N) or 0.227 pound (lb) pushing or moving a distance of 1 meter (m). Since 1 hour has 3600 seconds, 1 watt-hour (Wh) is equal to 3600 J. The stored energy in batteries is measured in either milliampere-hours if small or ampere-hours if large. Battery ratings are converted to energy if their average voltages are known during discharge. In other words, the average voltage of the battery is maintained relatively unchanged during the discharge cycle. The value in joules can also be converted into various other energy values as follows.

Battery Power Output

In each instance when power is discharged from a battery, the battery's energy is drained. The total quantity of energy drained equals the amount of power multiplied by the time the power flows. Energy has units of power and time, such as kilowatt-hours or watt-seconds. The stored battery energy is consumed until the available voltage and current to levels of the battery are exhausted. Upon depletion of stored energy, batteries are recharged over and over again until they deteriorate to a level where they must be replaced by new units. High-performance batteries in general have the following notable characteristics. First, they must be capable of meeting the power demand requirements of the connected loads by supplying the required current while maintaining a constant voltage, and they must have sufficient energy storage capacity to maintain the load power demand as long as required. In addition, they must be as inexpensive and economical as possible and be readily replaced and recharged as well as safe.

Foam Lead Acid Battery Technology

Historical Development

One of the major technology challenges that have been associated with conventional lead acid batteries has been corrosion of the battery's positive plate and sulfation of the battery's negative plate. The lead foam acid battery technology described in the following has been a result of extensive research and development by Caterpillar Inc., a world-renowned manufacturer of heavy equipment, which has produced a novel electrochemical battery manufacturing methodology that prevents corrosion by use of corrosion-resistant materials that displace much of the lead within a traditional battery and harness the battery's energy-producing chemistry in a more efficient and powerful manner. The research and development effort headed by Kurt Kelley at Caterpillar led into development of a patented Microcell™ foam battery technology, which has resulted in an electrical charge storage battery with superior performance and survivability in the harsh lead acid chemistry. The following foam lead acid battery description is based upon excerpts of a white paper by Firefly Energy Corporation, which manufactures foam batteries.

Battery Development Challenges and Objectives

In the design of any high-performance battery, there are always four major overriding objectives:

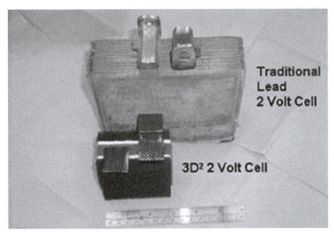

Figure 12.4. Conventional lead acid battery.
Source: Photo courtesy of Firefly Energy

1) Maximize specific energy (energy storage per unit of weight, measured in watt-hours per kilogram) over designated discharge scenarios.
2) Maximize the specific power (power per unit of weight, measured in watts per kilogram) over designated high rate discharge scenarios.
3) Maximize battery life, not only in environmental durability but most importantly in cycle life (number of possible charges and discharges).
4) Do it all at extremely low cost.

The final point has traditionally driven lead acid battery producers to the use of low-cost lead materials, which of course has limited the first three criteria. Given the long-standing use of lead metal, improvements in battery current collector (i.e., grid) design and advances in lead-alloys appear to be reaching a plateau. The lead acid battery industry has been working in obscure corners of the periodic table to find alloying elements that would help stabilize current collector behavior. Improved manufacturing techniques have allowed mechanically more delicate and thinner lead-based alloys to be employed in high-speed production. But without the development and insertion of new materials and processes, the traditional lead acid industry's gains appear to be approaching the limits for electrode thickness, alloy corrosion rates, and active material pellet structures. Figure 12.4 and 12.5 depict conventional lead acid battery configuration and lead acid grid system. Figure 12.6 depicts lead acid battery corroded grid.

Conventional Lead Acid Battery Failure Modes

Both corrosion on the positive plate and sulfation on the negative plate define two key failure modes of today's lead acid batteries. Although average industry battery life over the past decades has been extended over the past 20 years, the failure modes result in products that achieve only a fraction of the life associated with more advanced energy storage systems. Further, these life-limiting factors become exacerbated by the varying temperature environments of many applications.

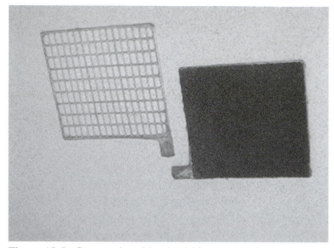

Figure 12.5. Conventional lead acid battery grid.
Source: Photo courtesy of Firefly Energy

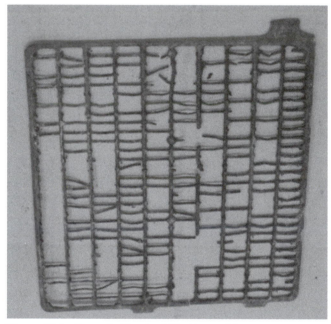

Figure 12.6. Corroded lead acid battery grid.
Source: Photo courtesy of Firefly Energy

Regarding corrosion failures, this failure mode begins to accelerate either as temperatures rise about 70°F and/or when the battery is left uncharged. To mitigate the effects of the corrosion process, most battery companies focus their research on developing more corrosion-resistant lead alloys and grid manufacturing processes that reduce the mechanical stresses in the as-manufactured grids. Regardless of the alloy or grid fabrication process, essentially all battery manufacturers engineer battery service life based on lead alloy and grid wire cross-sectional area. Normally this engineering translates as a change in grid thickness and corresponding plate thickness. Thicker grids provide longer life but usually sacrifice power density, cost, weight, and volume.

Regarding sulfation failures, when a lead acid battery is left on an open circuit stand or kept in a partially or fully discharged state for a period, the lead sulfate formed in the discharge reaction recrystallizes to form larger, low-surface-area lead sulfate crystals, which are often referred to as hard lead sulfate. This low surface area covered by nonconductive lead sulfate blocks the conductive path needed for recharging. These crystals, especially those farthest removed from the electrode grid, are difficult to convert back into the charged lead and lead dioxide active materials. Even a well-maintained battery will lose some capacity over time through the continued growth of large lead sulfate crystals that are not entirely recharged during each recharge. These sulfate crystals, at 6.287 grams per cubic centimeter (g/cc), are also larger in volume than the original paste, so they mechanically deform the plate and push material apart. Sulfation is the main problem in recreational applications during battery storage when the season ends. Boats, motorcycles, snowmobiles lie dormant in their off months and, left uncharged, discharge toward a 0% state of charge, leading to progressive sulfation of the battery. Thus, the battery cannot be recharged anymore, is irreversibly damaged, and must be replaced.

Cycle Life

Battery life cycle implies a number of discharges and charges. As users have come to know portable battery products in cell phones and laptop computers, they have correspondingly become comfortable with the process of taking a battery down to almost charge and then taking it back to full, complete charge and power capabilities within hours. Traditional lead acid batteries, because of their inherent design and active material utilization limitations, only provide relatively good cycle life when less than about 80% of the rated capacity is removed during each discharge event in an application. A battery of this type suffers a significant decrease in the number of times it can be discharged and recharged when 100% of the rated capacity is consumed during a single discharge in an application. Many new products that historically used lead acid batteries are requiring a significant jump in cycle life. The most notable examples are hybrid electric vehicles, which operate in a high rate partial-state-of-charge condition. This is a punishing application, which dramatically shortens the cycle life of a typical lead acid battery and has therefore prompted car companies to use much more expensive nickel-metal hydride batteries and experiment with lithium-ion batteries. Even though advanced performance of the technologies has superior current limitation performance characteristics compared to lead acid batteries, they are exceedingly expensive.

Recently many industries, such as transportation, have been forced to address new product designs with alternative battery systems employing expensive advanced materials and leaving their traditional long-term lead acid battery manufacturers behind. These conditions have led automotive manufacturers to embrace the new battery entrants such as those based in the Pacific Rim. As mentioned, unfortunately the cost penalty for these more exotic batteries can still be as high as 5–10 times the cost of lead-acid batteries.

Lead acid batteries have always faced severe trade-offs with respect to power, capacity, weight, life, and cost. Firefly technology breaks through these constraints and creates a much higher set of trade-offs, enabling new levels of performance, reliability, and affordability.

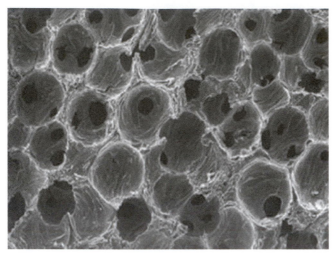

Figure 12.7. Foam-based battery cell structure.
Source: Photo courtesy of Firefly Energy

Foam-Based Battery Technology

Firefly's technology is an innovative material science that removes almost all limitations of current lead acid battery products. The materials also hold the promise of major simplification for manufacturing of lead acid batteries and will potentially deliver more flexible form factors or configurations, which may be the catalyst to change the entire distribution and profitability models of the battery industry.

During the research for a novel battery chemistry technology it was discovered that much of the lead in the grid structure of conventional batteries can be replaced by a totally new type of material. Once the basic material was determined to have the requisite physical and chemical properties, much subsequent research and testing were required to determine the optimum configuration and architecture within the battery itself, which subsequently led to securing two U.S. patents, which were granted in December 2005.

In the foam-based battery architectures, the composite foam "grids" are impregnated with a slurry of lead oxides, which are then formed up to the sponge lead and lead dioxide in the normal fashion. Because of the foam structure, the resultant negative and positive plates have enormous energy generating capacity advantages over conventional lead acid grid structures. This has resulted in much-improved active material utilization levels, as well as enhanced fast-recharge capability and greater high-rate over low-temperature discharge times. Figure 12.7 depicts foam-based battery cell structure.

The signal advantage of Firefly's Microcell™ technology is that it fundamentally changes the distribution of active materials within the lead acid cell because of its unique architecture. Overall, the composite foam electrode structure results in a redistribution of electrolyte from the smaller separator reservoir to the pores of the foam plates, resulting in a 70/30% to 30/70% reversal, respectively, relative to conventional lead acid technologies. Depending on the type of foam product, each foam plate contains hundreds or thousands of spherical microcells. This leads to enhanced active-material utilization levels, because each microcell has its full complement of sponge lead or lead dioxide and sulfuric acid electrolyte. Liquid diffusion distances are reduced from the traditional levels of millimeters over linear paths in the conventional two-dimensional (2D) diffusion mechanism found in the lead metal grid-based classic lead acid battery architecture to the level of micron diffusion path lengths in the three-dimensional space within the discrete microcells that collectively compose a totally new type of electrode structure referred to as a three-dimensional (3D) electrode. Such a structure results in much higher power and energy delivery and rapid recharge capabilities relative to conventional lead acid products.

Commercial Introduction

Firefly Energy has developed two significant technologies that will deliver advanced battery performance for an entire spectrum of uses served by lead acid–, nickel-, and lithium-based chemistries.

The two technologies, referred to as 3D & 3D^2, involve the use of a porous three-dimensional composite material to replace the lead metal grids in either flooded or VRLA lead acid battery designs. Consequently the technologies led into development of a novel lead acid chemistry for energy storage. The foam-based batteries are able to store considerably more charge per volume and have remarkably extended cycle life. The battery technology increases the cycle life of lead acid chemistry by a factor of 4, delivering performance very similar to that of advanced materials batteries such as lithium and nickel; however, the main advantage of the technology is that they can be built for cost comparable to that of conventional lead acid batteries, which amounts to about one-fifth of the cost of advanced material batteries.

3D Technology

As the initial implementation phase of its Microcell™ foam grid technology, the 3D cell architecture involves replacing the conventional negative lead metal–based plate composite with foam electrode. These products are configured in such a way as to be easily incorporated into the manufacturing processes used by all existing lead acid manufacturers. Figure 12.8 depicts foam-based battery architecture.

Because of this relatively seamless integration into established manufacturing techniques, Firefly will manufacture the pasted foam negative electrodes and furnish them to existing manufacturing partners who will incorporate them into finished battery products. To all outward appearances, these batteries will be indistinguishable from currently available products. Hydrogen overvoltage levels on the composite negative plates are comparable or slightly superior to conventional lead negatives, which mean that gassing levels and corresponding water loss rates are low. Self-discharge rates are very low at approximately 0.3 mV/cell-day, which equates to a shelf life of 2 years or more without recharge. Because of the use of an electrolyte compatible with conventional lead acid cell designs the open-circuit voltages and recharge/float voltages correspond to those for conventional lead acid batteries. This condition permits the use of conventional lead acid chargers with Firefly Energy's 3D batteries.

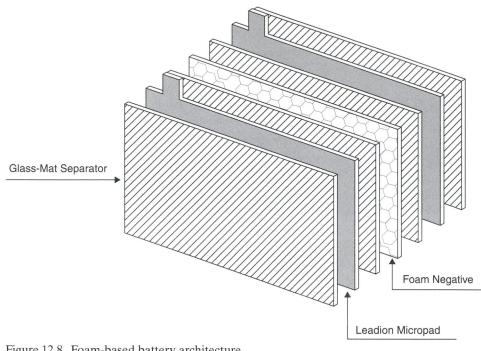

Figure 12.8. Foam-based battery architecture.
Source: Photo courtesy of Firefly Energy

3D Technology Energy and Power Performance Attributes

Surface Area

The real quest for performance improvements in lead acid batteries is all about surface area. Of course, selective enhancements have been applied that have made incremental gains over the last several decades, but the overwhelming restriction to lead acid battery efficiency has always been the lack of interface area between the active chemistry and the electrodes. Notably the 3D battery chemistry is such that it is capable of delivering approximately 170 watt-hours per kilogram (Whr/kg), compared to conventional lead acid batteries, which average around 30–50 Whr/kg. The main reason for such superior energy storage is that the battery electrodes have much larger surface areas. Figure 12.6 shows lead electrodes (on left side of picture) in lead acid batteries that corrode over time and decrease life. The foam-based battery technology shown in Figure 12.7 (right side of picture) has increased surface area.

Larger electrode surface area has added advantages of increasing the interface area between active chemistry and the electrodes to allow better and faster utilization of the chemistry. As a result of larger surface areas charge and discharge times decrease, since a higher percentage of active material becomes accessible, which in turn results in greater charge transfer efficiency. In foam-based batteries the utilization efficiencies can potentially rise more than 90%, resulting in gravimetric and volumetric energy and power values. As the electrode surface area increases, the real current density is reduced, and this causes a corresponding reduction in the battery's overpotential. In a practical sense, lower overpotential leads to higher efficiency because losses to heat are reduced, lower charging voltage is required, and there is less voltage drop on discharge. Discharge voltage curves are also flatter. It is also noted that low overpotential also results in lower open-circuit voltage (OCV).

Utilization and Spatial Efficiency

Conventional lead flat-plate and spiral wound electrodes shown in Figure 12.9 can be thought of as two-dimensional in terms of pore structure/reactivity and one-dimensional in terms of electrolyte diffusion. Roughly only one-half of the active materials are available for reaction to produce energy; this is a result of plate structure, where diffusion must take place through previously discharged material in order to sustain the energy-producing electrochemical reactions.

In traditional lead acid batteries the pasted lead plates result in linear diffusion of electrolyte into the plate pores from the separator or interplate reservoir, which delivers electrical energy during the discharge process. In lead acid batteries discharge begins at the plate surfaces and progresses into the interiors as long as electrolyte diffusion rates can support the load current being applied. In low-rate applications with lead acid technology, it is possible to achieve utilization levels [ratio of the amount of active material plates by about no more than 50–60%]. This is due to the long diffusion paths and the buildup of highly resistive lead sulfate in the plate pores. This effectively chokes off access of electrolyte to the deeper plate pores. The distances traveled by the electrolyte during full discharge are on the order of 1–3 millimeters to reach the interiors of the plates. In thick-plate deep-cycle batteries (discussed earlier in this chapter) it can be as much as 4–6 mm. Under higher rates of discharge, these relatively long diffusion path lengths become limiting in terms of total capacity output. Subsequently as a result, lead acid batteries for high-rate applications are fabricated from thin plates that have smaller plate spacing. However, such a design approach adds cost and weight to the battery since they require more grid material, which increases manufacturing scrap and shortens battery lifetimes through corrosion. Figure 12.8 shows a foam-based battery electrode system.

On the other hand, the foam-based battery architecture shown in Figure 12.7 deploys three-dimensional tubular plates, which result in a significant increase in active material utilization levels. The three-dimensional foam plate technologies allow increasing low-rate utilization levels that result in significantly higher-rate utilizations of about a 67% theoretical limit of conventional lead acid batteries. As mentioned earlier, conventional lead acid battery plates have a linear structure that requires electrolyte diffusion over relatively large distances (typical diffusion rate coefficients are on the order of 1×10^{-6} cm^2/sec, which means that in 1 second electrolyte will diffuse only a distance of 0.01 mm; in other words, it will take 100 seconds to diffuse 1 mm. On the other hand, foam-based battery diffusion path lengths are on the order of approximately 100 microns or less (0.1 mm). This type of battery architecture results in virtual use of the entire electrolyte within each tubular plate, resulting in complete depletion of charge in about 5 seconds under very high current loads. In addition, proper balance of active materials and electrolyte used in the foam battery design allows utilization levels well in excess of the practical limit of 67%, which results from the dispersion of lead sulfate ($PbSO_4$) buildup. Figure 12.9 shows foam-based battery electrode systems.

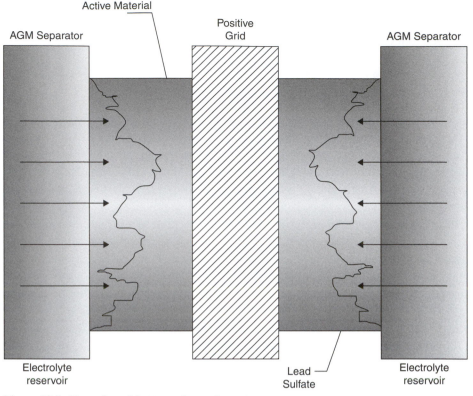

Figure 12.9. Foam-based battery electrode system.
Source: Photo courtesy of Firefly Energy

This feature prevents sulfate buildup, which is the principal cause that shuts down the discharge reaction by choking off electrolyte diffusion. Moreover, since electrolyte diffusion paths in the foam battery electrodes are on the order of microns, rather than millimeters, the unique tubular electrode design results in large increases in active-material utilizations and high-rate discharge capacities, as well as sharp reductions in recharge times.

Low- and High-Temperature Advantages

Cold-Temperature Operation

Though economical in many applications, conventional lead acid batteries have a relatively low specific energy and, similar to competitive batteries, are severely affected by cold temperatures. This effect, or increase in internal resistance, is due to the slowing down of the battery's chemical-reaction and ion-diffusion rates. In general, reaction rates in lead acid batteries are cut in half for each 10°C drop in temperature. Cold cranking is a discharge that needs a high current, and reaction rates are critical to sizing a battery. A high current implies a lot of active material conversion in a short time, and this is related to the amount of electrode surface area covered with active material that is available for conversion. Therefore, a starter battery needs a lot of surface area, which means a large number of lead plates.

Sizing a lead acid battery for starting applications at -18°C, for example, requires an approximate 200% size increase over room temperature operation. Because of an acknowledged corrosion rate for the positive lead grids in lead acid batteries, attempts to increase cold-temperature starting power by increasing electrode surface area without sizing up the overall battery result in short warm-temperature life.

In contrast, foam-based battery (3D and $3D^2$) products have outstanding discharge performance at low ambient temperatures relative to commercial flooded lead acid and VRLA batteries. This is due to the extremely high electrochemically available surface area of the composite foam coated with sponge lead. At high discharge rates and/or low temperatures, the discharge performance of a typical lead metal–based negative plate limits a cell's output, in large part because of the relatively low surface areas of conventional planar negative plates. The foam battery negative, with its hundreds or thousands of tiny microcells, each having its own complement of sponge lead and electrolyte, is ideal for discharge and charge conditions where electrolyte diffusion is limited by surface area, distance, or temperature. The distances for electrolyte diffusion in the foam battery microcells are on the order of tens of microns, while in conventional lead acid batteries path lengths are measured in millimeters. Diffusion rates at low temperatures are reduced in a 3D cell just as they are in conventional commercial products, but the distances traveled to react with the sponge lead are much smaller. This enhanced electrolyte supply also results in higher, flatter voltage-time curves on discharge, which means higher energy outputs when combined with the lower current densities that accrue to the high foam electrochemical surface area.

As the temperature is lowered, it takes more power to start the engine at the same time that the available power from the battery drops to only 40% of what can be provided at ambient when the car is started at –18°C. By comparison, foam-based batteries will provide 69% of their ambient temperature power at –18°C. This means that the battery could be smaller to have the same cold-crank amps, or it would be more powerful and last longer if its size were comparable to that of a commercial product.

This point is depicted in Figure 12.11, which compares the discharge performance of a foam battery with that of a typical commercial lead acid (VRLA) product at –20°C. It can be seen that at slow discharge rates there is a small but real advantage for a foam-based battery, which has a significantly greater rate of discharge. Although not shown, the foam battery will continue to operate at lower temperatures on both charge and discharge, even down to –40°C or lower where other lead acid products cease to function.

Hot-Temperature Operation

The optimum operating temperature for a lead acid battery is 25°C (77°F). In general, every 8–10°C (14–18°F) rise in temperature cuts the battery life in half. This is a result of chemical activity at higher temperatures. Under such elevated temperature conditions, lead grids corrode in the acidic electrolyte in the presence of lead dioxide, the positive plate active material. However, foam-based batteries, because of their inherent electrochemical and heat-transfer characteristics, have superior performance compared to lead acid and other batteries. Figure 12.10 shows foam-based battery

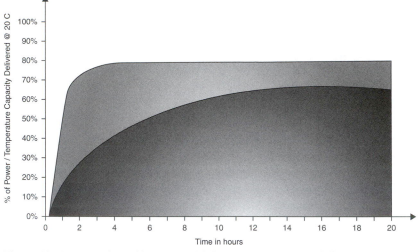

Figure 12.10. Foam-based battery at high temperature rate of discharge.

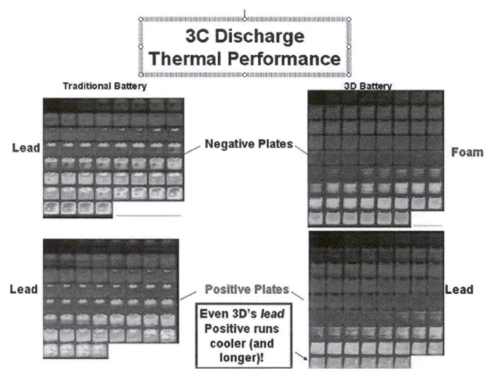

Figure 12.11. Foam-based battery discharge thermal performance.
Source: Graphic courtesy of Firefly Energy

power versus temperature performance. Figure 12.11 represents a graph of a foam-based battery high temperature rate of discharge. Figure 12.12 shows thermal images taken from foam-based cell and comparable commercial VRLA cell batteries. The gray and black shades correspond to ambient temperatures.

Average power	Temperature	Power required
from battery	°C	to crank engine
100%	27°	100%
65%	0°	155%
40%	−18o	210%

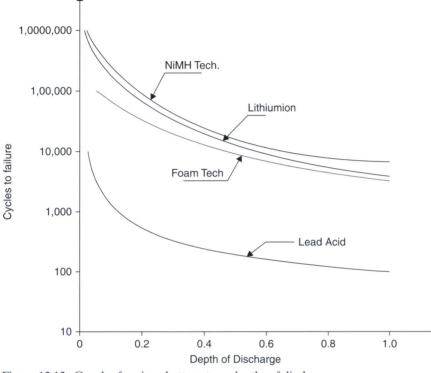

Figure 12.12. Graph of various battery type depths of discharge.

Even though the foam-based battery design utilizes a standard lead grid positive plate, the negative foam plate operates much cooler and generates a calming influence to reduce the temperature of the lead grid positive.

Dramatic Cycle Life Improvements

A full discharge of today's lead acid battery causes extra strain, and each cycle robs the battery of a small amount of capacity. In lead acid batteries, deeper discharges convert larger amounts of charged active material into lead sulfate. Lead sulfate has a significantly larger volume that is about 37% more than the charged material, and this volume change stresses the electrode structures. This expansion induces mechanical forces that deform the grid and ultimately result in the lead grid's disappearing into the paste. The resulting expansion and deformation of the plates also cause active material to separate from the electrodes with a commensurate loss of performance.

Figure 12.13. Foam-based battery used electrode.
Source: Photo courtesy of Firefly Energy

Additionally, over time, sulfate crystals can grow together, yielding large lead sulfate crystals that are difficult or impossible to convert back into the charged state. This wear-down characteristic also applies to other battery chemistries to varying degrees. To prevent the battery from being stressed through repetitive deep discharge, a larger lead acid battery and shallower discharge are typically recommended. Depending on the depth of discharge and operating temperature, the sealed lead acid battery provides 200 to 300 discharge/charge cycles. Short cycle life also results from grid corrosion of the positive electrode, which undergoes extensive oxidative stress during extended recharge conditions. These changes are exacerbated at higher operating temperatures. Figure 12.12 is a graph of foam-based battery depth of discharge.

In contrast, Microcell™ composite plate technology provides a design that fully accommodates the volume changes of the active material during charge and recharge. Within each Firefly plate is contained a full complement of active materials, electrolyte, and volume, which will allow complete discharge without causing physical stress on the plate itself. This results in an electrode plate that does not undergo volume change during deep discharges. Foam battery electrode material is not reactive in the chemistry and so does not corrode. This is in part due to a natural stability of the base material but is also due to the formation process used, which maximizes exposure of the most chemically resistive surfaces and minimizes exposure of chemically less-stable surfaces. Figure 12.13 is a photograph of a foam-based battery used electrode.

The growth of large sulfate crystals is also restricted, resulting in a low incidence of crystals that are too large to recharge. The strong resistance of the electrode material to corrosion also severely reduces the deleterious effects of long recharges.

Because of the removal of grid corrosion as a life-limiting factor, the Firefly approach offers significant improvements over conventional lead acid technologies in both float and deep-cycle applications.

Cycling in irregular applications such as partial-state-of-charge (PSoC) regimes used in hybrid vehicles and photovoltaic energy storage are also well suited to foam-based battery technology. This is because the conditions of partial or heavy sulfation of the negative plate, a process that can render present-generation lead acid products unrecoverable, are easily reversed in foam-based battery products even under long storage periods. Sulfation reversal is achieved because the nature of the lead sulfate deposits in foam-based cells is fundamentally different from those in traditional lead acid cells. In the latter, lead sulfate is deposited on the surfaces of the plates in dense layers of relatively large crystals, somewhat remote from the lead grid members. Because the sponge lead active material in a cell is deposited on the walls of the foam's many small pores in thin layers, and the surface characteristics of the foam result in relatively low current densities, the lead sulfate deposits are composed of small, porous crystal structures (on the order of 3–10 microns, much smaller than in commercial products using traditional manufacturing techniques) that are easily dissolved on the subsequent recharge. Moreover, these very small crystal sizes grow only slowly over time. A final factor that facilitates recharge is the proximity of the carbon-graphite foam (as well as residual sponge lead) that can act as an efficient current-carrying path during recharge for the small, local deposits of lead sulfate crystals. This resistance to the effects of sulfation makes the foam-based batteries ideal for seasonal applications where devices and their associated batteries such as electric lawn mowers, boats, RVs, and motorcycles may be unused for months on end, often in a partially or fully discharged state. Conventional batteries, on the other hand, are difficult or impossible to recover from these conditions and are often replaced far short of their potential life span.

The low self-discharge rate and easy recovery from sulfation also mean that foam-based batteries are not subject to the distribution chain and inventory time constraints of conventional lead acid products. The batteries can be subjected to much longer periods of inactivity without damaging effects. Conventional lead acid batteries are limited to storage times of 3–6 months at most before requiring a recharge.

To date, float and cycle lifetimes for foam-based batteries have yet to be fully determined, but it is anticipated that they will be superior to those of conventional lead acid batteries. Finally, because of the use of lightweight composite foam, their low mass makes the foam batteries highly resistant to vibration.

Performance Summary of Foam-Based Battery Technology

In summary, then, the foam-based battery cell architecture results in the following attributes:

- Instantaneous power (2 hours and faster run-time rates)
- Fast recharge capability
- Continuous power through discharge process
- Recovery to full capacity after off-season storage
- Excellent cold-temperature capacity utilization

- High temperature resiliency
- Recovery to full capacity after discharge
- Can be constructed in either flooded or VRLA configurations
- Extremely rapid recharge capability
- Superior discharge performance
- Longer cycle life compared to existing lead acid products, particularly in PSoC-type applications
- Longer float lifetimes, particularly in high-temperature usage

The remarkable attributes of the composite foam negative electrode noted make certain applications possible or more favorable, as well as improving many lead acid weak points, such as sulfation recovery and active-material utilization limitations. At slow discharge rates, modest weight and volume improvements over existing lead acid products (typically 15–20%) are achieved as well. For faster discharge rate applications, the weight and volume saving can approach 50–75% or more.

3D² Technology

At present, further research in composite foam-based batteries is being advanced to meet specific energy storage for the electrical grid technology. The technology currently referred to as 3D² deploys a unique architecture that is intended to achieve its fullest potential. While the 3D technology is a significant evolutionary improvement over existing lead acid technology, the true potential of the Microcell™ technology, it is currently limited by the conventional positive plate, which acts as a brake on the cell performance in fast charging and during high-rate/low-temperature discharges. In addition, the cell lifetime is still limited by the failure modes associated with the positive plate, which is subject to grid corrosion. Building on the 3D product design, the 3D² product continues to take advantage of the composite foam material for both the negative and positive plates alongside other components, replacing up to 70% of the lead utilized in traditional lead acid batteries. This delivers a formidable jump in power, energy, and cycle life beyond the 3D and approximates the performance of fully packaged lithium and nickel metal hydride batteries without the cost or safety factors. Figure 12.14 is a photograph of 3D² foam battery versus lead acid battery size.

Conversion of the conventional positive plate to a composite foam electrode eliminates positive grid growth and corrosion, but it also introduces a different set of challenges. Various foams used in the positive plate are affected to varying degrees when exposed to extreme overcharge conditions. As a matter of thermodynamics, composite foams can corrode (oxidize) in the positive potential region similarly to the way lead grids corrode; however, electrochemical data show that foams of certain grades react differently in the upper potential ranges commonly experienced during recharge of the positive plate. Currently Firefly is actively refining and stabilizing foam chemistry as well as increasing the robustness of the foams used in positive plates through manipulation of combinations of both foam chemistry and processing, as well as methods of plate preparation. At present, these chemistries and methods are trade-secret intellectual property, for which Firefly is actively pursuing additional patent protection.

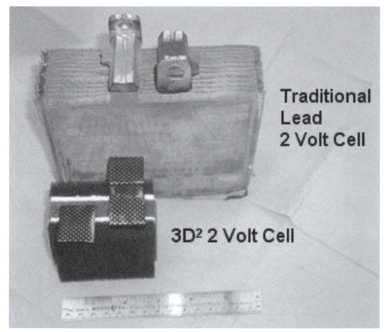

Figure 12.14. $3D^2$ foam battery versus lead acid battery size.
Source: Photo courtesy of Firefly Energy

In summary, $3D^2$ technology in the near future is expected to be a breakthrough in large energy storage systems in that it will offer the promise of allowing the use of lead acid chemistry in new applications, some of which are currently being served by NiCad, NiMH, and/or Li-Ion, as well as improving performance and life in existing lead acid duty cycles. For further information readers may refer to the following Web link: www.fireflyenergy.com/environmental.

Lead Carbon Battery

In recent years advancements in lead acid battery research have led to development of a novel battery technology known as the lead-carbon battery. In this technology electrode structure of the battery is constructed from lead carbon material, which enables much higher volumetric energy density. The battery construction involves charge and discharge double layered electrodes. Lead carbon batteries in addition to durability have achieved significant extension of the operational life cycle.

In this technology carbon is added to the negative electrodes. While the carbon does not affect the chemical nature of the charge-transfer reaction, it augments power storage capacity and significantly reduces sulfation during charging cycles, which is the principal cause of battery failure in conventional lead acid batteries.

Lithium-Ion Batteries

Lithium-ion batteries, which have within the last couple of decades been used in consumer products, have also made a gradual transition into hybrid vehicles and large-scale energy storage; however, at present, because of increased cost of production, the charge and discharge characteristics, as well as charge capacity issues, the technology will perhaps within the near future have a prospect for use in large-scale energy storage. In view of this technology's future potential for use as large energy storage technology, in this section we will have a brief discussion about lithium-ion batteries.

Lithium-Ion Battery Electrochemistry

Lithium (Li) is a soft, silvery white alkali metal, which has atomic number 3 in the periodic table and is the lightest metallic substance. Lithium has a single electron in its valance orbit and is highly reactive and flammable. Because of its high reactivity, lithium never occurs as a free element in nature but appears as an ionic compound in a number of minerals. Water soluble material lithium also is commonly found in the form of brines in seawater and clays. Commercial production of lithium involves an electrolysis process formed from a mixture of lithium chloride and potassium chloride.

Lithium-ion batteries (LIB) are a class of rechargeable battery technologies in which lithium ions migrate from the negative electrode to the positive during discharge and reverse the process of ion migration during the charging phase. Unlike common disposable batteries, lithium-ion electrochemical cell positive and negative electrodes are constructed from nonmetallic lithium compounds that allow capture or insertion of ions, referred to as *intercalation*. Intercalations are a chemical phenomenon when a molecule or group of molecules undergoes temporary inclusion and becomes an integral part of other molecular structures, which can be released and are reconstituted to their original state under certain chemical or charge reversal processes.

Commercially lithium-ion battery electrode materials are graphite compounds. In general, cathodes are fabricated from a number of compounds such as lithium–cobalt oxide, lithium–iron phosphate, or lithium–manganese oxide.

The electrolytes are commonly a mixture of nonaqueous organic carbonates such as diethyl-carbonate, which contain complex lithium Li_6 (stable with 3 neutrons) and Li_7 (stable with 4 neutrons) ions. The nonaqueous electrolytes used are lithium salts such as lithium phosphate fluoride ($LiPF_6$), lithium arsenic fluoride ($LiAsF_6$), lithium perchlorate ($LiClO_4$), or lithium boron fluoride ($LiBF_4$). Depending on the chemical compounds used, lithium-ion battery specific electrical performance characteristics as well as life expectancy can be altered drastically. Because of the current cost of lithium and compounds used in fabrication, lithium ion batteries are considerably more expensive per unit of charge holding capacity per volume; however, they operate over a wider range of energy densities, are small in volume, and operate at wider temperature ranges. Table 12.1 represents characteristic electrical properties of lithium-ion battery electrode materials.

Lithium-Ion Battery Charge Transfer Electrochemistry

The main elements of electrochemistry ion transfer in lithium-ion batteries are the anode, cathode, and electrolyte. Electrical charge storage takes place by electron flow through an external circuit. Chemical charge flow equations representing the

Table 12.1. Characteristic electrical properties of lithium-ion battery electrode materials

Positive electrode

Material	Potential (volts)	Specific capacity	Specific energy
$LiCoO_2$	3.7 V	140 ma-h/g	0.518 kW-h/kg
$LiMn_2O_4$	4.0 V	100 ma-h/g	0.400 kW-h/kg
$LiNiO_2$	3.5 V	180 ma-h/g	0.630 kW-h/kg
$LiFePO_4$	3.3 V	150 ma-h/g	0.495 kW-h/kg
$LiFePO_4F$	3.6 V	115 ma-h/g	0.414 kW-h/kg
Negative electrode			
Graphite (LiC_6)	0.1–0.2 V	372 ma-h/g	0.037–0.07 kW-h/kg
Titanate ($Li_4Ti_5O_{12}$)	1–2 V	160 ma-h/g	0.16–0.32 kW-h/kg
Si (Li_4Si)	0.5–1 V	4212 ma-h/g	2.1–4.2 kW-h/kg
Ge (Li_4Ge)	0.7–1.2 V	1624 ma-h/g	1.137–1.949 kW-h/kg

ion insertion or intercalation are shown in the following formulas. The notation x designates multiples of moles.

Positive electrode during half-reaction charge period:

$$LiCoO_2 \leftrightarrow Li_{1-x}CoO_2 + xLi+ + xe^-$$

Negative electrode during half-reaction charge period:

$$xLi+ + xe^- + 6C \leftrightarrow Li_xC_6$$

Lithium-ion battery ions as described previously are transported from the cathode or the anode during charging and discharging.

At present 12-MW lithium-ion battery storage manufactured by AES, installed in the Atacama Desert, Chile, is deployed in frequency regulation of the local grid. Figure 12.16 is a photograph of a sodium sulfur battery.

Operational Characteristics of Various Lithium-Ion Base Battery Technologies

Lithium Iron Phosphate ($LiFePO_4$)

Lithium iron phosphate ($LiFePO_4$), also known as LFP, is one of the most common compounds used in manufacture of lithium-ion batteries, which are commonly used in electric driven power tools as well as car batteries (Figure 12.15). Other types of lithium batteries used in consumer electronics, such as computers and cell phones, are constructed from lithium manganese oxide ($LiMn_2O_4$), lithium nickel oxide ($LiNiO_2$), or lithium cobalt oxide ($LiCoO_2$). In general anodes of lithium-ion batteries described previously are made from carbon materials.

Among the preceding technologies lithium cobalt oxide ($LiCoO_2$) is the most common and by far the most expensive battery. $LiCoO_2$ in addition are quite toxic and because of formation of gases within the battery they overheat and result in battery explosion, which is a significant shortcoming. On the other hand, $LiFePO_4$ are fabricated from less expensive materials and are not susceptible to overheating; however, they have 25% lower charge storage capacity. Cost of fabrication of the LFP

Figure 12.15. Lithium-ion battery.
Source: Photo courtesy of VARAT battery

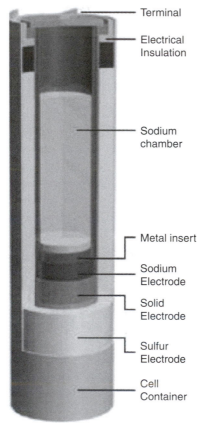

Figure 12.16. Sodium sulfur battery cell.
Source: Courtesy of NGK Japan

batteries due to use of abundant materials such as iron and phosphate and lithium carbonate is considerably less expensive. Moreover, LFP batteries as a result of use of nontoxic materials are considered to be environmentally friendly and appear to have a great prospect for use in electric hybrid cars.

Even though LFP batteries have 25% less charge storage capacity than other lithium-ion batteries, because of their specific material use and construction they have 70% higher performance and operational safety and stability.

Sodium Sulfur Batteries

The sodium sulfur (Na/S) battery is a technology classified as a high-temperature battery that utilizes liquid metallic sodium as the negative electrode and a beta-alumina ceramic separator that functions as the electrolyte. This technology was introduced in the market in the mid-1970s and has since undergone considerable advancements. Essentially sodium sulfur batteries have significantly higher energy density and more efficiency as well as longer cycle life compared to lead acid type batteries. Moreover, they are fabricated from inexpensive and plentiful sulfur and sodium materials. Beta aluminum has a fast ion conduction property, which is also used as a separator in several types of molten salt electrochemical batteries.

Major drawbacks of these batteries are high operational temperature of 300–350°C and the excessively corrosive nature of polysulfide, which is discharged from the battery. Sodium sulfur batteries because of their high charge storage density are suitable for use in stationary applications such as grid energy storage. An additional disadvantage of the Na/S batteries is their thermal management, which is required to maintain the ceramic separator function.

An advanced molten salt battery called the sulfur-metal-chloride (SMC) was developed in the early 1980s and offers less expensive fabrication and a longer operational life cycle. Even though similar in construction to sodium sulfur battery technology, the SMC cells include a secondary electrolyte of molten sodium tetrachloroaluminate ($NaAlCl_4$) in the positive electrode, which is an insoluble transition metal chloride ($FeCl_2$) of nickel chloride ($NiCl_2$) used as a positive electrode, which prevents cell corrosion and allows the battery to operate at a higher temperature range that results in a higher output voltage.

One of the critical components of molten salt type batteries is their thermal management control system, which ensures that each cell is maintained at relatively constant high temperature. Other important and sensitive technology components include a secure internal cell interconnection, thermal cell enclosures, management and maintenance of internal cell electrical heater circuitry, and the battery voltage regulation system. Figure 12.16 depicts a sodium sulfur battery graphic diagram. Figure 12.17 depicts a graphic diagram of sodium sulfur battery assembly- Courtesy of NGK Japan. Figure 12.18 – is a graphic representation of sodium sulfur battery installation.

In order to maintain cell temperature at 300–350°C, electrical heaters are installed within enclosures initially to warm the cells and maintain internal temperature while the batteries are in idle, noncharging, or operational state. Upon battery charge and discharge, because of internal ohmic heating, thermal heaters are deactivated.

Sodium sulfur batteries with their small footprint and high energy charge storage density and exceptional efficiency (90%) are considered to be excellent candidates for use in grid level energy storage. Additional advantages of the Na/S battery technology are minimal maintenance, charge cycling flexibility, and cell modularity. At present, many manufacturers offer self-contained battery modules with 10 to 50

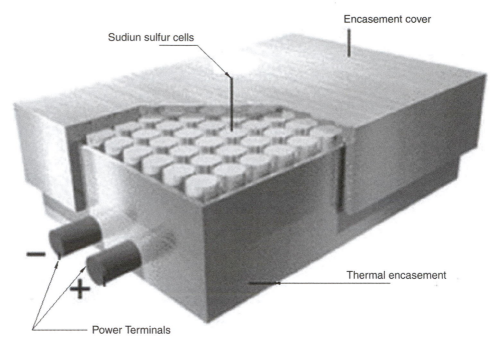

Figure 12.17. Sodium sulfur battery assembly.
Source: Courtesy of NGK Japan

Figure 12.18. Sodium sulfur battery installation.
Source: Courtesy of NGK Japan

kW of energy storage capacity that can be connected in series and parallel to obtain desired voltage, energy, and power requirements. Figure 12.19 is a photograph of sodium sulfur battery installation.

A proof of concept project built in 1990 by a Japanese company, Yuasa deployed 26,880 sodium sulfur batteries, which store 8 megawatt-hours (MWh) of energy. The proof concept project, which was named Moonlight, was terminated after a few

Table 12.2. *Worldwide deployment of Na/S battery projects as of 2009*

Project	Country	Kilowatts	Date of operation
TEPCO	Japan	200,000	2008
HOKKAIDO ELECTRIC	Japan	1,500	2008
JAPAN WIND ELECTRIC	Japan	60,000	2008
AEP	USA	11,000	2009
NEW YORK POWER	USA	1,000	2008
PACIFIC GAS & ELECTRIC	USA	6,000	2009
XCEL	USA	1,000	2008
YOUNICUS	Germany	1,000	2008
ENERCON	Germany	800	2008
EDF	France	1,000	2009
ABU DHABI WATER	Abu Dhabi	48,000	2010

years of operation for technical and economic reasons. At present the technology is still considered in developmental stage. Total combined capacity installed worldwide is estimated to be about 400 MWh. Table 12.2 lists Na/S battery projects deployed worldwide.

Flow Batteries

Flow batteries are a class of electrical energy storage systems unlike conventional batteries in that the power conversion and the electrolyte storage are combined in the same compartment. In these types of technology, power as well as the energy rating are fixed and decoupled from one another. In flow batteries the electrolytes flow through an electrochemical cell in which chemical energy is converted into electricity. Flow batteries are also called redox flow batteries. In flow type batteries the electrolytes are separately stored in large storage tanks outside the battery stacks. As such, separation of electrolyte and the battery stacks allows the amount of the electrolyte storage tanks to define the energy capacity of the system. As a result, the cost of flow battery systems can be divided into two separate components, namely, the electrochemical reactor and the electrolyte storage system. It should be noted that increasing the storage capacity of the electrolyte tanks and their associated cost does not change the cost of the electrochemical reactor. As a result, the cost per kilowatt hour (kWh) of stored energy can be drastically reduced. Figure 12.19(a) is a schematic diagram of a flow battery cell storage system. Figure 12.19(b) is a graphic diagram representing the charge and discharge cycle of flow battery systems. Figure 12.20 is a photograph of a 200-kW vanadium redox battery installation.

The Reduction-Oxidation Process (Redox)

To understand the electrochemical energy generation and charge transfer process it is important to understand the reduction-oxidation (REDOX) reaction. Redox is a chemical reaction in which atoms have their oxidation number changed. In flow battery cells two electrolytes are separated from each other by a semipermeable membrane (SPM). The SPM allows ion flow through the membrane; however, it prevents ion flow across the membrane, as such flow of electrons results in electrical current across conductive buses between the positive and negative electrodes.

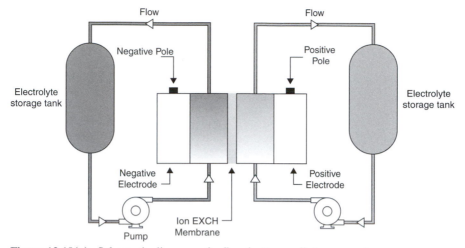

Figure 12.19(a). Schematic diagram of a flow battery cell storage system.

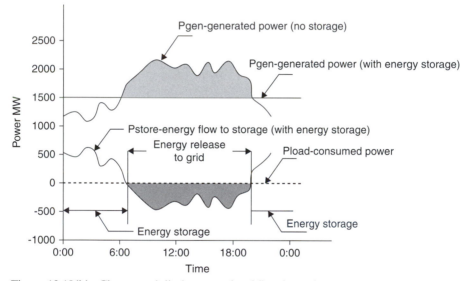

Figure 12.19(b). Charge and discharge cycle of flow batteries.

As mentioned, redox reactions describe all chemical reactions in which atoms have their oxidation state changed. For instance, oxidation of carbon yields carbon dioxide (CO_2), and the reduction of carbon by hydrogen yields methane (CH_4). Redox reactions, or oxidation-reduction reactions, also take place in acid-base reactions. Essentially, redox reactions refer to transfer of electrons between various elements. It should be noted that redox reactions are a matched set; in other words, there cannot be an oxidation reaction without a reduction reaction that occurs simultaneously. Oxidation refers to the loss of electrons, while reduction refers to the gain of electrons. Each reaction by itself is called a "half-reaction," because there must be two half-reactions to form a whole reaction. As such, in notating redox reactions, gain and loss of electrons are shown in numerals beside the element. In summary, the redox reaction is considered as the following:

Figure 12.20. A 200-kW vanadium redox flow battery installation.
Source: Photo courtesy of VRB-ESS, Utah, United States

- Oxidation is the **loss** of electrons or an increase in oxidation state by a molecule, atom, or ion. Oxidation is also defined as an increase in oxidation state
- Reduction is the **gain** of electrons or a *decrease* in oxidation state by a molecule, atom, or ion. Reduction is also referred to as a change in oxidation state

An exception to the process is the transfer of electrons in reactions of covalent bonding, when even though referred to as "redox reactions," they involve no electron transfer.

Electron charge transfer in a redox process is expressed and formulated as the following:

Oxidation process

Oxidant + e⁻ → product (electron gained; oxidation number reduced)
Reduction → product + e⁻ (electron lost; oxidation number increased)

In short, in redox processes, the *reductant* transfers electrons to the *oxidant*. During the reaction, the reductant or *reducing agent* loses electrons and is oxidized, while the oxidant or *oxidizing agent* gains electrons and is reduced. The pair of an oxidizing and reducing agent involved in a particular reaction is called a *redox pair*.

Likewise oxidizing agents or oxidizers remove electrons from another substance and themselves by accepting electrons. They are also referred to as *electron acceptor*s. Table 12.3 shows the redox reaction that takes place between negative and positive cells of flow batteries that use various redox pairs.

In this section we will discuss two of the most promising redox couple flow battery cells, the vanadium redox battery (VRB) and the zinc bromide battery (ZnBr), which have been developed over the past few years.

Vanadium Redox Battery (VRB) Technology

The key feature of the vanadium redox battery is based on exploitation of the ability of vanadium to exist in solution in four different oxidation states. This unique

Table 12.3. *Redox reaction process between flow battery cells*

Redox pair	Process at negative cell	Process at positive cell
Fe/Ti	$Ti^{3+} + e^- \rightarrow Ti^{2+}$	$Fe^{3+} + e^- \rightarrow Fe^{2+}$
V/V	$V^{3+} + e^- \rightarrow V^{2+}$	$V^{3+} + e^- \rightarrow V^{4+}$
Zn/Br	$Zn^{2+} + 2e^- \rightarrow Zn$	$Br_2 + 2e^- \rightarrow 2Br$

property is used to make a battery that uses one type of electroactive element instead of two used in other flow type batteries. Vanadium redox batteries use two different vanadium electrolytes, namely, V^{2+}/V^{3+} and V^{4+}/V^{5+}, both of which are mildly acid solutions with four varying ionization levels. During the charge and discharge cycles, hydrogen ions H+ are exchanged between the two electrolytes through a permeable polymer separator membrane, which results in the following reaction during charge and discharge cycles:

Charge cycle reaction:

$$V^{3+} + e^- - \rightarrow V^{2+}$$
$$V^{4+} \rightarrow V^{5+} + e^-$$

Discharge cycle reaction:

$$V2^+ \rightarrow V^{3+} + e^-$$
$$V4^+ + e^- \rightarrow V^{4+}$$

In both of the half-cells vanadium is used as an electrolyte, which results in elimination of what is known as cross-contamination by diffusion of ions, which may occur across the ion exchange membrane.

Vanadium battery technology development was spearheaded by the National Aeronautics and Space Administration (NASA) in the 1970s. Subsequently, the first commercial development of the technology was that by Maria Skyllas-Kazacoss and coworkers at the University of New South Wales, Australia, in the 1980s. At present there are a number of manufacturers of vanadium redox battery systems in Australia, the United States, Germany, Thailand, and elsewhere. Figure 12.21 is a photograph of a 200-kW vanadium redox flow battery installation in Utah, in the United States.

Construction of a vanadium redox battery consists of power cells that deploy two electrolytes that are separated by a proton exchange membrane. Electrolytes used in the battery are prepared by dissolving vanadium peroxide (V_2O_5) in sulfuric acid (H_2SO_4). The solution is therefore very acidic when in use. In the ion exchange process in vanadium flow batteries, the electrolytes of both half-cells are connected to storage tanks within the vicinity of the battery and are circulated in large volumes by two separate pumps. Separation of the electrolytes limits vanadium batteries strictly to stationary operation and prevents mobile use.

The main advantages of vanadium redox flow batteries are fast response to charging loads and vast capacity to deliver power to large loads, which makes them very suitable for grid energy charge storage. Under discharge condition, vanadium batteries are capable of sustaining 400% overload for periods of up to 10 seconds. Efficiency of vanadium flow batteries as a result of energy use of circulating pumps and control electronics is limited to 60–75%. A more recent advancement

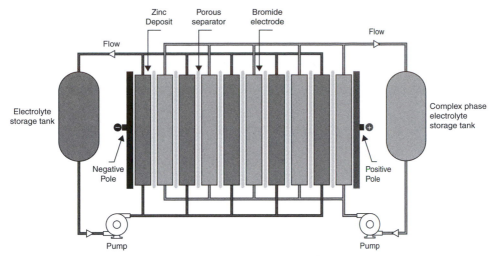

Figure 12.21. Zinc-bromine battery system configuration diagram.

in vanadium flow battery design uses vanadium-polyhalide electrolyte, which apparently augments charge density by 100% and increases battery operational temperature to a higher range.

In view of large charge storage capacity, extended life cycle, usability of electrolyte, and fast discharge of large amounts of stored power, vanadium is an excellent candidate for grid storage applications, as well as wind and solar power energy production projects. The following is a listing of typical vanadium redox battery characteristics.

Specific energy storage capacity		
Energy density	10–20 watt-hours / kilogram	15–25 watt-hours/ liter
Charge/discharge efficiency	60–80%	
Battery life cycle		10–20 years
Cycle durability		Less than 10,000 cycles
Nominal cell voltage		1.15–1.55 volts

Zinc-Bromine Battery (ZnBr) Technology

The zinc-bromine (ZnBr) battery is one of the most attractive flow battery technologies and perhaps because of its specific performance characteristics may become one of the preferred utility-scale energy storage systems in the near future. The zinc-bromine batteries consist of a zinc-negative electrode and a bromine-positive electrode, which are separated by an ion exchange barrier fabricated from microporous noncorrosive material. As in all flow type battery technologies, the electrochemical reactions in ZnBr battery technology that store electrical energy are realized through the system configuration described, which consists of the bipolar electrodes and an aqueous electrolyte, which are contained in a reservoir. The electrolyte is an aqueous solution of zinc-bromide, circulated with pumps, that passes through the electrode surfaces. As mentioned earlier, the electrode surfaces

Figure 12.22. Zinc-bromine battery installation.
Source: Photo courtesy of Redflow

of zinc and bromine are separated by a plastic ion exchange film. The electrolyte flow takes place in two streams, with two separate pumps, one on the positive electrode side and another on the negative side. Figure 12.22 depicts the zinc-bromide battery system configuration.

During charging, zinc is deposited at the negative electrode, and bromine is collected at the positive electrode. During discharge, zinc and bromide ions are formed at each electrode. The ion exchange porous separator placed between the electrodes prevents diffusion of bromine to reach the zinc deposit and blocks the chemical reaction that otherwise would discharge the cell.

Chemical electrolyte during the charge process forms complex bromine and zinc ion pairs that have various oxidation levels. The electrolyte also contains complexing chemical agents such as N-methyl-N-ethylmorpholinium (MEMBr) that reduce the amount of bromine in each cell that prevents the self-discharge reaction. Upon formation of the aqueous complexed polybromine, the liquid is pumped out and stored in large container tanks, while zinc is deposited on the negative electrode. The ZnBr battery therefore is considered an electroplating process. The electrodes of ZnBr batteries are fabricated from a nonmetallic composite carbon-plastic material. Figure 21.23 is a photograph of zinc-bromine battery installation.

As mentioned, during the charging process, complexed polybromide is formed, and during discharge complex polybromide is circulated back to the electrodes, at which time the electrolyte reverts to its original chemical composition as a zinc-bromide solution. The following is the chemical reaction representing the ZnBr battery technology. The (aq) in the formula designates aqueous solution and (s) solid material:

Charge phase – At negative electrode
$Zn^2 + (aq) + 2e^- \rightarrow Zn^0 (s)$

Figure 12.23. Zinc-bromine battery installation in solar power system application.
Source: Photo courtesy of Redflow

Charge phase – At positive electrode
$2Br^- (aq) \rightarrow Br_2 (aq) + 2e^-$

Battery cell combine reaction
$Zn^{2}+ (aq) + 2Br (aq) \rightarrow Zn (s) + Br_2 (aq)$

During the discharge, the reactions are reversed. During the charge process zinc also reacts with water and results in production of small amounts of hydrogen. There is a slight negative effect on the chemical reaction, however; since the cells do not contain metallic components or impurities, a trace component of hydrogen does not diminish battery energy storage capacity. Performance efficiency of zinc-bromine batteries varies with discharge rate. At higher discharge rates battery efficiency decreases and battery temperature increases. During the charge and discharge operation, a significant portion of the energy stored is diverted to auxiliary systems such as circulating pumps, control valves, thermal management, and controls. Mean performance efficiency of energy output in ZnBr battery technologies ranges from 60% to 70%. A minor amount of energy loss, approximately 1%/per hour, also occurs during standby time. Upon depletion of bromine from the cell stacks, the battery self-discharge ceases. Table 12.4 shows the effect of battery discharge rates and temperature increase. Table 12.5 is an overview of characteristics of various types of energy storage technologies.

It is interesting to know that the concept of the zinc bromide technology was patented more than 100 years ago; however, because of its inherent properties, development of a commercial product did not materialize. The main obstacle for maturation of the technology was that during the electroplating or charging process, zinc has a tendency to form dendrites, which upon deposition on the negative electrode cause the electrochemical process to deteriorate. Figure 12.23 is a photograph of zinc-bromine battery installation in a solar power system application.

Table 12.4. *Characteristic performance variations of ZnBr energy storage technologies*

Discharge current (A)	Discharge time (hours)	Maximum temp.	Energy (°C)	Output (kWh)
35.5	5.61	30.6	19.83	
42.8	4.67	31.6	19.71	
53.3	3.75	33.2	19.43	
71.2	2.82	35.0	19.15	
104.9	1.87	39.5	17.86	
209.9	0.82	50.9	13,54	

Source: Clark and Eidler and Lex.

Table 12.5. *Overview of characteristics of various energy storage systems*

Technology	Typical power	Typical energy	Duration of discharge	
Lead acid, NiCad, Li-ion	Up to 500 kW	Up to 100 MWh	1–8 hours	
Pumped hydro	500 kW–1 MW	100 kWh–15 GWh	< 5 minutes	
CAES		25–3000 MW	200 MWh–10 GWh	1–20 hours
NaS	1 MW	1 MWh	1 hour	
Flow batteries	100 kW–10 MW	1–100 MWh	10 hours	

Flow Battery System Cost Comparison

Three types of flow batteries discussed previously have an approximately equivalent range of system cost per kilowatt. At present, the approximate cost of commercially available flow batteries ranges from $250.00 to $300.00/kWh of storage capacity. Among the three technologies upscaling system size, which translates into large electrolyte tank capacity, ZnBr has the lowest electrolyte cost. However, vanadium system (V/V) batteries offer greater scalability of electrolyte storage. In the near future it is expected that cost per kilowatt-hour of storage will be reduced to $150.00 to $200.00/kWh.

Compared to lead acid or lithium-ion battery technologies, flow batteries are relatively inexpensive and offer longer life and considerable scalability for use in grid connected systems. Table 12.6 is an overview of various types of storage systems applications.

Pumped Hydro

The storage system technologies discussed offer substantial promise for deployment in large-scale grid connected applications. However, to date, most demonstration projects of the technologies are in progress. In order for the technologies to mature into large-scale commercial production, in the coming years extensive field testing is required.

Viability of Flow Battery Use in Solar Power Systems

As discussed, all three major flow battery technologies use external electrolyte storage tanks and associated circulating pumps, electronic controls, and temperature

Table 12.6. *Overview of various types of storage systems applications*

Storage system	Application	Specification
Na/S, V/V, ZnBr Pumped hydro	Long-term energy management	10–100 MW, 1–10 hours
Flow batteries, Na/S Lead acid	Integration of renewable energy	0.1–100 MW, 1–10 minutes Sources such as wind and solar
Flow batteries, Na/S	Backup renewable for energy	Less than 1 MW, 1–20 hours
Flow batteries, Na/S	Peak power generation support	1–100 MW, 1 second

regulation, as well as DC to AC power conversion equipment to store electrical charge. In view of the significant energy required for supporting the auxiliary equipment, energy efficiency of the overall flow battery technologies is reduced by approximately 30–40%. In other words, when storing electrical energy from various flow-based sources, there is a considerable loss, which must be taken into account.

In some applications such as wind power, most often electrical energy is usually produced in off-peak hours, in the evening and nighttime. Even though the off-peak energy produced may not be suitable for grid loading, however, storage of energy and disbursement during peak demand hours, such as midday, render use of large-scale power storage quite viable; otherwise, wind energy produced during off-peak hours will have minimal economic value.

In the case of solar power energy production, in order to validate financial viability of electrical energy storage systems, regardless of the cost per kilowatt of storage, low-performance efficiency of the flow battery system is such that losses of 30–40% could significantly burden and invalidate use of flow battery storage, since solar power systems at best operate at a maximum efficiency of 80%. By superimposing a loss of 30–40%, the overall efficiency of a solar power system could be drastically reduced to about 55–60%, which would otherwise be economically unacceptable.

In view of the degradation of efficiency, the only viable alternative large-scale energy storage technology available to date is the sodium sulfur (Na/s) battery system, which has operational efficiency of approximately 89%.

Unit Conversion and Design Reference Tables

Renewable Energy Tables and Important Solar Power Facts

1 Recent analysis by the Department of Energy (DOE) shows that by year 2025, one-half of the new U.S. electricity generation could come from the sun.
2 The United States has generated only 4 GW (1 GW is 1000 MW) of solar power. By the year 2030, it is estimated to be 200 GW.
3 A typical nuclear power plant generates about 1 GW of electric power, which is equal to 5 GW of solar power (daily power generation is limited to an average of 5 to 6 hours per day).
4 Global sales of solar power systems have been growing at a rate of 45% in the past few years.
5 It is projected that by the year 2020, the United States will be producing about 7.2 GW of solar power.
6 Shipment of U.S. solar power systems has fallen by 10% annually but has increased by 45% throughout Europe.
7 Annual sales growth globally has been 35%.
8 Present cost of solar power modules on the average is $2.33/W. By 2030 it should be about $0.38/W.
9 World production of solar power is 1 GW/year.
10 Germany has a $0.50/W grid feed incentive that will be valid for the next 20 years. The incentive is to be decreased by 5% per year.
11 In the past few years, Germany installed 130 MW of solar power per year.
12 Japan has a 50% subsidy for solar power installations of 3- to 4-kW systems and has about 800 MW of grid-connected solar power systems. Solar power in Japan has been used since 1994.
13 California, in 1996, set aside $540 million for renewable energy, which has provided a $4.50/W to $3.00/W buyback as a rebate.
14 In the years 2015 through 2024, it is estimated that California could produce an estimated $40 billion of solar power sales.
15 In the United States, 20 states have a solar rebate program. Nevada and Arizona have set aside a state budget for solar programs.
16 Total U.S. production has been just about 18% of global production.
17 For each megawatt of solar power produced, we employ 32 people.

18 A solar power collector, sized 100 bv 00 mi, in the southwestern United States could produce sufficient electric power to satisfy the country's yearly energy needs.

19 For every kilowatt of power produced by nuclear or fossil fuel plants, 1/2 gal of water is used for scrubbing, cleaning, and cooling. Solar power requires practically no water usage.

20 Significant impact of solar power cogeneration:

a. Boosts economic development.
b. Lowers cost of peak power.
c. Provides greater grid stability.
d. Lowers air pollution.
e. Lowers greenhouse gas emissions.
f. Lowers water consumption and contamination.

21 A mere 6.7 mi/gal efficiency increase in cars driven in the United States could offset our share of imported Saudi oil.

22 State of solar power technology at present:

Crystalline
Polycrystalline
Amorphous
Thin- and thick-film technologies

23 State of solar power technology in future:

Plastic solar cells
Nano-structured materials
Dye-synthesized cells

Energy Conversion Table

Energy units

1 J (joule) = 1 Ws = 4.1868 cal
1 GJ (gigajoule) = 10 E9 J
1 TJ (terajoule) = 10 E12 J
1 PJ (petajoule) = 10 E15 J
1 kWh (kilowatt-hour) = 3,600,000 J
1 toe (ton oil equivalent) = 7.4 barrels of crude oil in primary energy = 7.8 barrels in total final consumption = 1270 m3 of natural gas = 2.3 metric tons of coal

Mt (million ton oil equivalent) = 41.868 PJ

Power – Electric power is usually measured in watts (W), kilowatts (kW), megawatts (MW), and so forth. Power is energy transfer per unit of time.

1 kW = 1000 W
1 MW = 1,000,000
1 GW = 1000 MW
1 TW = 1,000,000 MW

Power (e.g., in W) may be measured at any point in time, whereas energy (e.g., in kWh) has to be measured during a certain period, for example, a second, an hour, or a year.

Unit Abbreviations

m = meter = 3.28 feet (ft)
s = second
h = hour
W = watt
hp = horsepower
J = joule
cal = calorie
toe = tons of oil equivalent
Hz = hertz (cycles per second)
10 E–12 = pico (p) = 1/1,000,000,000,000,000
10 E–9 = nano (n) = 1/1,000,000,000
10 E–6 = micro (μ) = 1/1,000,000
10 E–3 = milli (m) = 1/1,000
10 E–3 = kilo (k) = 1,000 = thousands
10 E–6 = mega (M) = 1,000,000 = millions
10 E–9 = giga (G) = 1,000,000,000
10 E–12 = tera (T) = 1,000,000,000,000
10 E–15 = peta (P) = 1,000,000,000,000,000

Wind Speeds

1 m/s = 3.6 km/h = 2.187 mi/h = 1.944 knots
1 knot = 1 nautical mile per hour = 0.5144 m/s = 1.852 km/h = 1.125 mi/h

voltage Drop Calculation for Copper Wires:Solar Photovoltaic Module Tilt Angle Correction Table:SOLAR PANEL ORIENTATION TILT CORRECTION FACTOR SOLAR TILT ANGLE FROM HORIZONTAL

Tilt angle/facing	0	15	30	45	60	90
South	0.89	0.97	1.00	0.97	0.88	0.56
SSE or SSW	0.89	0.97	0.99	0.96	0.87	0.57
SE or SW	0.89	0.95	0.96	0.93	0.85	0.59
ESE or WSW	0.89	0.92	0.91	0.87	0.79	0.57
East or West	0.89	0.88	0.84	0.78	0.7	0.51

TILT ANGLE EFFICIENCY MULTIPLIER TABLE: COLLECTOR TILT ANGLE FROM HORIZONTAL (DEGREES)

0	15	30	45	60	90	
Fresno						
South	0.90	0.98	1.00	0.96	0.87	0.55
SSE, SSW	0.90	0.97	0.99	0.96	0.87	0.56

(continued)

0	15	30	45	60	90	
SE, SW	0.90	0.95	0.96	0.92	0.84	0.68
ESE, WSW	0.90	0.92	0.91	0.87	0.79	0.57
E, W	0.90	0.88	0.86	0.78	0.70	0.51
Daggett						
South	0.88	0.97	1.00	0.97	0.88	0.56
SSE, SSW	0.88	0.96	0.99	0.96	0.87	0.58
SE, SW	0.88	0.94	0.96	0.93	0.85	0.59
ESE, WSW	0.88	0.91	0.91	0.86	0.78	0.57
E, W	0.88	0.87	0.83	0.77	0.69	0.51
Santa Maria						
South	0.89	0.97	1.00	0.97	0.88	0.57
SSE, SSW	0.89	0.97	0.99	0.96	0.87	0.58
SE, SW	0.89	0.95	0.96	0.93	0.86	0.59
ESE, WSW	0.89	0.92	0.91	0.87	0.79	0.67
E, W	0.89	0.88	0.84	0.78	0.70	0.52
Los Angeles						
South	0.89	0.97	1.00	0.97	0.88	0.57
SSE, SSW	0.89	0.97	0.99	0.96	0.87	0.58
SE, SW	0.89	0.95	0.96	0.93	0.85	0.69
ESE, WSW	0.89	0.92	0.91	0.87	0.79	0.57
E, W	0.89	0.88	0.85	0.78	0.70	0.51
San Diego						
South	0.89	0.98	1.00	0.97	0.88	0.57
SSE, SSW	0.89	0.97	0.99	0.96	0.87	0.58
SE, SW	0.89	0.95	0.96	0.92	0.54	0.59
ESE, WSW	0.89	0.92	0.91	0.87	0.79	0.57
E, W	0.89	0.88	0.85	0.78	0.70	0.51

Solar Insolation Table for Major Cities in the United States

State	City	High	Low	Avg.	State	City	High	Low	Avg.
AK	Fairbanks	5.87	2.12	3.99	GA	Griffin	5.41	4.26	4.99
AK	Matanuska	5.24	1.74	3.55	III	Honolulu	6.71	5.59	6.02
AL	Montgomery	4.69	3.37	4.23	IA	Ames	4.80	3.73	4.40
AR	Bethel	6.29	2.37	3.81	ill	Boise	5.83	3.33	4.92
AR	Little Rock	5.29	3.88	4.69	ill	Twin Falls	5.42	3.42	4.70
AZ	Tucson	7.42	6.01	6.57	IL	Chicago	4.08	1.47	3.14
AZ	Page	7.30	5.65	6.36	IN	Indianapolis	5.02	2.55	4.21
AZ	Phoenix	7.13	5.78	6.58	KS	Manhattan	5.08	3.62	4.57
CA	Santa Maria	6.52	5.42	5.94	KS	Dodge City	4.14	5.28	5.79
CA	Riverside	6.35	5.35	5.87	KY	Lexington	5.97	3.60	4.94
CA	Davis	6.09	3.31	5.10	LA	Lake Charles	5.73	4.29	4.93
CA	Fresno	6.19	3.42	5.38	LA	New Orleans	5.71	3.63	4.92
CA	Los Angeles	6.14	5.03	5.62	LA	Shreveport	4.99	3.87	4.63
CA	Soda Springs	6.47	4.40	5.60	MA	E. Wareham	4.48	3.06	3.99
CA	La Jolla	5.24	4.29	4.77	MA	Boston	4.27	2.99	3.84
CA	Inyokern	8.70	6.87	7.66	MA	Blue Hill	4.38	3.33	4.05
CO	Grandbaby	7.47	5.15	5.69	MA	Natick	4.62	3.09	4.10
CO	Grand Lake	5.86	3.56	5.08	MA	Lynn	4.60	2.33	3.79
CO	Grand Junction	6.34	5.23	5.85	MD	Silver Hill	4.71	3.84	4.47
CO	Boulder	5.72	4.44	4.87	ME	Caribou	5.62	2.57	4.19

State	City	High	Low	Avg.	State	City	High	Low	Avg.
DC	Washington	4.69	3.37	4.23	ME	Portland	5.23	3.56	4.51
FL	Apalachicola	5.98	4.92	5.49	MI	Sault Ste. Marie	4.83	2.33	4.20
FL	Belie Is.	5.31	4.58	4.99	MI	E. Lansing	4.71	2.70	4.00
FL	Miami	6.26	5.05	5.62	MN	St. Cloud	5.43	3.53	4.53
FL	Gainesville	5.81	4.71	5.27	MO	Columbia	5.50	3.97	4.73
FL	Tampa	6.16	5.26	5.67	MO	St. Louis	4.87	3.24	4.38
GA	Atlanta	5.16	4.09	4.74	MS	Meridian	4.86	3.64	4.43
MT	Glasgow	5.97	4.09	5.15	PA	Pittsburg	4.19	1.45	3.28
MT	Great Falls	5.70	3.66	4.93	PA	State College	4.44	2.79	3.91
MT	Summit	5.17	2.36	3.99	RI	Newport	4.69	3.58	4.23
NM	Albuquerque	7.16	6.21	6.77	SC	Charleston	5.72	4.23	5.06
NB	Lincoln	5.40	4.38	4.79	SD	Rapid City	5.91	4.56	5.23
NB	N. Omaha	5.28	4.26	4.90	1N	Nashville	5.2	3.14	4.45
NC	Cape Hatteras	5.81	4.69	5.31	1N	Oak Ridge	5.06	3.22	4.37
NC	Greensboro	5.05	4.00	4.71	TX	San Antonio	5.88	4.65	5.3
ND	Bismarck	5.48	3.97	5.01	TX	Brownsville	5.49	4.42	4.92
NJ	Sea Brook	4.76	3.20	4.21	TX	EI Paso	7.42	5.87	6.72
NV	Las Vegas	7.13	5.84	6.41	TX	Midland	6.33	5.23	5.83
NV	Ely	6.48	5.49	5.98	TX	Fort Worth	6.00	4.80	5.43
NY	Binghamton	3.93	1.62	3.16	UT	Salt Lake City	6.09	3.78	5.26
NY	Ithaca	4.57	2.29	3.79	UT	Flaming Gorge	6.63	5.48	5.83
NY	Schenectady	3.92	2.53	3.55	VA	Richmond	4.50	3.37	4.13
NY	Rochester	4.22	1.58	3.31	WA	Seattle	4.83	1.60	3.57
NY	New York City	4.97	3.03	4.08	WA	Richland	6.13	2.01	4.44
OH	Columbus	5.26	2.66	4.15	WA	Pullman	6.07	2.90	4.73
OH	Cleveland	4.79	2.69	3.94	WA	Spokane	5.53	1.16	4.48
OK	Stillwater	5.52	4.22	4.99	WA	Prosser	6.21	3.06	5.03
OK	Oklahoma City	6.26	4.98	5.59	WI	Madison	4.85	3.28	4.29
OR	Astoria	4.76	1.99	3.72	WV	Charleston	4.12	2.47	3.65
OR	Corvallis	5.71	1.90	4.03	WY	Lander	6.81	5.50	6.06
OR	Medford	5.84	2.02	4.51					

Note: Values are given in kilowatt-hours per square meter per day.

Longitude and latitude tables

Alabama	Latitude	Longitude
Alexander City	32° 57 N	85° 57 W
Anniston	33° 35 N	85° 51 W
Auburn	32° 36 N	85° 30 W
Birmingham	33° 34 N	86° 45 W
Decatur	34° 37 N	86° 59 W
Dothan	31° 19 N	85° 27 W
Florence	34° 48 N	87° 40 W
Gadsden	34° 1 N	86° 0 W
Huntsville	34° 42 N	86° 35 W
Mobile	30° 41 N	88° 15 W
Mobile Co	30° 40 N	88° 15 W

(continued)

Alabama	Latitude	Longitude
Montgomery	32° 23 N	86° 22 W
Selma-Craig AFB	32° 20 N	87° 59 W
Talladega	33° 27 N	86° 6 W
Tuscaloosa	33° 13 N	87° 37 W
Alaska		
Anchorage	61° 10 N	150° 1 W
Barrow (S)	71° 18 N	156° 47 W
Fairbanks (S)	64° 49 N	147° 52 W
Juneau	58° 22 N	134° 35 W
Kodiak	57° 45 N	152° 29 W
Nome	64° 30 N	165° 26 W
Barstow	34° 51 N	116° 47 W
Blythe	33° 37 N	114° 43 W
Burbank	34° 12 N	118° 21 W
Chico	39° 48 N	121° 51 W
Concord	37° 58 N	121° 59 W
Covina	34° 5 N	117° 52 W
Crescent City	41° 46 N	124° 12 W
Downey	33° 56 N	118° 8 W
El Cajon	32° 49 N	116° 58 W
El Cerrito (S)	32° 49 N	115° 40 W
Escondido	33° 7 N	117° 5 W
Eureka/Arcata	40° 59 N	124° 6 W
Fairfield-Trafis AFB	38° 16 N	121° 56 W
Fresno (S)	36° 46 N	119° 43 W
Hamilton AFB	38° 4 N	122° 30 W
Laguna Beach	33° 33 N	117° 47 W
Livermore	37° 42 N	121° 57 W
Lompoc, Vandenberg AFB	34° 43 N	120° 34 W
Long Beach	33° 49 N	118° 9 W
Los Angeles (S)	33° 56 N	118° 24 W
Los Angeles CO (S)	34° 3 N	118° 14 W
Merced-Castle AFB	37° 23 N	120° 34 W
Modesto	37° 39 N	121° 0 W
Arizona		
Douglas	31° 27 N	109° 36 W
Flagstaff	35° 8 N	111° 40 W
Fort Huachuca (S)	31° 35 N	110° 20 W
Kingman	35° 12 N	114° 1 W
Nogales	31° 21 N	110° 55 W
Phoenix (S)	33° 26 N	112° 1 W
Prescott	34° 39 N	112° 26 W
Tucson (S)	32° 7 N	110° 56 W
Winslow	35° 1 N	110° 44 W
Yuma	32° 39 N	114° 37 W
Arkansas		
Blytheville AFB	35° 57 N	89° 57 W
Camden	33° 36 N	92° 49 W
El Dorado	33° 13 N	92° 49 W
Fayetteville	36° 0 N	94° 10 W
Fort Smith	35° 20 N	94° 22 W
Hot Springs	34° 29 N	93° 6 W

Alabama	Latitude	Longitude
Jonesboro	35° 50 N	90° 42 W
Little Rock (S)	34° 44 N	92° 14 W
Pine Bluff	34° 18 N	92° 5 W
Texarkana	33° 27 N	93° 59 W
California		
Bakersfield	35° 25 N	119° 3 W
Monterey	36° 36 N	121° 54 W
Napa	38° 13 N	122° 17 W
Needles	34° 36 N	114° 37 W
Oakland	37° 49 N	122° 19 W
Oceanside	33° 14 N	117° 25 W
Ontario	34° 3 N	117° 36 W
Oxnard	34° 12 N	119° 11 W
Palmdale	34° 38 N	118° 6 W
Palm Springs	33° 49 N	116° 32 W
Pasadena	34° 9 N	118° 9 W
Petaluma	38° 14 N	122° 38 W
Pomona Co	34° 3 N	117° 45 W
Redding	40° 31 N	122° 18 W
Redlands	34° 3 N	117° 11 W
Richmond	37° 56 N	122° 21 W
Riverside-March AFB (S)	33° 54 N	117° 15 W
Sacramento	38° 31 N	121° 30 W
Salinas	36° 40 N	121° 36 W
San Bernadino, Norton AFB	34°8 N	117° 16 W
San Diego	32° 44 N	117° 10 W
San Fernando	34° 17 N	118° 28 W
San Francisco	37° 37 N	122° 23 W
San Francisco Co	37° 46 N	122° 26 W
San Jose	37° 22 N	121° 56 W
San Louis Obispo	35° 20 N	120° 43 W
Santa Ana	33° 45 N	117° 52 W
Santa Barbara MAP	34° 26 N	119° 50 W
Santa Cruz	36° 59 N	122° 1 W
Santa Maria (S)	34° 54 N	120° 27 W
Santa Monica CIC	34° 1 N	118° 29 W
Santa Paula	34° 21 N	119° 5 W
Santa Rosa	38° 31 N	122° 49 W
Stockton	37° 54 N	121° 15 W
Ukiah	39° 9 N	123° 12 W
Visalia	36° 20 N	119° 18 W
Yreka	41° 43 N	122° 38 W
Yuba City	39° 8 N	121° 36 W
Colorado		
Alamosa	37° 27 N	105° 52 W
Boulder	40° 0 N	105° 16 W
Colorado Springs	38° 49 N	104° 43 W
Denver	39° 45 N	104° 52 W
Durango	37° 17 N	107° 53 W
Fort Collins	40° 45 N	105° 5 W
Grand Junction (S)	39° 7 N	108° 32 W

(continued)

Alabama	Latitude	Longitude
Greeley	40° 26 N	104° 38 W
Lajunta	38° 3 N	103° 30 W
Fort Myers	26° 35 N	81° 52 W
Fort Pierce	27° 28 N	80° 21 W
Gainesville (S)	29° 41 N	82° 16 W
Jacksonville	30° 30 N	81° 42 W
Key West	24° 33 N	81° 45 W
Lakeland Co (S)	28° 2 N	81° 57 W
Miami (S)	25° 48 N	80° 16 W
Miami Beach Co	25° 47 N	80° 17 W
Ocala	29° 11 N	82° 8 W
Orlando	28° 33 N	81° 23 W
Panama City, Tyndall AFB	30° 4 N	85° 35 W
Pensacola Co	30° 25 N	87° 13 W
St. Augustine	29° 58 N	81° 20 W
St. Petersburg	27° 46 N	82° 80 W
Stanford	28° 46 N	81° 17 W
Sarasota	27° 23 N	82° 33 W
Tallahassee (S)	30° 23 N	84° 22 W
Tampa (S)	27° 58 N	82° 32 W
West Palm Beach	26° 41 N	80° 6 W
Georgia		
Albany, Turner AFB	31° 36 N	84° 5 W
Americus	32° 3 N	84° 14 W
Athens	33° 57 N	83° 19 W
Atlanta (S)	33° 39 N	84° 26 W
Leadville	39° 15 N	106° 18 W
Pueblo	38° 18 N	104° 29 W
Sterling	40° 37 N	103° 12 W
Trinidad	37° 15 N	104° 20 W
Connecticut		
Bridgeport	41° 11 N	73° 11 W
Hartford, Brainard Field	41° 44 N	72° 39 W
New Haven	41° 19 N	73° 55 W
New London	41° 21 N	72° 6 W
Norwalk	41° 7 N	73° 25 W
Norwick	41° 32 N	72° 4 W
Waterbury	41° 35 N	73° 4 W
Widsor Locks, Bradley Fld	41° 56 N	72° 41 W
Delaware		
Dover AFB	39° 8 N	75° 28 W
Wilmington	39° 40 N	75° 36 W
District of Columbia		
Andrews AFB	38° 5 N	76° 5 W
Washington, National	38° 51 N	77° 2 W
Florida		
Belle Glade	26° 39 N	80° 39 W
Cape Kennedy	28° 29 N	80° 34 W
Daytona Beach	29° 11 N	81° 3 W
E Fort Lauderdale	26° 4 N	80° 9 W
Augusta	33° 22 N	81° 58 W
Brunswick	31° 15 N	81° 29 W

Alabama	Latitude	Longitude
Columbus, Lawson AFB	32° 31 N	84° 56 W
Dalton	34° 34 N	84° 57 W
Dublin	32° 20 N	82° 54 W
Gainesville	34° 11 N	83° 41 W
Griffin	33° 13 N	84° 16 W
LaGrange	33° 1 N	85° 4 W
Macon	32° 42 N	83° 39 W
Marietta, Dobbins AFB	33° 55 N	84° 31 W
Savannah	32° 8 N	81° 12 W
Valdosta-Moody AFB	30° 58 N	83° 12 W
Waycross	31° 15 N	82° 24 W
Hawaii		
Hilo (S)	19° 43 N	155° 5 W
Honolulu	21° 20 N	157° 55 W
Kaneohe Bay MCAS	21° 27 N	157° 46 W
Wahiawa	21° 3 N	158° 2 W
Idaho		
Boise (S)	43° 34 N	116° 13 W
Burley	42° 32 N	113° 46 W
Coeur D'Alene	47° 46 N	116° 49 W
Idaho Falls	43° 31 N	112° 4 W
Lewiston	46° 23 N	117° 1 W
Moscow	46° 44 N	116° 58 W
Mountain Home AFB	43° 2 N	115° 54 W
Pocatello	42° 55 N	112° 36 W
Twin Falls (S)	42° 29 N	114° 29 W
Illinois		
Aurora	41° 45 N	88° 20 W
Belleville, Scott AFB	38° 33 N	89° 51 W
Bloomington	40° 29 N	88° 57 W
Carbondale	37° 47 N	89° 15 W
Champaign/Urbana	40° 2 N	88° 17 W
Chicago, Midway	41° 47 N	87° 45 W
Chicago, O'Hare	41° 59 N	87° 54 W
Chicago Co	41° 53 N	87° 38 W
Danville	40° 12 N	87° 36 W
Decatur	39° 50 N	88° 52 W
Dixon	41° 50 N	89° 29 W
Elgin	42° 2 N	88° 16 W
Freeport	42° 18 N	89° 37 W
Galesburg	40° 56 N	90° 26 W
Greenville	38° 53 N	89° 24 W
Joliet	41° 31 N	88° 10 W
Kankakee	41° 5 N	87° 55 W
La Salle/Peru	41° 19 N	89° 6 W
La Porte	41° 36 N	86° 43 W
Marion	40° 29 N	85° 41 W
Peru, Grissom AFB	40° 39 N	86° 9 W
Richmond	39° 46 N	84° 50 W
Shelbyville	39° 31 N	85° 47 W
South Bend	41° 42 N	86° 19 W

(continued)

Alabama	Latitude	Longitude
Terre Haute	39° 27 N	87° 18 W
Valparaiso	41° 31 N	87° 2 W
Vincennes	38° 41 N	87° 32 W
Iowa		
Ames (S)	42° 2 N	93° 48 W
Burlington	40° 47 N	91° 7 W
Cedar Rapids	41° 53 N	91° 42 W
Clinton	41°50 N	90° 13 W
Council Bluffs	41° 20 N	95° 49 W
Des Moines	41° 32 N	93° 39 W
Dubuque	42° 24 N	90° 42 W
Fort Dodge	42° 33 N	94° 11 W
Iowa City	41° 38 N	91° 33 W
Keokuk	40° 24 N	91° 24 W
Marshalltown	42° 4 N	92° 56 W
Mason City	43° 9 N	93° 20 W
Newton	41° 41 N	93° 2 W
Macomb	40° 28 N	90° 40 W
Moline	41° 27 N	90° 31 W
Mt Vernon	38° 19 N	88° 52 W
Peoria	40° 40 N	89° 41 W
Quincy	39° 57 N	91° 12 W
Rantoul, Chanute AFB	40° 18 N	88° 8 W
Rockford	42° 21 N	89° 3 W
Springfield	39° 50 N	89° 40 W
Waukegan	42° 21 N	87° 53 W
Indiana		
Anderson	40° 6 N	85° 37 W
Bedford	38° 51 N	86° 30 W
Bloomington	39° 8 N	86° 37 W
Columbus, Bakalar AFB	39° 16 N	85° 54 W
Crawfordsville	40° 3 N	86° 54 W
Evansville	38° 3 N	87° 32 W
Fort Wayne	41° 0 N	85° 12 W
Goshen	41° 32 N	85° 48 W
Hobart	41° 32 N	87° 15 W
Huntington	40° 53 N	85° 30 W
Indianapolis	39° 44 N	86° 17 W
Jeffersonville	38° 17 N	85° 45 W
Kokomo	40° 25 N	86° 3 W
Lafayette	40° 2 N	86° 5 W
Ottumwa	41° 6 N	92° 27 W
Sioux City	42° 24 N	96° 23 W
Waterloo	42° 33 N	92° 24 W
Kansas		
Atchison	39° 34 N	95° 7 W
Chanute	37° 40 N	95° 29 W
Dodge City (S)	37° 46 N	99° 58 W
El Dorado	37° 49 N	96° 50 W
Emporia	38° 20 N	96° 12 W
Garden City	37° 56 N	100° 44 W
Goodland	39° 22 N	101° 42 W
Great Bend	38° 21 N	98° 52 W

Alabama	Latitude	Longitude
Hutchinson	38° 4 N	97° 52 W
Liberal	37° 3 N	100° 58 W
Manhattan, Ft Riley	39° 3 N	96° 46 W
Parsons	37° 20 N	95° 31 W
Russell	38° 52 N	98° 49 W
Salina	38° 48 N	97° 39 W
Topeka	39° 4 N	95° 38 W
Wichita	37° 39 N	97° 25 W
Kentucky		
Ashland	38° 33 N	82° 44 W
Bowling Green	35° 58 N	86° 28 W
Corbin	36° 57 N	84° 6 W
Covington	39° 3 N	84° 40 W
Hopkinsville, Ft Campbell	36° 40 N	87° 29 W
Lexington (S)	38° 2 N	84° 36 W
Louisville	38° 11 N	85° 44 W
Madisonville	37° 19 N	87° 29 W
Owensboro	37° 45 N	87° 10 W
Paducah	37° 4 N	88° 46 W
Louisiana		
Alexandria	31° 24 N	92° 18 W
Baton Rouge	30° 32 N	91° 9 W
Bogalusa	30° 47 N	89° 52 W
Houma	29° 31 N	90° 40 W
Lafayette	30° 12 N	92° 0 W
Lake Charles (S)	30° 7 N	93° 13 W
Minden	32° 36 N	93° 18 W
Monroe	32° 31 N	92° 2 W
Natchitoches	31° 46 N	93° 5 W
New Orleans	29° 59 N	90° 15 W
Shreveport (S)	32° 28 N	93° 49 W
Maine		
Augusta	44° 19 N	69° 48 W
Bangor, Dow AFB	44° 48 N	68° 50 W
Caribou (S)	46° 52 N	68° 1 W
Lewiston	44° 2 N	70° 15 W
Michigan		
Adrian	41° 55 N	84° 1 W
Alpena	45° 4 N	83° 26 W
Battle Creek	42° 19 N	85° 15 W
Benton Harbor	42° 8 N	86° 26 W
Detroit	42° 25 N	83° 1 W
Escanaba	45° 44 N	87° 5 W
Flint	42° 58 N	83° 44 W
Grand Rapids	42° 53 N	85° 31 W
Holland	42° 42 N	86° 6 W
Jackson	42° 16 N	84° 28 W
Kalamazoo	42° 17 N	85° 36 W
Lansing	42° 47 N	84° 36 W
Marquette Co	46° 34 N	87° 24 W
Mt Pleasant	43° 35 N	84° 46 W

(continued)

Alabama	Latitude	Longitude
Muskegon	43° 10 N	86° 14 W
Pontiac	42° 40 N	83° 25 W
Port Huron	42° 59 N	82° 25 W
Saginaw	43° 32 N	84° 5 W
Sault Ste. Marie (S)	46° 28 N	84° 22 W
Traverse City	44° 45 N	85° 35 W
Ypsilanti	42° 14 N	83° 32 W
Minnesota		
Albert Lea	43° 39 N	93° 21 W
Alexandria	45° 52 N	95° 23 W
Millinocket	45° 39 N	68° 42 W
Portland (S)	43° 39 N	70° 19 W
Waterville	44° 32 N	69° 40 W
Maryland		
Baltimore	39° 11 N	76° 40 W
Baltimore Co	39° 20 N	76° 25 W
Cumberland	39° 37 N	78° 46 W
Frederick	39° 27 N	77° 25 W
Hagerstown	39° 42 N	77° 44 W
Salisbury (S)	38° 20 N	75° 30 W
Massachusetts		
Boston	42° 22 N	71° 2 W
Clinton	42° 24 N	71° 41 W
Fall River	41° 43 N	71° 8 W
Framingham	42° 17 N	71° 25 W
Gloucester	42° 35 N	70° 41 W
Greenfield	42° 3 N	72° 4 W
Lawrence	42° 42 N	71° 10 W
Lowell	42° 39 N	71° 19 W
New Bedford	41° 41 N	70° 58 W
Pittsfield	42° 26 N	73° 18 W
Springfield, Westover AFB	42° 12 N	72° 32 W
Taunton	41° 54 N	71° 4 W
Worcester	42° 16 N	71° 52 W
Minnesota		
Bemidji	47° 31 N	94° 56 W
Brainerd	46° 24 N	94° 8 W
Duluth	46° 50 N	92° 11 W
Fairbault	44° 18 N	93° 16 W
Fergus Falls	46° 16 N	96° 4 W
International Falls	48° 34 N	93° 23 W
Mankato	44° 9 N	93° 59 W
Minneapolis/St. Paul	44° 53 N	93° 13 W
Rochester	43° 55 N	92° 30 W
St. Cloud (S)	45° 35 N	94° 11 W
Virginia	47° 30 N	92° 33 W
Willmar	45° 7 N	95° 5 W
Winona	44° 3 N	91° 38 W
Mississippi		
Biloxi – Keesler AFB	30° 25 N	88° 55 W
Clarksdale	34° 12 N	90° 34 W
Columbus AFB	33° 39 N	88° 27 W

Alabama	Latitude	Longitude
Greenville AFB	33° 29 N	90° 59 W
Greenwood	33° 30 N	90° 5 W
Hattiesburg	31° 16 N	89°15 W
Jackson	32° 19 N	90° 5 W
Laurel	31° 40 N	89° 10 W
McComb	31° 15 N	90° 28 W
Meridian	32° 20 N	88° 45 W
Natchez	31° 33 N	91° 23 W
Tupelo	34° 16 N	88° 46 W
Vicksburg Co	32° 24 N	90° 47 W
Missouri		
Cape Girardeau	37° 14 N	89° 35 W
Columbia (S)	38° 58 N	92° 22 W
Farmington	37° 46 N	90° 24 W
Hannibal	39° 42 N	91° 21 W
Jefferson City	38° 34 N	92° 11 W
Joplin	37° 9 N	94° 30 W
Kansas City	39° 7 N	94° 35 W
Kirksville	40° 6 N	92° 33 W
Mexico	39° 11 N	91° 54 W
Moberly	39° 24 N	92° 26 W
Poplar Bluff	36° 46 N	90° 25 W
Rolla	37° 59 N	91° 43 W
St. Joseph	39° 46 N	94° 55 W
St. Louis	38° 45 N	90° 23 W
St. Louis CO	38° 39 N	90° 38 W
Sikeston	36° 53 N	89° 36 W
Sedalia – Whiteman AFB	38° 43 N	93° 33 W
Sikeston	36° 53 N	89° 36 W
Springfield	37° 14 N	93° 23 W
McCook	40° 12 N	100° 38 W
Norfolk	41° 59 N	97° 26 W
North Platte (S)	41° 8 N	100° 41 W
Omaha	41° 18 N	95° 54 W
Scottsbluff	41° 52 N	103° 36 W
Sidney	41° 13 N	103° 6 W
Nevada		
Carson City	39° 10 N	119° 46 W
Elko	40° 50 N	115° 47 W
Ely (S)	39° 17 N	114° 51 W
Las Vegas (S)	36° 5 N	115° 10 W
Lovelock	40° 4 N	118° 33 W
Reno (S)	39° 30 N	119° 47 W
Reno Co	39° 30 N	119° 47 W
Tonopah	38° 4 N	117° 5 W
Winnemucca	40° 54 N	117° 48 W
New Hampshire		
Berlin	44° 3 N	71° 1 W
Claremont	43° 2 N	72° 2 W
Concord	43° 12 N	71° 30 W
Keene	42° 55 N	72° 17 W

(continued)

Alabama	Latitude	Longitude
Laconia	43° 3 N	71° 3 W
Manchester, Grenier AFB	42° 56 N	71° 26 W
Portsmouth, Pease AFB	43° 4 N	70° 49 W
Montana		
Billings	45° 48 N	108° 32 W
Bozeman	45° 47 N	111° 9 W
Butte	45° 57 N	112° 30 W
Cut Bank	48° 37 N	112° 22 W
Glasgow (S)	48° 25 N	106° 32 W
Glendive	47° 8 N	104° 48 W
Great Falls (S)	47° 29 N	111° 22 W
Havre	48° 34 N	109° 40 W
Helena	46° 36 N	112° 0 W
Kalispell	48° 18 N	114°16 W
Lewiston	47° 4 N	109° 27 W
Livingstown	45° 42 N	110° 26 W
Miles City	46° 26 N	105° 52 W
Missoula	46° 55 N	114° 5 W
Nebraska		
Beatrice	40° 16 N	96° 45 W
Chadron	42° 50 N	103° 5 W
Columbus	41° 28 N	97° 20 W
Fremont	41° 26 N	96° 29 W
Grand Island	40° 59 N	98° 19 W
Hastings	40° 36 N	98° 26 W
Kearney	40° 44 N	99° 1 W
Lincoln Co (S)	40° 51 N	96° 45 W
New Jersey		
Atlantic City CO	39° 23 N	74° 26 W
Long Branch	40° 19 N	74° 1 W
Newark	40° 42 N	74° 10 W
New Brunswick	40° 29 N	74° 26 W
Paterson	40° 54 N	74° 9 W
Phillipsburg	40° 41 N	75° 11 W
Trenton Co	40° 13 N	74° 46 W
Vineland	39° 29 N	75° 0 W
New Mexico		
Alamagordo° Holloman AFB	32° 51 N	106° 6 W
Albuquerque (S)	35° 3 N	106° 37 W
Artesia	32° 46 N	104° 23 W
Carlsbad	32° 20 N	104° 16 W
Clovis	34° 23 N	103° 19 W
Farmington	36° 44 N	108° 14 W
Gallup	35° 31 N	108° 47 W
Grants	35° 10 N	107° 54 W
Hobbs	32° 45 N	103° 13 W
Las Cruces	32° 18 N	106° 55 W
Los Alamos	35° 52 N	106° 19 W
Raton	36° 45 N	104° 30 W
Roswell, Walker AFB	33° 18 N	104° 32 W
Santa Fe CO	35° 37 N	106° 5 W
Silver City	32° 38 N	108° 10 W

Alabama	Latitude	Longitude
Socorro	34° 3 N	106° 53 W
Tucumcari	35° 11 N	103° 36 W
New York		
Albany (S)	42° 45 N	73° 48 W
Albany Co	42° 39 N	73° 45 W
Auburn	42° 54 N	76° 32 W
Batavia	43° 0 N	78° 11 W
Binghamton	42° 13 N	75° 59 W
Buffalo	42° 56 N	78° 44 W
Cortland	42° 36 N	76° 11 W
Dunkirk	42° 29 N	79° 16 W
Elmira	42° 10 N	76° 54 W
Geneva (S)	42° 45 N	76° 54 W
Glens Falls	43° 20 N	73° 37 W
Gloversville	43° 2 N	74° 21 W
Hornell	42° 21 N	77° 42 W
Ithaca (S)	42° 27 N	76° 29 W
Jamestown	42° 7 N	79° 14 W
Kingston	41° 56 N	74° 0 W
Lockport	43° 9 N	79° 15 W
Massena	44° 56 N	74° 51 W
Newburgh, Stewart AFB	41° 30 N	74° 6 W
Greenville	35° 37 N	77° 25 W
Henderson	36° 22 N	78° 25 W
Hickory	35° 45 N	81° 23 W
Jacksonville	34° 50 N	77° 37 W
Lumberton	34° 37 N	79° 4 W
New Bern	35° 5 N	77° 3 W
Raleigh/Durham (S)	35° 52 N	78° 47 W
Rocky Mount	35° 58 N	77° 48 W
Wilmington	34° 16 N	77° 55 W
Winston-Salem	36° 8 N	80° 13 W
North Dakota		
Bismarck (S)	46° 46 N	100° 45 W
Devils Lake	48° 7 N	98° 54 W
Dickinson	46° 48 N	102° 48 W
Fargo	46° 54 N	96° 48 W
Grand Forks	47° 57 N	97° 24 W
Jamestown	46° 55 N	98° 41 W
Minot	48° 25 N	101° 21 W
Williston	48° 9 N	103° 35 W
Ohio		
Akron-Canton	40° 55 N	81° 26 W
Ashtabula	41° 51 N	80° 48 W
Athens	39° 20 N	82° 6 W
Bowling Green	41° 23 N	83° 38 W
NYC-Central Park (S)	40° 47 N	73° 58 W
NYC-Kennedy	40° 39 N	73° 47 W
NYC-La Guardia	40° 46 N	73° 54 W
Niagara Falls	43° 6 N	79° 57 W
Olean	42° 14 N	78° 22 W

(continued)

Alabama	Latitude	Longitude
Oneonta	42° 31 N	75° 4 W
Oswego Co	43° 28 N	76° 33 W
Plattsburg AFB	44° 39 N	73° 28 W
Poughkeepsie	41° 38 N	73° 55 W
Rochester	43° 7 N	77° 40 W
Rome, Griffiss AFB	43° 14 N	75° 25 W
Schenectady (S)	42° 51 N	73° 57 W
Suffolk County AFB	40° 51 N	72° 38 W
Syracuse	43° 7 N	76° 7 W
Utica	43° 9 N	75° 23 W
Watertown	43° 59 N	76° 1 W
North Carolina		
Asheville	35° 26 N	82° 32 W
Charlotte	35° 13 N	80° 56 W
Durham	35° 52 N	78° 47 W
Elizabeth City	36° 16 N	76° 11 W
Fayetteville, Pope AFB	35° 10 N	79° 1 W
Goldsboro, Seymour-Johnson	35° 20 N	77° 58 W
Greensboro (S)	36° 5 N	79° 57 W
Cambridge	40° 4 N	81° 35'W
Chillicothe	39° 21 N	83° 0 W
Cincinnati Co	39° 9 N	84° 31 W
Cleveland (S)	41° 24 N	81° 51 W
Columbus (S)	40° 0 N	82° 53 W
Dayton	39° 54 N	84° 13 W
Defiance	41° 17 N	84° 23 W
Findlay	41° 1 N	83° 40 W
Fremont	41° 20 N	83° 7 W
Hamilton	39° 24 N	84° 35 W
Lancaster	39° 44 N	82° 38 W
Lima	40° 42 N	84° 2 W
Mansfield	40° 49 N	82° 31 W
Marion	40° 36 N	83° 10 W
Middletown	39° 31 N	84° 25 W
Newark	40° 1 N	82° 28 W
Norwalk	41° 16 N	82° 37 W
Portsmouth	38° 45 N	82° 55 W
Sandusky Co	41° 27 N	82° 43 W
Springfield	39° 50 N	83° 50 W
Steubenville	40° 23 N	80° 38 W
Toledo	41° 36 N	83° 48 W
Warren	41° 20 N	80° 51 W
Wooster	40° 47 N	81° 55 W
Youngstown	41° 16 N	80° 40 W
Zanesville	39° 57 N	81° 54 W
Oklahoma		
Ada	34° 47 N	96° 41 W
Altus AFB	34° 39 N	99° 16 W
Ardmore	34° 18 N	97° 1 W
Bartlesville	36° 45 N	96° 0 W
Chickasha	35° 3 N	97° 55 W
Enid, Vance AFB	36° 21 N	97° 55 W
Lawton	34° 34 N	98° 25 W

Alabama	Latitude	Longitude
McAlester	34° 50 N	95° 55 W
Muskogee	35° 40 N	95° 22 W
Norman	35° 15 N	97° 29 W
Oklahoma City (S)	35° 24 N	97° 36 W
Ponca City	36° 44 N	97° 6 W
Seminole	35° 14 N	96° 40 W
Stillwater (S)	36° 10 N	97° 5 W
Tulsa	36° 12 N	95° 54 W
Woodward	36° 36 N	99° 31 W
Oregon		
Albany	44° 38 N	123° 7 W
Astoria (S)	46° 9 N	123° 53 W
Baker	44° 50 N	117° 49 W
Pittsburgh	40° 30 N	80° 13 W
Pittsburgh Co	40° 27 N	80° 0 W
Reading Co	40° 20 N	75° 38 W
Scranton/Wilkes-Barre	41° 20 N	75° 44 W
State College (S)	40° 48 N	77° 52 W
Sunbury	40° 53 N	76° 46 W
Uniontown	39° 55 N	79° 43 W
Warren	41° 51 N	79° 8 W
West Chester	39° 58 N	75° 38 W
Williamsport	41° 15 N	76° 55 W
York	39° 55 N	76° 45 W
Rhode Island		
Newport (S)	41° 30 N	71° 20 W
Providence	41° 44 N	71° 26 W
South Carolina		
Anderson	34° 30 N	82° 43 W
Charleston AFB (S)	32° 54 N	80° 2 W
Charleston Co	32° 54 N	79° 58 W
Columbia	33° 57 N	81° 7 W
Florence	34° 11 N	79° 43 W
Georgetown	33° 23 N	79° 17 W
Greenville	34° 54 N	82° 13 W
Greenwood	34° 10 N	82° 7 W
Orangeburg	33° 30 N	80° 52 W
Bend	44° 4 N	121° 19 W
Corvallis (S)	44° 30 N	123° 17 W
Eugene	44° 7 N	123° 13 W
Grants Pass	42° 26 N	123° 19 W
Klamath Falls	42° 9 N	121° 44 W
Medford (S)	42° 22 N	122° 52 W
Pendleton	45° 41 N	118° 51 W
Portland	45° 36 N	122° 36 W
Portland Co	45° 32 N	122° 40 W
Roseburg	43° 14 N	123° 22 W
Salem	44° 55 N	123° 1 W
The Dalles	45° 36 N	121° 12 W
Pennsylvania		
Allentown	40° 39 N	75° 26 W

(continued)

Alabama	Latitude	Longitude
Altoona Co	40° 18 N	78° 19 W
Butler	40° 52 N	79° 54 W
Chambersburg	39° 56 N	77° 38 W
Erie	42° 5 N	80° 11'W
Harrisburg	40° 12 N	76° 46 W
Johnstown	40° 19 N	78° 50 W
Lancaster	40° 7 N	76° 18'W
Meadville	41° 38 N	80° 10 W
New Castle	41° 1 N	80° 22'W
Philadelphia	39° 53 N	75° 15 W
Rock Hill	34° 59 N	80° 58 W
Spartanburg	34° 58 N	82° 0 W
Sumter, Shaw AFB	33° 54 N	80° 22 W
South Dakota		
Aberdeen	45° 27 N	98° 26 W
Brookings	44° 18 N	96° 48 W
Huron	44° 23 N	98° 13 W
Mitchell	43° 41 N	98° 1 W
Pierre	44° 23 N	100° 17 W
Rapid City (S)	44° 3 N	103° 4 W
Sioux Falls	43° 34 N	96° 44 W
Watertown	44° 55 N	97° 9 W
Yankton	42° 55 N	97° 23 W
Tennessee		
Athens	35° 26 N	84° 35 W
Bristol-Tri City	36° 29 N	82° 24 W
Chattanooga	35° 2 N	85° 12'W
Clarksville	36° 33 N	87° 22 W
Columbia	35° 38 N	87° 2 W
Dyersburg	36° 1 N	89° 24'W
Greenville	36° 4 N	82° 50'W
Jackson	35° 36 N	88° 55 W
Knoxville	35° 49 N	83° 59 W
Memphis	35° 3 N	90° 0 W
Murfreesboro	34° 55 N	86° 28 W
Nashville (S)	36° 7 N	86° 41'W
Tullahoma	35° 23 N	86° 5 W
Texas		
Abilene	32° 25 N	99° 41 W
Alice	27° 44 N	98° 2 W
Amarillo	35° 14 N	100° 42 W
Austin	30° 18 N	97° 42 W
Bay City	29° 0 N	95° 58 W
Beaumont	29° 57 N	94° 1 W
Beeville	28° 22 N	97° 40 W
Big Spring (S)	32° 18 N	101° 27 W
Brownsville (S)	25° 54 N	97° 26 W
Brownwood	31° 48 N	98° 57 W
Bryan	30° 40 N	96° 33 W
Corpus Christi	27° 46 N	97° 30 W
Corsicana	32° 5 N	96° 28 W
Dallas AP	32° 51 N	96° 51 W
Del Rio, Laughlin AFB	29° 22 N	100° 47 W

Alabama	Latitude	Longitude
Denton	33° 12 N	97° 6 W
Eagle Pass	28° 52 N	100° 32 W
El Paso (S)	31° 48 N	106° 24 W
Fort Worth (S)	32° 50 N	97° 3 W
Snyder	32° 43 N	100° 55 W
Temple	31° 6 N	97° 21 W
Tyler	32° 21 N	95° 16 W
Vernon	34° 10 N	99° 18 W
Victoria	28° 51 N	96° 55 W
Waco	31° 37 N	97° 13 W
Wichita Falls	33° 58 N	98° 29 W
Utah		
Cedar City	37° 42 N	113° 6 W
Logan	41° 45 N	111° 49 W
Moab	38° 36 N	109° 36 W
Ogden	41° 12 N	112° 1 W
Price	39° 37 N	110° 50 W
Provo	40° 13 N	111° 43 W
Richfield	38° 46 N	112° 5 W
St George Co	37° 2 N	113° 31 W
Salt Lake City (S)	40° 46 N	111° 58 W
Vernal	40° 27 N	109° 31 W
Vermont		
Barre	44° 12 N	72° 31 W
Burlington (S)	44° 28 N	73° 9 W
Rutland	43° 36 N	72° 58 W
Virginia		
Charlottesville	38° 2 N	78° 31 W
Galveston	29° 18 N	94° 48 W
Greenville	33° 4 N	96° 3 W
Harlingen	26° 14 N	97° 39 W
Harrisonburg	38° 27 N	78° 54 W
Houston	29° 58 N	95° 21 W
Houston Co	29° 59 N	95° 22 W
Huntsville	30° 43 N	95° 33 W
Killeen, Robert Gray AAF	31° 5 N	97° 41W
Lamesa	32° 42 N	101° 56 W
Laredo AFB	27° 32 N	99° 27 W
Longview	32° 28 N	94° 44 W
Lubbock	33° 39 N	101° 49 W
Lufkin	31° 25 N	94° 48 W
Mcallen	26° 12 N	98° 13 W
Midland (S)	31° 57 N	102° 11 W
Mineral Wells	32° 47 N	98° 4 W
Palestine Co	31° 47 N	95° 38 W
Pampa	35° 32 N	100° 59 W
Pecos	31° 25 N	103° 30 W
Plainview	34° 11 N	101° 42 W
Port Arthur	29° 57 N	94° 1 W
Goodfellow AFB	31° 26 N	100° 24 W
San Antonio (S)	29° 32 N	98° 28 W

(continued)

Alabama	Latitude	Longitude
Sherman, Perrin AFB	33° 43 N	96° 40 W
Danville	36° 34 N	79° 20 W
Fredericksburg	38° 18 N	77° 28 W
Lynchburg	37° 20 N	79° 12 W
Norfolk	36° 54 N	76° 12 W
Petersburg	37° 11 N	77° 31 W
Richmond	37° 30 N	77° 20 W
Roanoke	37° 19 N	79° 58 W
Staunton	38° 16 N	78° 54 W
Winchester	39° 12 N	78° 10 W
Washington		
Aberdeen	46° 59 N	123° 49 W
Bellingham	48° 48 N	122° 32 W
Bremerton	47° 34 N	122° 40 W
Ellensburg	47° 2 N	120° 31'W
Everett, Paine AFB	47° 55 N	122° 17 W
Kennewick	46° 13 N	119° 8 W
Longview	46° 10 N	122° 56 W
Moses Lake, Larson AFB	47° 12 N	119° 19 W
Olympia	46° 58 N	122° 54 W
Port Angeles	48° 7 N	123° 26 W
Seattle-Boeing Field	47° 32 N	122° 18 W
Seattle Co (S)	47° 39 N	122° 18 W
Seattle-Tacoma (S)	47° 27 N	122° 18 W
Spokane (S)	47° 38 N	117° 31 W
Tacoma, McChord AFB	47° 15 N	122° 30 W
Walla Walla	46° 6 N	118° 17 W
Wenatchee	47° 25 N	120° 19 W
Yakima	46° 34 N	120° 32 W
West Virginia		
Beckley	37° 47 N	81°7 W
Bluefield	37° 18 N	81° 13 W
Charleston	38° 22 N	81° 36 W
Clarksburg	39° 16 N	80° 21 W
Elkins	38° 53 N	79° 51 W
Huntington Co	38° 25 N	82° 30 W
Martinsburg	39° 24 N	77° 59 W
Morgantown	39° 39 N	79° 55 W
Parkersburg Co	39° 16 N	81° 34 W
Wheeling	40° 7 N	80° 42 W
Wisconsin		
Appleton	44° 15 N	88° 23 W
Ashland	46° 34 N	90° 58 W
Beloit	42° 30 N	89° 2 W
Eau Claire	44° 52 N	91° 29 W
Fond Du Lac	43° 48 N	88° 27 W
Green Bay	44° 29 N	88° 8 W
La Crosse	43° 52 N	91° 15 W
Madison (S)	43° 8 N	89° 20 W
Manitowoc	44° 6 N	87° 41 W
Marinette	45° 6 N	87° 38 W
Milwaukee	42° 57 N	87° 54 W
Racine	42° 43 N	87° 51 W

Sheboygan	43° 45 N	87° 43 W
Stevens Point	44° 30 N	89° 34 W
Waukesha	43° 1 N	88° 14 W
Wausau	44° 55 N	89° 37 W
Wyoming		
Casper	42° 55 N	106° 28 W
Cheyenne	41° 9 N	104° 49 W
Cody	44° 33 N	109° 4 W
Evanston	41° 16 N	110° 57 W
Lander (S)	42° 49 N	108° 44 W
Laramie (S)	41° 19 N	105° 41 W
Newcastle	43° 51 N	104° 13 W
Rawlins	41° 48 N	107° 12 W
Rock Springs	41° 36 N	109° 0 W
Sheridan	44° 46 N	106° 58 W
Torrington	42° 5 N	104° 13 W

Canada longitudes and latitudes

Alberta	Latitude	Longitude
Calgary	51° 6 N	114° 1 W
Edmonton	53° 34 N	113° 31 W
Grande Prairie	55° 11 N	118° 53 W
Jasper	52° 53 N	118° 4 W
Lethbridge (S)	49° 38 N	112° 48 W
McMurray	56° 39 N	111° 13 W
Medicine Hat	50° 1 N	110° 43 W
Red Deer	52° 11 N	113° 54 W
British Columbia		
Dawson Creek	55° 44 N	120° 11 W
Fort Nelson (S)	58° 50 N	122° 35 W
Kamloops Co	50° 43 N	120° 25 W
Nanaimo (S)	49° 11 N	123° 58 W
New Westminster	49° 13 N	122° 54 W
Penticton	49° 28 N	119° 36 W
Prince George (S)	53° 53 N	122° 41 W
Prince Rupert Co	54° 17 N	130° 23 W
Trail	49° 8 N	117° 44 W
Vancouver (S)	49° 11 N	123° 10 W
Victoria Co	48° 25 N	123° 19 W
Manitoba		
Brandon	49° 52 N	99° 59 W
Churchill (S)	58° 45 N	94° 4 W
Dauphin	51° 6 N	100° 3 W
Flin Flon	54° 46 N	101° 51 W
Portage La Prairie	49° 54 N	98° 16 W
The Pas (S)	53° 58 N	101° 6 W
Winnipeg (S)	49° 54 N	97° 14 W
New Brunswick		
Campbellton Co	48° 0 N	66° 40 W

(continued)

Alabama	Latitude	Longitude
Chatham	47° 1 N	65° 27 W
Edmundston Co	47° 22 N	68° 20 W
Fredericton (S)	45° 52 N	66° 32 W
Moncton (S)	46° 7 N	64° 41 W
Saint John	45° 19 N	65° 53 W
Newfoundland		
Corner Brook	48° 58 N	57° 57 W
Gander	48° 57 N	54° 34 W
Goose Bay (S)	53° 19 N	60° 25 W
St John's (S)	47° 37 N	52° 45 W
Stephenville	48° 32 N	58° 33 W
Northwest Territories		
Fort Smith (S)	60° 1 N	111° 58 W
Frobisher (S)	63° 45 N	68° 33 W
Inuvik (S)	68° 18 N	133° 29 W
Resolute (S)	74° 43 N	94° 59 W
Yellowknife	62° 28 N	114° 27 W
Nova Scotia		
Amherst	45° 49 N	64°13 W
Halifax (S)	44° 39 N	63° 34 W
Kentville (S)	45° 3 N	64° 36 W
New Glasgow	45° 37 N	62° 37 W
Sydney	46° 10 N	60° 3 W
Truro Co	45° 22 N	63° 16 W
Yarmouth	43° 50 N	66° 5 W
Ontario		
Belleville	44° 9 N	77° 24 W
Chatham	42° 24 N	82° 12 W
Cornwall	45° 1 N	74° 45 W
Hamilton	43° 16 N	79° 54 W
Kapuskasing (S)	49° 25 N	82° 28 W
Kenora	49° 48 N	94° 22 W
Kingston	44° 16 N	76° 30 W
Kitchener	43° 26 N	80° 30 W
London	43° 2 N	81° 9 W
Toronto (S)	43° 41 N	79° 38 W
Windsor	42° 16 N	82° 58 W
Prince Edward Island		
Charlottetown (S)	46° 17 N	63° 8 W
Summerside	46° 26 N	63° 50 W
Quebec		
Bagotville	48° 20 N	71° 0 W
Chicoutimi	48° 25 N	71° 5 W
Drummondville	45° 53 N	72° 29 W
Granby	45° 23 N	72° 42 W
Hull	45° 26 N	75° 44 W
Megantic	45° 35 N	70° 52 W
Montreal (S)	45° 28 N	73° 45 W
Quebec	46° 48 N	71° 23 W
Rimouski	48° 27 N	68° 32 W
St Jean	45° 18 N	73° 16 W
St Jerome	45° 48 N	74° 1 W

Alabama	Latitude	Longitude
Sept. Iles (S)	50° 13 N	66° 16 W
Shawinigan	46° 34 N	72° 43 W
Sherbrooke Co	45° 24 N	71° 54 W
Thetford Mines	46° 4 N	71° 19 W
Trois Rivieres	46° 21 N	72° 35 W
Val D'or	48° 3 N	77° 47 W
Valleyfield	45° 16 N	74° 6 W
Ontario		
North Bay	46° 22 N	79° 25 W
Oshawa	43° 54 N	78° 52 W
Ottawa (S)	45° 19 N	75° 40 W
Owen Sound	44° 34 N	80° 55 W
Peterborough	44° 17 N	78° 19 W
St Catharines	43° 11 N	79° 14 W
Sarnia	42° 58 N	82° 22 W
Sault Ste Marie	46° 32 N	84° 30 W
Sudbury	46° 37 N	80° 48 W
Thunder Bay	48° 22 N	89° 19 W
Timmins	48° 34 N	81° 22 W
Saskatchewan		
Estevan	49° 4 N	103° 0 W
Moose Jaw	50° 20 N	105° 33 W
North Battleford	52° 46 N	108° 15 W
Prince Albert	53° 13 N	105° 41 W
Regina	50° 26 N	104° 40 W
Saskatoon (S)	52° 10 N	106° 41 W
Swift Current (S)	50° 17 N	107° 41 W
Yorkton	51° 16 N	102° 28 W
Yukon Territory		
Whitehorse (S)	60° 43 N	135° 4 W
Afghanistan		
Kabul	34° 35 N	69° 12 E
Algeria		
Algiers	36° 46 N	30° 3 E
Argentina		
Buenos Aires	34° 35 S	58° 29 W
Cordoba	31° 22 S	64° 15 W
Tucuman	26° 50 S	65° 10 W
Australia		
Adelaide	34° 56 S	138° 35 E
Alice Springs	23° 48 S	133° 53 E
Brisbane	27° 28 S	153° 2 E
Darwin	12° 28 S	130° 51 E
Melbourne	37° 49 S	144° 58 E
Perth	31°57 S	115°51 E
Sydney	33° 52 S	151° 12 E
Austria		
Vienna	48° 15 N	16° 22 E
Azores		
Lajes (Terceira)	38° 45 N	27° 5 W

(continued)

Alabama	Latitude	Longitude
Bahamas		
Nassau	25° 5 N	77° 21 W
Burma		
Mandalay	21° 59 N	96° 6 E
Rangoon	16° 47 N	96° 9 E
Cambodia		
Phnom Penh	11° 33 N	104° 51 E
Chile		
Punta Arenas	53° 10 S	70° 54 W
Santiago	33° 27 S	70° 42 W
Valparaiso	33° 1 S	71° 38 W
China		
Chongquing	29° 33 N	106° 33 E
Shanghai	31° 12 N	121° 26 E
Colombia		
Baranquilla	10° 59 N	74° 48 W
Bogota	4° 36 N	74° 5 W
Cali	3° 25 N	76° 30 W
Medellin	6° 13 N	75° 36 W
Congo		
Brazzaville	4°15 S	15° 15 E
Cuba		
Guantanamo Bay	19° 54 N	75° 9 W
Havana	23° 8 N	82° 21 W
Bangladesh		
Chittagong	22° 21 N	91° 50 E
Belgium		
Brussels	50° 48 N	4°21 E
Bermuda		
Kindley AFB	33° 22 N	64° 41 W
Bolivia		
La Paz	16° 30 S	68° 9 W
Brazil		
Belem	1° 27 S	48° 29 W
Belo Horizonte	19° 56 S	43° 57 W
Brasilia	15° 52 S	47° 55 W
Curitiba	25° 25 S	49° 17 W
Fortaleza	3° 46 S	38° 33 W
Porto Alegre	30° 2 S	51° 13 W
Recife	8° 4 S	34° 53 W
Rio de Janeiro	22° 55 S	43° 12 W
Salvador	13° 0 S	38° 30 W
Sao Paulo	23° 33 S	46° 38 W
Belize		
Belize	17° 31 N	88° 11 W
Bulgaria		
Sofia	42° 42 N	23° 20 E
Strasbourg	48° 35 N	7° 46 E
Czechoslovakia		

Alabama	Latitude	Longitude
Prague	50° 5 N	14° 25 E
Denmark		
Copenhagen	55° 41 N	12° 33 E
Dominican Republic		
Santo Domingo	18° 29 N	69° 54 W
Egypt		
Cairo	29° 52 N	31° 20 E
El Salvador		
San Salvador	13° 42 N	89° 13 W
Equador		
Guayaquil	2°0 S	79° 53 W
Quito	0°13 S	78° 32 W
Ethiopia		
Addis Ababa	90° 2 N	38° 45 E
Asmara	15° 17 N	38° 55 E
Finland		
Helsinki	60° 10 N	24° 57 E
France		
Lyon	45° 42 N	4° 47 E
Marseilles	43° 18 N	5° 23 E
Nantes	47° 15 N	1° 34 W
Nice	43° 42 N	7° 16 E
Paris	48° 49 N	2° 29 E
French Guiana		
Cayenne	4° 56 N	52° 27 W
Germany		
Berlin (West)	52° 27 N	13° 18 E
Hamburg	53° 33 N	9° 58 E
Hannover	52° 24 N	9° 40 E
Mannheim	49° 34 N	8° 28 E
Munich	48° 9 N	11° 34 E
Ghana		
Accra	5° 33 N	0° 12 W
Gibraltar		
Gibraltar	36° 9 N	5° 22 W
Greece		
Athens	37° 58 N	23° 43 E
Thessaloniki	40° 37 N	22° 57 E
Greenland		
Narsarssuaq	61° 11 N	45° 25 W
Guatemala		
Guatemala City	14° 37 N	90° 31 W
Guyana		
Georgetown	6° 50 N	58° 12 W
Haiti		
Port Au Prince	18° 33 N	72° 20 W

(continued)

Alabama	Latitude	Longitude
Iran		
Abadan	30° 21 N	48° 16 E
Meshed	36° 17 N	59° 36 E
Tehran	35° 41 N	51° 25 E
Iraq		
Baghdad	33° 20 N	44° 24 E
Mosul	36° 19 N	43° 9 E
Ireland		
Dublin	53° 22 N	6° 21 W
Shannon	52° 41 N	8° 55 W
Irian Barat		
Manokwari	0° 52 S	134° 5 E
Israel		
Jerusalem	31° 47 N	35° 13 E
Tel Aviv	32° 6 N	34° 47 E
Italy		
Milan	45° 27 N	9° 17 E
Naples	40° 53 N	14° 18 E
Rome	41° 48 N	12° 36 E
Ivory Coast		
Abidjan	5° 19 N	4° 1 W
Japan		
Fukuoka	33° 35 N	130° 27 E
Honduras		
Tegucigalpa	14° 6 N	87° 13 W
Hong Kong		
Hong Kong	22° 18 N	114° 10 E
Hungary		
Budapest	47° 31 N	19° 2 E
Iceland		
Reykjavik	64°8 N	21°56 E
India		
Ahmedabad	23° 2 N	72° 35 E
Bangalore	12° 57 N	77° 37 E
Bombay	18° 54 N	72° 49 E
Calcutta	22° 32 N	88° 20 E
Madras	13° 4 N	80° 15 E
Nagpur	21°9 N	79°7 E
New Delhi	28° 35 N	77° 12 E
Indonesia		
Djakarta	6° 11 S	106° 50 E
Kupang	10° 10 S	123° 34 E
Makassar	5° 8 S	119° 28 E
Medan	3° 35 N	98° 41 E
Palembang	3° 0 S	104° 46 E
Surabaya	7°13 S	112° 43 E
Japan		
Sapporo	43° 4 N	141° 21 E
Tokyo	35° 41 N	139° 46 E
Jordan		

Alabama	Latitude	Longitude
Amman	31° 57 N	35° 57 E
Kenya		
Nairobi	1° 16 S	36° 48 E
Korea		
Pyongyang	39° 2 N	125° 41 E
Seoul	37° 34 N	126° 58 E
Lebanon		
Beirut	33° 54 N	35° 28 E
Liberia		
Monrovia	6° 18 N	10° 48 W
Libya		
Benghazi	32° 6 N	20° 4 E
Tananarive	18° 55 S	47° 33 E
Malaysia		
Kuala Lumpur	3° 7 N	101° 42 E
Penang	5° 25 N	100° 19 E
Martinique		
Fort De France	14° 37 N	61° 5 W
Mexico		
Guadalajara	20° 41 N	103° 20 W
Merida	20° 58 N	89° 38 W
Mexico City	19° 24 N	99° 12 W
Monterrey	25° 40 N	100° 18 W
Vera Cruz	19° 12 N	96° 8 W
Morocco		
Casablanca	33° 35 N	7° 39 W
Nepal		
Katmandu	27° 42 N	85° 12 E
Netherlands		
Amsterdam	52° 23 N	4° 55 E
New Zealand		
Auckland	36° 51 S	174° 46 E
Christchurch	43° 32 S	172° 37 E
Wellington	41° 17 S	174° 46 E
Nicaragua		
Managua	12° 10 N	86° 15 W
Nigeria		
Lagos	6° 27 N	3° 24 E
Norway		
Bergen	60° 24 N	5° 19 E
Oslo	59° 56 N	10° 44 E
Pakistan		
Karachi	24° 48 N	66° 59 E
Lahore	31° 35 N	74° 20 E
Peshwar	34° 1 N	71° 35 E
Kiev	50° 27 N	30° 30 E
Kharkov	50° 0 N	36° 14 E

(continued)

Alabama	Latitude	Longitude
Kuibyshev	53° 11 N	50° 6 E
Leningrad	59° 56 N	30° 16 E
Minsk	53° 54 N	27° 33 E
Moscow	55° 46 N	37° 40 E
Odessa	46° 29 N	30° 44 E
Petropavlovsk	52° 53 N	158° 42 E
Rostov on Don	47° 13 N	39° 43 E
Sverdlovsk	56° 49 N	60° 38 E
Tashkent	41° 20 N	69° 18 E
Tbilisi	41° 43 N	44° 48 E
Vladivostok	43° 7 N	131° 55 E
Volgograd	48° 42 N	44° 31 E
Saudi Arabia		
Dhahran	26° 17 N	50° 9 E
Jedda	21° 28 N	39° 10 E
Riyadh	24° 39 N	46° 42 E
Senegal		
Dakar	14° 42 N	17° 29 W
Singapore		
Singapore	1° 18 N	103° 50 E
Somalia		
Mogadiscio	2° 2 N	49° 19 E
Panama		
Panama City	8° 58 N	79° 33 W
Papua New Guinea		
Port Moresby	9° 29 S	147° 9 E
Paraguay		
Ascuncion	25° 17 S	57° 30 W
Peru		
Lima	12° 5 S	77° 3 W
Philippines		
Manila	14° 35 N	120° 59 E
Poland		
Krakow	50° 4 N	19° 57 E
Warsaw	52° 13 N	21° 2 E
Portugal		
Lisbon	38° 43 N	9° 8 W
Puerto Rico		
San Juan	18° 29 N	66° 7 W
Rumania		
Bucharest	44° 25 N	26° 6 E
Russia		
Alma Ata	43° 14 N	76° 53 E
Archangel	64° 33 N	40° 32 E
Kaliningrad	54° 43 N	20° 30 E
Krasnoyarsk	56° 1 N	92° 57 E
South Africa		
Cape Town	33° 56 S	18° 29 E
Johannesburg	26° 11 S	28° 3 E
Pretoria	25° 45 S	28° 14 E

Alabama	Latitude	Longitude
South Yemen		
Aden	12° 50 N	45° 2 E
Spain		
Barcelona	41° 24 N	2° 9 E
Madrid	40° 25 N	3° 41 W
Valencia	39° 28 N	0° 23 W
Sri Lanka		
Colombo	6° 54 N	79° 52 E
Sudan		
Khartoum	15° 37 N	32° 33 E
Surinam		
Paramaribo	5° 49 N	55° 9 W
Sweden		
Stockholm	59° 21 N	18° 4 E
Switzerland		
Zurich	47° 23 N	8° 33 E
Syria		
Damascus	33° 30 N	36° 20 E
Taiwan		
Tainan	22° 57 N	120° 12 E
Taipei	25°2 N	121°31 E
Tanzania		
Dar Es Salaam	6° 50 S	39° 18 E
Thailand		
Bangkok	13° 44 N	100° 30 E
Trinidad		
Port Of Spain	10° 40 N	61° 31 W
Tunisia		
Tunis	36° 47 N	10° 12 E
Turkey		
Adana	36° 59 N	35° 18 E
Ankara	39° 57 N	32° 53 E
Istanbul	40° 58 N	28° 50 E
Izmir	38° 26 N	27° 10 E
United Kingdom		
Belfast	54° 36 N	5° 55 W
Birmingham	52° 29 N	1° 56 W
Cardiff	51° 28 N	3°10 W
Edinburgh	55° 55 N	3° 11 W
Glasgow	55° 52 N	4° 17 W
London	51° 29 N	0° 0 W
Uruguay		
Montevideo	34° 51 S	56° 13 W
Venezuela		
Caracas	10° 30 N	66° 56 W
Maracaibo	10° 39 N	71° 36 W
Vietnam		

(continued)

Alabama	Latitude	Longitude
Da Nang	16° 4 N	108° 13 E
Hanoi	21° 2 N	105° 52 E
Ho Chi Minh City (Saigon)	10° 47 N	106° 42 E
Yugoslavia		
Belgrade	44° 48 N	20° 28 E
Zaire		
Kinshasa (Leopoldville)	4° 20 S	15° 18 E
Kisangani (Stanleyville)	0° 26 S	15° 14 E

Energy Systems

The following narrative is a summary in the Wikipedia Web encyclopedia. For broader details refer to www.wikipedia.com. In order to differentiate forms of alternative energy sources, it is important to understand the various definitions of energy. Energy in physics, chemistry, and nature occurs in numerous forms, all of which imply similar connotations to the ability to perform work. In physics and other sciences, energy is a scalar quantity that is a property of objects and systems and is conserved by nature.

Several different forms of energy, including kinetic energy, potential energy, thermal energy, gravitational energy, electromagnetic radiation energy, chemical energy, and nuclear energy have been defined to explain all known natural phenomena.

Conservation of Energy

Energy is transformation from one form to another, but it is never created or destroyed. This principle, the law of conservation of energy, was first postulated in the early nineteenth century and applies to any isolated system. The total energy of a system does not change over time, but its value may depend on the frame of reference. For example, a seated passenger in a moving vehicle has zero kinetic energy relative to the vehicle but does indeed have kinetic energy relative to Earth.

The Concept of Energy in Various Scientific Fields

- In chemistry, the energy differences between chemical substances determine whether, and to what extent, they can be converted into, or react with, other substances.
- In biology, the chemical bonding are often broken and made during metabolism. Energy is often stored by the body in the form of carbohydrates s and lipids, both of which release energy when reacting with Oxygen.
- In Earth sciences, continental drift, volcanic activity, and earthquakes are phenomena that can be explained in terms of energy transformation, in the Earth's interior. Meteorological phenomena like wind, rain, hail, snow, lightning, tornado, and hurricanes are all a result of energy transformations brought about by solar energy.

Energy transformations in the universe are characterized by various kinds of potential energy that have been available since the Big Bang, later "released" to be transformed into more active types of energy.

Nuclear Decay

Examples of such processes include those in which energy that was originally "stored" in heavy isotopes such as uranium and thorium are released by nucleosynthesis. In this process, gravitational potential energy, released from the gravitational collapse of supernovae, is used to store energy in the creation of these heavy elements before their incorporation into the solar system and Earth. This energy is triggered and released in the nuclear fission bomb.

Fusion

In a similar chain of transformations at the dawn of the universe, the nuclear fusion of hydrogen in the Sun released another store of potential energy that was created at the time of the Big Bang. Space expanded and the universe cooled too rapidly for hydrogen to fuse completely into heavier elements. Hydrogen thus represents a store of potential energy that can be released by nuclease fusion.

Sunlight Energy Storage

Light from our Sun may again be stored as gravitational potential energy after it strikes the Earth. After being released at a hydroelectric dam, this water can be used to drive turbines and generators to produce electricity. Sunlight also drives all weather phenomena, including such events as hurricanes, in which large unstable areas of warm ocean, heated over months, suddenly give up some of their thermal energy to power intense air movement.

Kinetic versus Potential Energy

An important distinction should be made between kinetic and potential energy before continuing. Potential energy is the energy of matter due to its position or arrangement. This stored energy can be found in any lifted objects, which have the force of gravity bringing them down to their original positions. Kinetic energy is the energy that an object possesses due to its motion. A great example of this is seen with a ball that falls under the influence of gravity. As it accelerates downward, its potential energy is converted into kinetic energy. When it hits the ground and deforms, the kinetic energy converts into elastic potential energy. Upon bouncing back up, this potential energy once again becomes kinetic energy.

The two forms, though seemingly very different, play important roles in complementing each other.

Gravitational Potential Energy

The gravitational force near the Earth's surface is equal to the mass, m, multiplied by the gravitational acceleration, $g = 9.81$ m/s^2.

Temperature

On the macroscopic scale, temperature is the unique physical property that determines the direction of heat flow between two objects placed in thermal contact.

If no heat flow occurs, the two objects have the same temperature, as heat flows from the hotter object to the colder object. These two basic principles are stated in the zeroth law of thermodynamics and the second law of thermodynamics of thermodynamics, respectively. For a solid, these microscopic motions are principally the vibrations of its atoms about their sites in the solid.

In most of the world (except for the United States, Jamaica, and a few other countries), the degree Celsius scale is used for most temperature measuring purposes. The global scientific community, with the United States included, measures temperature using the Celsius scale and thermodynamic temperature using the Kelvin scale, in which $0 \text{ K} = -273.15°\text{C}$, or absolute zero.

The United States is the last major country in which the degree Fahrenheit scale is popularly used in everyday life.

Specific Heat Capacity

Specific heat capacity, also known as specific heat, is a measure of the energy that is needed to raise the temperature of a quantity of a substance by a certain temperature.

Chemical Energy

Chemical energy is defined as the work done by electric forces during the rearrangement of electric charges, electrons, and protons in the process of aggregation.

If the chemical energy of a system decreases during a chemical reaction, it is transferred to the surroundings in some form of energy (often heat). On the other hand, if the chemical energy of a system increases as a result of a chemical reaction, it is from the conversion of another form of energy from its surroundings.

Moles are the typical units used to describe change in chemical energy, and values can range from tens to hundreds of kJ/mol.

Radiant Energy

Radiant energy is the energy of electromagnetic wave, or sometimes of other forms of radiation. Like all forms of energy, its unit is the joule. The term is used especially when radiation is emitted by a source into the surrounding environment.

As electromagnetic (EM) radiation can be conceptualized as a stream of photons, radiant energy can be seen as the energy carried by these photons. EM radiation can also be seen as an electromagnetic wave that carries energy in its oscillating electric and magnetic fields. Quantum field theory reconciles these two views.

EM radiation can have a range of frequencies. From the viewpoint of photons, the energy carried by each photon is proportional to its frequency. From the wave viewpoint, the energy of a monochromatic wave is proportional to its intensity.

Thus, it can be implied that if two EM waves have the same intensity, but different frequencies, the wave with the higher frequency contains fewer photons.

When EM waves are absorbed by an object, their energy is typically converted to heat. This is an everyday phenomenon, seen, for example, when sunlight warms the surfaces it irradiates. This is often associated with infrared radiation, but any kind of EM radiation will warm the object that absorbs it. EM waves can also be reflected or scattered, causing their energy to be redirected or redistributed.

Energy can enter or leave an open system in the form of radiant energy. Such a system can be man-made, as with a solar energy collector, or natural, as with the Earth's atmosphere. Green house gases trap the Sun's radiant energy at certain wavelengths, allowing it to penetrate deep into the atmosphere or all the way to the Earth's surface, where it is reemitted as longer wavelengths. Radiant energy is produced in the Sun due to the phenomenon of nuclear fusion.

Glossary of Solar Energy Power Terms

Absorber – In a photovoltaic device, the material that readily absorbs photons to generate charge carriers (free electrons or holes).

AC – See alternating current.

Activated Shelf Life – The period of time, at a specified temperature, that a charged battery can be stored before its capacity falls to an unusable level.

Activation Voltage(s) – The voltage(s) at which a charge controller will take action to protect the batteries.

Adjustable Set Point – A feature allowing the user to adjust the voltage levels at which a charge controller will become active.

Alternating Current (AC) – A type of electrical current, the direction of which is reversed at regular intervals or cycles. In the United States, the standard is 120 reversals or 60 cycles per second. Electricity transmission networks use AC because voltage can be controlled with relative ease.

Acceptor – A dopant material, such as boron, which has fewer outer shell electrons than required in an otherwise balanced crystal structure, providing a hole, which can accept a free electron.

AIC – See amperage interrupt capability.

Air mass (sometimes called air mass ratio) – Equal to the cosine of the zenith angle – that angle from directly overhead to a line intersecting the Sun. The air mass is an indication of the length of the path solar radiation travels through the atmosphere. An air mass of 1.0 means the Sun is directly overhead and the radiation travels through one atmosphere (thickness).

Ambient Temperature – The temperature of the surrounding area.

Amorphous Semiconductor – A noncrystalline semiconductor material that has no long-range order.

Amorphous Silicon – A thin-film, silicon photovoltaic cell having no crystalline structure. Manufactured by depositing layers of doped silicon on a substrate. See also single-crystal silicon and poly-crystalline silicon.

Amperage Interrupt Capability (AIC) – direct current fuses should be rated with a sufficient AIC to interrupt the highest possible current.

Ampere (amp) – A unit of electrical current or rate of flow of electrons. One volt across 1 ohm of resistance causes a current flow of 1 ampere.

Ampere-Hour (Ah/AH) – A measure of the flow of current (in amperes) over 1 hour; used to measure battery capacity.

Ampere Hour Meter – An instrument that monitors current with time. The indication is the product of current (in amperes) and time (in hours).

Angle of Incidence – The angle that a ray of Sun makes with a line perpendicular to the surface. For example, a surface that directly faces the Sun has a solar angle of incidence of zero, but if the surface is parallel to the Sun (for example, sunrise striking a horizontal rooftop), the angle of incidence is 90°.

Annual Solar Savings – The annual solar savings of a solar building is the energy savings attributable to a solar feature relative to the energy requirements of a nonsolar building.

Anode – The positive electrode in an electrochemical cell (battery). Also, the earth or ground in a cathodic protection system. Also, the positive terminal of a diode.

Antireflection Coating – A thin coating of a material applied to a solar cell surface that reduces the light reflection and increases light transmission.

Array – See photovoltaic (PV) array.

Array Current – The electrical current produced by a photovoltaic array when it is exposed to sunlight.

Array Operating Voltage – The voltage produced by a photovoltaic array when exposed to sunlight and connected to a load.

Autonomous System – See stand-alone system.

Availability – The quality or condition of a photovoltaic system being available to provide power to a load. Usually measured in hours per year. One minus availability equals downtime.

Azimuth Angle – The angle between true south and the point on the horizon directly below the Sun.

Balance-of-System – Represents all components and costs other than the photovoltaic modules/array. It includes design costs, land, site preparation, system installation, support structures, power conditioning, operation and maintenance costs, indirect storage, and related costs.

Band Gap – In a semiconductor, the energy difference between the highest valence band and the lowest conduction band.

Band Gap Energy (Eg) – The amount of energy (in electron volts) required to free an outer shell electron from its orbit about the nucleus to a free state, and thus promote it from the valence to the conduction level.

Barrier Energy – The energy given up by an electron in penetrating the cell barrier; a measure of the electrostatic potential of the barrier.

Base Load – The average amount of electric power that a utility must supply in any period.

Battery – Two or more electrochemical cells enclosed in a container and electrically interconnected in an appropriate series/parallel arrangement to provide the required operating voltage and current levels. Under common usage, the term "battery" also applies to a single cell if it constitutes the entire electrochemical storage system.

Battery Available Capacity – The total maximum charge, expressed in ampere-hours, that can be withdrawn from a cell or battery under a specific set of operating conditions including discharge rate, temperature, initial state of charge, age, and cut-off voltage.

Battery Capacity – The maximum total electrical charge, expressed in ampere-hours, which a battery can deliver to a load under a specific set of conditions.

Battery Cell – The simplest operating unit in a storage battery. It consists of one or more positive electrodes or plates, an electrolyte that permits ionic conduction, one or more negative electrodes or plates, separators between plates of opposite polarity, and a container for all the above.

Battery Cycle Life – The number of cycles, to a specified depth of discharge, that a cell or battery can undergo before failing to meet its specified capacity or efficiency performance criteria.

Battery Energy Capacity – The total energy available, expressed in watt-hours (kilowatt-hours), which can be withdrawn from a fully charged cell or battery. The energy capacity of a given cell varies with temperature, rate, age, and cut-off voltage. This term is more common to system designers than it is to the battery industry where capacity usually refers to ampere-hours.

Battery Energy Storage – Energy storage using electrochemical batteries. The three main applications for battery energy storage systems include spinning reserve at generating stations, load leveling at substations, and peak shaving on the customer side of the meter.

Battery Life – The period during which a cell or battery is capable of operating above a specified capacity or efficiency performance level. Life may be measured in cycles and/or years, depending on the type of service for which the cell or battery is intended.

BIPV (Building-Integrated Photovoltaics) – A term for the design and integration of photovoltaic (PV) technology into the building envelope, typically replacing conventional building materials. This integration may be in vertical facades, replacing view glass, spandrel glass, or other facade material; into semitransparent skylight systems; into roofing systems, replacing traditional roofing materials; into shading "eyebrows" over windows; or other building envelope systems.

Blocking Diode – A semiconductor connected in series with a solar cell or cells and a storage battery to keep the battery from discharging through the cell when there is no output, or low output, from the solar cell. It can be thought of as a one-way valve that allows electrons to flow forward, but not backward.

Boron (B) – The chemical element commonly used as the dopant in photovoltaic device or cell material.

Boule – A sausage-shaped, synthetic single-crystal mass grown in a special furnace, pulled and turned at a rate necessary to maintain the single-crystal structure during growth.

Btu (British Thermal Unit) – The amount of heat required to raise the temperature of one pound of water 1 degree Fahrenheit; equal to 252 calories.

Bypass Diode – A diode connected across one or more solar cells in a photovoltaic module such that the diode will conduct if the cell(s) become reverse biased. It protects these solar cells from thermal destruction in case of total or partial shading of individual solar cells while other cells are exposed to full light.

Cadmium (Cd) – A chemical element used in making

certain types of solar cells and batteries.

Cadmium Telluride (CdTe) – A polycrystalline thin-film photovoltaic material.

Capacity (C) – See battery capacity.

Capacity Factor – The ratio of the average load on (or power output of) an electricity generating unit or system to the capacity rating of the unit or system over a specified period.

Captive Electrolyte Battery – A battery having an immobilized electrolyte (gelled or absorbed in a material).

Cathode – The negative pole or electrode of an electrolytic cell, vacuum tube, etc., where electrons enter (current leaves) the system; the opposite of an anode.

Cathodic Protection – A method of preventing oxidation of the exposed metal in structures by imposing a small electrical voltage between the structure and the ground.

Cd – See cadmium.

CdTe – See cadmium telluride.

Cell (battery) – A single unit of an electrochemical device capable of producing direct voltage by converting chemical energy into electrical energy. A battery usually consists of several cells electrically connected together to produce higher voltages. (Sometimes the terms "cell" and "battery" are used interchangeably). Also see photovoltaic (PV) cell.

Cell Barrier – A very thin region of static electric charge along the interface of the positive and negative layers in a photovoltaic cell. The barrier inhibits the movement of electrons from one layer to the other, so that higher-energy electrons from one side diffuse preferentially through it in one direction, creating a current and thus a voltage across the cell. Also called depletion zone or space charge.

Cell Junction – The area of immediate contact between two layers (positive and negative) of a photovoltaic cell. The junction lies at the center of the cell barrier or depletion zone.

Charge – The process of adding electrical energy to a battery.

Charge Carrier – A free and mobile conduction electron or hole in a semiconductor.

Charge Controller – A component of a photovoltaic system that controls the flow of current to and from the battery to protect it from overcharge and overdischarge. The charge controller may also indicate the system operational status.

Charge Factor – A number representing the time in hours during which a battery can be charged at a constant current without damage to the battery. Usually expressed in relation to the total battery capacity, i.e., C/5 indicates a charge factor of 5 hours. Related to charge rate.

Charge Rate – The current applied to a cell or battery to restore its available capacity. This rate is commonly normalized by a charge control device with respect to the rated capacity of the cell or battery.

Chemical Vapor Deposition (CVD) – A method of depositing thin semiconductor films used to make certain types of photovoltaic devices. With this method, a substrate is exposed to one or more vaporized compounds, one or more of which contain desirable constituents. A chemical reaction is initiated, at or near the substrate surface, to produce the desired material that will condense on the substrate.

Cleavage of Lateral Epitaxial Films for Transfer (CLEFT) – A process for making inexpensive Gallium Arsenide (GaAs) photovoltaic cells in which a thin film of GaAs is grown atop a thick, single-crystal GaAs (or other suitable material) substrate and then is cleaved from the substrate and incorporated into a cell, allowing the substrate to be reused to grow more thin-film GaAs.

Cloud Enhancement – The increase in solar intensity caused by reflected irradiance from nearby clouds.

Combined Collector – A photovoltaic device or module that provides useful heat energy in addition to electricity.

Concentrator – A photovoltaic module, which includes optical components such as lenses (Fresnel lens) to direct and concentrate sunlight onto a solar cell of smaller area. Most concentrator arrays must directly face or track the sun. They can increase the power flux of sunlight hundreds of times.

Conduction Band (or conduction level) – An energy band in a semiconductor in which electrons can move freely in a solid, producing a net transport of charge.

Conductor – The material through which electricity is transmitted, such as an electrical wire, or transmission or distribution line.

Contact Resistance – The resistance between metallic contacts and the semiconductor.

Conversion Efficiency – See photovoltaic (conversion) efficiency.

Converter – A unit that converts a direct current (DC) voltage to another dc voltage.

Copper Indium Diselenide (CuInSe2, or CIS) – A polycrystalline thin-film photovoltaic material (sometimes incorporating gallium (CIGS) and/or sulfur).

Crystalline Silicon – A type of photovoltaic cell made from a slice of single-crystal silicon or polycrystalline silicon.

Current – See electric current.

Current at Maximum Power (Imp) – The current at which maximum power is available from a module.

Cutoff Voltage – The voltage levels (activation) at which the charge controller disconnects the photovoltaic array from the battery or the load from the battery.

Cycle – The discharge and subsequent charge of a battery.

Czochralski Process – A method of growing large-size, high-quality semiconductor crystal by slowly lifting a seed crystal from a molten bath of the material under careful cooling conditions.

Dangling Bonds – A chemical bond associated with an atom on the surface layer of a crystal. The bond does not join with another atom of the crystal, but extends in the direction of exterior of the surface.

Days of Storage – The number of consecutive days the stand-alone system will meet a defined load without solar energy input. This term is related to system availability.

DC – See direct current.

DC-to-DC Converter – Electronic circuit to convert direct current voltages (e.g., photovoltaic module voltage) into other levels (e.g., load voltage). Can be part of a maximum power point tracker.

Deep-Cycle Battery – A battery with large plates that can withstand many discharges to a low state of charge.

Deep Discharge – Discharging a battery to 20% or less of its full charge capacity.

Depth of Discharge (DOD) – The ampere-hours removed from a fully charged cell or battery, expressed as a percentage of rated capacity. For example, the removal of 25 ampere-hours from a fully charged 100 ampere-hour rated cell results in a 25% depth of discharge. Under certain conditions, such as discharge rates lower than that used to rate the cell, depth of discharge can exceed 100%.

Dendrite – A slender threadlike spike of pure crystalline material, such as silicon.

Dendritic Web Technique – A method for making sheets of polycrystalline silicon in which silicon dendrites are slowly withdrawn from a melt of silicon whereupon a web of silicon forms between the dendrites and solidifies as it rises from the melt and cools.

Depletion Zone – Same as cell barrier. The term derives from the fact that this microscopically thin region is depleted of charge carriers (free electrons and hole).

Design Month – The month having the combination of insolation and load that requires the maximum energy from the photovoltaic array.

Diffuse Insolation – Sunlight received indirectly as a result of scattering due to clouds, fog, haze, dust, or other obstructions in the atmosphere. Opposite of direct insolation.

Diffuse Radiation – Radiation received from the Sun after reflection and scattering by the atmosphere and ground.

Diffusion Furnace – Furnace used to make junctions in semiconductors by diffusing dopant atoms into the surface of the material.

Diffusion Length – The mean distance a free electron or hole moves before recombining with another hole or electron.

Diode – An electronic device that allows current to flow in one direction only. See blocking diode and bypass diode.

Direct Beam Radiation – Radiation received by direct solar rays. Measured by a pyrheliometer with a solar aperture of 5.7° to transcribe the solar disc.

Direct Current (DC) – A type of electricity transmission and distribution by which electricity flows in one direction through the conductor, usually relatively low voltage and high current. To be used for typical 120 volt or 220 volt household appliances, DC must be converted to alternating current, its opposite.

Direct Insolation – Sunlight falling directly upon a collector. Opposite of diffuse insolation.

Discharge – The withdrawal of electrical energy from a battery.

Discharge Factor – A number equivalent to the time in hours during which a battery is discharged at constant current usually expressed as a percentage of the total battery capacity, i.e., C/5 indicates a discharge factor of 5 hours. Related to discharge rate.

Discharge Rate – The rate, usually expressed in amperes or time, at which electrical current is taken from the battery.

Disconnect – Switch gear used to connect or disconnect components in a photovoltaic system.

Distributed Energy Resources (DER) – A variety of small, modular power-generating technologies that can be combined with energy management and storage systems and used to improve the operation of the electricity delivery system, whether or not those technologies are connected to an electricity grid.

Distributed Generation – A popular term for localized or on-site power generation.

Distributed Power – Generic term for any power supply located near the point where the power is used. Opposite of central power. See stand-alone systems.

Distributed Systems – Systems that are installed at or near the location where the electricity is used, as opposed to central systems that supply electricity to grids. A residential photovoltaic system is a distributed system.

Donor – In a photovoltaic device, an n-type dopant, such as phosphorus, that puts an additional electron into an energy level very near the conduction band; this electron is easily exited into the conduction band where it increases the electrical conductivity over than of an undoped semiconductor.

Donor Level – The level that donates conduction electrons to the system.

Dopant – A chemical element (impurity) added in small amounts to an otherwise pure semiconductor material to modify the electrical properties of the material. An n-dopant introduces more electrons. A p-dopant creates electron vacancies (holes).

Doping – The addition of dopants to a semiconductor.

Downtime – Time when the photovoltaic system cannot provide power for the load. Usually expressed in hours per year or that percentage.

Dry Cell – A cell (battery) with a captive electrolyte. A primary battery that cannot be recharged.

Duty Cycle – The ratio of active time to total time. Used to describe the operating regime of appliances or loads in photovoltaic systems.

Duty Rating – The amount of time an inverter (power conditioning unit) can produce at full rated power.

Edge-Defined Film-Fed Growth (EFG) – A method for making sheets of polycrystalline silicon for photovoltaic devices in which molten silicon is drawn upward by capillary action through a mold.

Electric Circuit – The path followed by electrons from a power source (generator or battery), through an electrical system, and returning to the source.

Electric Current – The flow of electrical energy (electricity) in a conductor, measured in amperes.

Electrical grid – An integrated system of electricity distribution, usually covering a large area.

Electricity – Energy resulting from the flow of charge particles, such as electrons or ions.

Electrochemical Cell – A device containing two conducting electrodes, one positive and the other negative, made of dissimilar materials (usually metals) that are immersed in a chemical solution (electrolyte) that transmits positive ions from the negative to the positive electrode and thus forms an electrical charge. One or more cells constitute a battery.

Electrode – A conductor that is brought in conducting contact with a ground.

Electrodeposition – Electrolytic process in which a metal is deposited at the cathode from a solution of its ions.

Electrolyte – A nonmetallic (liquid or solid) conductor that carries current by the movement of ions (instead of electrons) with the liberation of matter at the electrodes of an electrochemical cell.

Electron – An elementary particle of an atom with a negative electrical charge and a mass of 1/1837 of a proton; electrons surround the positively charged nucleus of an atom and determine the chemical properties of an atom. The movement of electrons in an electrical conductor constitutes an electric current.

Electron Volt (eV) – The amount of kinetic energy gained by an electron when accelerated through an electric potential difference of 1 Volt; equivalent to 1.603×10^{-19}; a unit of energy or work.

Energy – The capability of doing work; different forms of energy can be converted to other forms, but the total amount of energy remains the same.

Energy Audit – A survey that shows how much energy used in a home, which helps find ways to use less energy.

Energy Contribution Potential – Recombination occurring in the emitter region of a photovoltaic cell.

Energy Density – The ratio of available energy per pound; usually used to compare storage batteries.

Energy Levels – The energy represented by an electron in the band model of a substance.

Epitaxial Growth – The growth of one crystal on the surface of another crystal. The growth of the deposited crystal is oriented by the lattice structure of the original crystal.

Equalization – The process of restoring all cells in a battery to an equal state-of-charge. Some battery types may require a complete discharge as a part of the equalization process.

Equalization Charge – The process of mixing the electrolyte in batteries by periodically overcharging the batteries for a short time.

Equalizing Charge – A continuation of normal battery charging, at a voltage level slightly higher than the normal end-of-charge voltage, in order to provide cell equalization within a battery.

Equinox – The two times of the year when the sun crosses the equator and night and day are of equal length; usually occurs on March 21 (spring equinox) and September 23 (fall equinox).

Extrinsic Semiconductor – The product of doping a pure semiconductor.

Fermi Level – Energy level at which the probability of finding an electron is one-half. In a metal, the Fermi level is very near the top of the filled levels in the partially filled valence band. In a semiconductor, the Fermi level is in the band gap.

Fill Factor – The ratio of a photovoltaic cell's actual power to its power if both current and voltage were at their maxima. A key characteristic in evaluating cell performance.

Fixed Tilt Array – A photovoltaic array set in at a fixed angle with respect to horizontal.

Flat-Plate Array – A photovoltaic (PV) array that consists of non-concentrating PV modules.

Flat-Plate Module – An arrangement of photovoltaic cells or material mounted on a rigid flat surface with the cells exposed freely to incoming sunlight.

Flat-Plate Photovoltaics (PV) – A PV array or module that consists of nonconcentrating elements. Flat-plate arrays and modules use direct and diffuse sunlight, but if the array is fixed in position, some portion of the direct sunlight is lost because of oblique sun-angles in relation to the array.

Float Charge – The voltage required to counteract the self-discharge of the battery at a certain temperature.

Float Life – The number of years that a battery can keep its stated capacity when it is kept at float charge.

Float Service – A battery operation in which the battery is normally connected to an external current source; for instance, a battery charger that supplies the battery load under normal conditions, while also providing enough energy input to the battery to make up for its internal quiescent losses, thus keeping the battery always up to full power and ready for service.

Float-Zone Process – A method of growing a large-size, high-quality crystal whereby coils heat a polycrystalline ingot placed atop a single-crystal seed. As the coils are slowly raised the molten interface beneath the coils becomes single crystal.

Float-Zone Process – In reference to solar photovoltaic cell manufacture, a method of growing a large-size, high-quality crystal whereby coils heat a polycrystalline ingot placed atop a single-crystal seed. As the coils are slowly raised the molten interface beneath the coils becomes a single crystal.

Frequency – The number of repetitions per unit time of a complete waveform, expressed in hertz (Hz).

Frequency Regulation – This indicates the variability in the output frequency. Some loads will switch off or not operate properly if frequency variations exceed 1%.

Fresnel Lens – An optical device that focuses light like a magnifying glass; concentric rings are faced at slightly different angles so that light falling on any ring is focused to the same point.

Full Sun – The amount of power density in sunlight received at the earth's surface at noon on a clear day (about 1000 watts/square meter).

Ga – See gallium.

GaAs – See gallium arsenide.

Gallium (Ga) – A chemical element, metallic in nature, used in making certain kinds of solar cells and semiconductor devices.

Gallium Arsenide (GaAs) – A crystalline, high-efficiency compound used to make certain types of solar cells and semiconductor material.

Gassing – The evolution of gas from one or more of the electrodes in the cells of a battery. Gassing commonly results from local action self-discharge or from the electrolysis of water in the electrolyte during charging.

Gassing Current – The portion of charge current that goes into electrolytical production of hydrogen and oxygen from the electrolytic liquid. This current increases with increasing voltage and temperature.

Gel-Type Battery – Lead acid battery in which the electrolyte is composed of a silica gel matrix.

Gigawatt (GW) – A unit of power equal to 1 billion watts; 1 million kilowatts, or 1000 megawatts.

Grid – See electrical grid.

Grid-Connected System – A solar electric or photovoltaic (PV) system in which the PV array acts like a central generating plant, supplying power to the grid.

Grid-Interactive System – Same as grid-connected system.

Grid Lines – Metallic contacts fused to the surface of the solar cell to provide a low resistance path for electrons to flow out to the cell interconnect wires.

Harmonic Content – The number of frequencies in the output waveform in addition to the primary frequency (50 or 60 Hz.). Energy in these harmonic frequencies is lost and may cause excessive heating of the load.

Heterojunction – A region of electrical contact between two different materials.

High Voltage Disconnect – The voltage at which a charge controller will disconnect the photovoltaic array from the batteries to prevent overcharging.

High Voltage Disconnect Hysteresis – The voltage difference between the high voltage disconnect set point and the voltage at which the full photovoltaic array current will be reapplied.

Hole – The vacancy where an electron would normally exist in a solid; behaves like a positively charged particle.

Homojunction – The region between an n-layer and a p-layer in a single material, photovoltaic cell.

Hybrid System – A solar electric or photovoltaic system that includes other sources of electricity generation, such as wind or diesel generators.

Hydrogenated Amorphous Silicon – Amorphous silicon with a small amount of incorporated hydrogen. The hydrogen neutralizes dangling bonds in the amorphous silicon, allowing charge carriers to flow more freely.

Incident Light – Light that shines onto the face of a solar cell or module.

Indium Oxide – A wide band gap semiconductor that can be heavily doped with tin to make a highly conductive, transparent thin film. Often used as a front contact or one component of a heterojunction solar cell.

Infrared Radiation – Electromagnetic radiation whose wavelengths lie in the range from 0.75 micrometer to 1000 micrometers; invisible long wavelength radiation (heat) capable of producing a thermal or photovoltaic effect, though less effective than visible light.

Input Voltage – This is determined by the total power required by the alternating current loads and the voltage of any direct current loads. Generally, the larger the load, the higher the inverter input voltage. This keeps the current at levels where switches and other components are readily available.

Insolation – The solar power density incident on a surface of stated area and orientation, usually expressed as Watts per square meter or Btu per square foot per hour. See diffuse insolation and direct insolation.

Interconnect – A conductor within a module or other means of connection that provides an electrical interconnection between the solar cells.

Intrinsic Layer – A layer of semiconductor material, used in a photovoltaic device, whose properties are essentially those of the pure, undoped, material.

Intrinsic Semiconductor – An undoped semiconductor.

Inverter – A device that converts direct current electricity to alternating current either for stand-alone systems or to supply power to an electricity grid.

Ion – An electrically charged atom or group of atoms that has lost or gained electrons; a loss makes the resulting particle positively charged; a gain makes the particle negatively charged.

Irradiance – The direct, diffuse, and reflected solar radiation that strikes a surface. Usually expressed in kilowatts per square meter. Irradiance multiplied by time equals insolation.

ISPRA Guidelines – Guidelines for the assessment of photovoltaic power plants, published by the Joint Research Centre of the Commission of the European Communities, Ispra, Italy.

I-Type Semiconductor – Semiconductor material that is left intrinsic, or undoped so that the concentration of charge carriers is characteristic of the material itself rather than of added impurities.

I-V Curve – A graphical presentation of the current versus the voltage from a photovoltaic device as the load is increased from the short circuit (no load) condition to the open circuit (maximum voltage) condition. The shape of the curve characterizes cell performance.

Joule – A metric unit of energy or work; 1 joule per second equals 1 watt or 0.737 foot-pounds; 1 Btu equals 1,055 joules.

Junction – A region of transition between semiconductor layers, such as a p/n junction, which goes from a region that has a high concentration of acceptors (p-type) to one that has a high concentration of donors (n-type).

Junction Box – A photovoltaic (PV) generator junction box is an enclosure on the module where PV strings are electrically connected and where protection devices can be located, if necessary.

Junction Diode – A semiconductor device with a junction and a built-in potential that passes current better in one direction than the other. All solar cells are junction diodes.

Kilowatt (kW) – A standard unit of electrical power equal to 1000 watts, or to the energy consumption at a rate of 1000 joules per second.

Kilowatt-Hour (kWh) – 1000 watts acting over a period of 1 hour. The kWh is a unit of energy. 1 kWh =3 600 kJ.

Langley (L) – Unit of solar irradiance. One gram calorie per square centimeter. 1 L = 85.93 kWh/cm².

Lattice – The regular periodic arrangement of atoms or molecules in a crystal of semiconductor material.

Lead Acid Battery – A general category that includes batteries with plates made of pure lead, lead-antimony, or lead-calcium immersed in an acid electrolyte.

Life – The period during which a system is capable of operating above a specified performance level.

Life-Cycle Cost – The estimated cost of owning and operating a photovoltaic system for the period of its useful life.

Light-Induced Defects – Defects, such as dangling bonds, induced in an amorphous silicon semiconductor upon initial exposure to light.

Light Trapping – The trapping of light inside a semiconductor material by refracting and reflecting the light at critical angles; trapped light will travel further in the material, greatly increasing the probability of absorption and hence of producing charge carriers.

Line-Commutated Inverter – An inverter that is tied to a power grid or line. The commutation of power (conversion from direct current to alternating current) is controlled by the power line, so that, if there is a failure in the power grid, the photovoltaic system cannot feed power into the line.

Liquid Electrolyte Battery – A battery containing a liquid solution of acid and water. Distilled water may be added to these batteries to replenish the electrolyte as necessary. Also called a flooded battery because the plates are covered with the electrolyte.

Load – The demand on an energy producing system; the energy consumption or requirement of a piece or group of equipment. Usually expressed in terms of amperes or watts in reference to electricity.

Load Circuit – The wire, switches, fuses, etc., that connect the load to the power source.

Load Current (A) – The current required by the electrical device.

Load Resistance – The resistance presented by the load. See resistance.

Low Voltage Cutoff (LVC) – The voltage level at which a charge controller will disconnect the load from the battery.

Low Voltage Disconnect – The voltage at which a charge controller will disconnect the load from the batteries to prevent over-discharging.

Low Voltage Disconnect Hysteresis – The voltage difference between the low voltage disconnect set point and the voltage at which the load will be reconnected.

Low Voltage Warning – A warning buzzer or light that indicates the low battery voltage set point has been reached.

Maintenance-Free Battery – A sealed battery to which water cannot be added to maintain electrolyte level.

Majority Carrier – Current carriers (either free electrons or holes) that are in excess in a specific layer of a semiconductor material (electrons in the n-layer, holes in the p-layer) of a cell.

Maximum Power Point (MPP) – The point on the current-voltage (I-V) curve of a module under illumination, where the product of current and voltage is maximum. For a typical silicon cell, this is at about 0.45 volt.

Maximum Power Point Tracker (MPPT) – Means of a power conditioning unit that automatically operates the photovoltaic generator at its maximum power point under all conditions.

Maximum Power Tracking – Operating a photovoltaic array at the peak power point of the array's I-V curve where maximum power is obtained. Also called peak power tracking.

Megawatt (MW) – 1000 kilowatts, or 1 million watts; standard measure of electric power plant generating capacity.

Megawatt-Hour – 1000 kilowatt-hours or 1 million watt-hours.

Microgroove – A small groove scribed into the surface of a solar cell, which is filled with metal for contacts.

Minority Carrier – A current carrier, either an electron or a hole, that is in the minority in a specific layer of a semiconductor material; the diffusion of minority carriers under the action of the cell junction voltage is the current in a photovoltaic device.

Minority Carrier Lifetime – The average time a minority carrier exists before recombination.

Modified Sine Wave – A waveform that has at least three states (i.e., positive, off, and negative). Has less harmonic content than a square wave.

Modularity – The use of multiple inverters connected in parallel to service different loads.

Module – See photovoltaic (PV) module.

Module Derate Factor – A factor that lowers the photovoltaic module current to account for field operating conditions such as dirt accumulation on the module.

Monolithic – Fabricated as a single structure.

Movistor – Metal oxide varistor. Used to protect electronic circuits from surge currents such as those produced by lightning.

Multicrystalline – A semiconductor (photovoltaic) material composed of variously oriented, small, individual crystals. Sometimes referred to as polycrystalline or semicrystalline.

Multijunction Device – A high-efficiency photovoltaic device containing two or more cell junctions, each of which is optimized for a particular part of the solar spectrum.

Multistage Controller – A charging controller unit that allows different charging currents as the battery nears full state_of_charge.

National Electrical Code (NEC) – Contains guidelines for all types of electrical installations. The 1984 and later editions of the NEC contain Article 690, "Solar Photovoltaic Systems," which should be followed when installing a PV system.

National Electrical Manufacturers Association (NEMA) – This organization sets standards for some nonelectronic products like junction boxes.

NEC – See National Electrical Code.

NEMA – See National Electrical Manufacturers Association.

Nickel-Cadmium Battery – A battery containing nickel and cadmium plates and an alkaline electrolyte.

Nominal Voltage – A reference voltage used to describe batteries, modules, or systems (i.e., a 12-volt or 24-volt battery, module, or system).

Normal Operating Cell Temperature (NOCT) – The estimated temperature of a photovoltaic module when operating under 800 W/m^2 irradiance, 20°C ambient temperature and wind speed of 1 meter per second. NOCT is used to estimate the nominal operating temperature of a module in its working environment.

N-Type – Negative semiconductor material in which there are more electrons than holes; current is carried through it by the flow of electrons.

N-Type Semiconductor – A semiconductor produced by doping an intrinsic semiconductor with an electron-donor impurity (e.g., phosphorus in silicon).

N-Type Silicon – Silicon material that has been doped with a material that has more electrons in its atomic structure than does silicon.

Ohm – A measure of the electrical resistance of a material equal to the resistance of a circuit in which the potential difference of 1 volt produces a current of 1 ampere.

One-Axis Tracking – A system capable of rotating about one axis.

Open-Circuit Voltage (Voc) – The maximum possible voltage across a photovoltaic cell; the voltage across the cell in sunlight when no current is flowing.

Operating Point – The current and voltage that a photovoltaic module or array produces when connected to a load. The operating point is dependent on the load or the batteries connected to the output terminals of the array.

Orientation – Placement with respect to the cardinal directions, N, S, E, W; azimuth is the measure of orientation from north.

Outgas – See gassing.

Overcharge – Forcing current into a fully charged battery. The battery will be damaged if overcharged for a long period.

Packing Factor – The ratio of array area to actual land area or building envelope area for a system; or, the ratio of total solar cell area to the total module area, for a module.

Panel – See photovoltaic (PV) panel.

Parallel Connection – A way of joining solar cells or photovoltaic modules by connecting positive leads together and negative leads together; such a configuration increases the current, but not the voltage.

Passivation – A chemical reaction that eliminates the detrimental effect of electrically reactive atoms on a solar cell's surface.

Peak Demand/Load – The maximum energy demand or load in a specified time period.

Peak Power Current – Amperes produced by a photovoltaic module or array operating at the voltage of the I-V curve that will produce maximum power from the module.

Peak Power Point – Operating point of the I-V (current-voltage) curve for a solar cell or photovoltaic module where the product of the current value times the voltage value is a maximum.

Peak Power Tracking – See maximum power tracking.

Peak Sun Hours – The equivalent number of hours per day when solar irradiance averages 1000 w/m². For example, 6 peak Sun hours means that the energy received during total daylight hours equals the energy that would have been received had the irradiance for 6 hours been 1000 w/m².

Peak Watt – A unit used to rate the performance of solar cells, modules, or arrays; the maximum nominal output of a photovoltaic device, in watts (Wp) under standardized test conditions, usually 1000 watts per square meter of sunlight with other conditions, such as temperature specified.

Phosphorus (P) – A chemical element used as a dopant in making n-type semiconductor layers.

Photocurrent – An electric current induced by radiant energy.

Photoelectric Cell – A device for measuring light intensity that works by converting light falling on, or reach it, to electricity, and then measuring the current; used in photometers.

Photoelectrochemical Cell – A type of photovoltaic device in which the electricity induced in the cell is used immediately within the cell to produce a chemical, such as hydrogen, which can then be withdrawn for use.

Photon – A particle of light that acts as an individual unit of energy.

Photovoltaic(s) (PV) – Pertaining to the direct conversion of light into electricity.

Photovoltaic (PV) Array – An interconnected system of PV modules that function as a single electricity-producing unit. The modules are assembled as a discrete structure, with common support or mounting. In smaller systems, an array can consist of a single module.

Photovoltaic (PV) Cell – The smallest semiconductor element within a PV module to perform the immediate conversion of light into electrical energy (direct current voltage and current). Also called a solar cell.

Photovoltaic (PV) Conversion Efficiency – The ratio of the electric power produced by a photovoltaic device to the power of the sunlight incident on the device.

Photovoltaic (PV) Device – A solid-state electrical device that converts light directly into direct current electricity of voltage-current characteristics that are a function of the characteristics of the light source and the materials in and design of the device. Solar photovoltaic devices are made of various semiconductor materials including silicon, cadmium sulfide, cadmium telluride, and gallium arsenide, and in single crystalline, multicrystalline, or amorphous forms.

Photovoltaic (PV) Effect – The phenomenon that occurs when photons, the "particles" in a beam of light, knock electrons loose from the atoms they strike. When this property of light is combined with the properties of semiconductors, electrons flow in one direction across a junction, setting up a voltage. With the addition of circuitry, current will flow and electric power will be available.

Photovoltaic (PV) Generator – The total of all PV strings of a PV power supply system, which are electrically interconnected.

Photovoltaic (PV) Module – The smallest environmentally protected, essentially planar assembly of solar cells and ancillary parts, such as interconnections, terminals, [and protective devices such as diodes] intended to generate direct current power under unconcentrated sunlight. The structural (load carrying) member of a module can either be the top layer (superstrate) or the back layer (substrate).

Photovoltaic (PV) Panel – often used interchangeably with PV module (especially in one-module systems), but more accurately used to refer to a physically connected collection of modules (i.e., a laminate string of modules used to achieve a required voltage and current).

Photovoltaic (PV) System – A complete set of components for converting sunlight into electricity by the photovoltaic process, including the array and balance-of-system components.

Photovoltaic-Thermal (PV/T) System – A photovoltaic system that, in addition to converting sunlight into electricity, collects the residual heat energy and delivers both heat and electricity in usable form. Also called a total energy system.

Physical Vapor Deposition – A method of depositing thin semiconductor photovoltaic films. With this method, physical processes, such as thermal evaporation or bombardment of ions, are used to deposit elemental semiconductor material on a substrate.

P-I-N – A semiconductor photovoltaic (PV) device structure that layers an intrinsic semiconductor between a p-type semiconductor and an n-type semiconductor; this structure is most often used with amorphous silicon PV devices.

Plates – A metal plate, usually lead or lead compound, immersed in the electrolyte in a battery.

P/N – A semiconductor photovoltaic device structure in which the junction is formed between a p-type layer and an n-type layer.

Pocket Plate – A battery in which active materials are held in a perforated metal pocket.

Point-Contact Cell – A high efficiency silicon photovoltaic concentrator cell that employs light trapping techniques and point-diffused contacts on the rear surface for current collection.

Polycrystalline – See Multicrystalline.

Polycrystalline Silicon – A material used to make photovoltaic cells, which consist of many crystals unlike single-crystal silicon.

Power Conditioning – The process of modifying the characteristics of electrical power (for e.g., inverting direct current to alternating current).

Power Conditioning Equipment – Electrical equipment, or power electronics, used to convert power from a photovoltaic array into a form suitable for subsequent use. A collective term for inverter, converter, battery charge regulator, and blocking diode.

Power Conversion Efficiency – The ratio of output power to input power of the inverter.

Power Density – The ratio of the power available from a battery to its mass (W/kg) or volume (W/l).

Power Factor (PF) – The ratio of actual power being used in a circuit, expressed in watts or kilowatts, to the power that is apparently being drawn from a power source, expressed in volt-amperes or kilovolt-amperes.

Primary Battery – A battery whose initial capacity cannot be restored by charging.

Projected Area – The net south-facing glazing area projected on a vertical plane.

P-Type Semiconductor – A semiconductor in which holes carry the current; produced by doping an intrinsic semiconductor with an electron acceptor impurity (e.g., boron in silicon).

Pulse-Width-Modulated (PWM) Wave Inverter – A type of power inverter that produce a high quality (nearly sinusoidal) voltage, at minimum current harmonics.

PV – See photovoltaic(s).

Pyranometer – An instrument used for measuring global solar irradiance.

Pyrheliometer – An instrument used for measuring direct beam solar irradiance. Uses an aperture of 5.7° to transcribe the solar disc.

Quad – One quadrillion Btu (1,000,000,000,000,000 Btu).

Qualification Test – A procedure applied to a selected set of photovoltaic modules involving the application of defined electrical, mechanical, or thermal stress in a prescribed manner and amount. Test results are subject to a list of defined requirements.

Rated Battery Capacity – The term used by battery manufacturers to indicate the maximum amount of energy that can be withdrawn from a battery under specified discharge rate and temperature. See battery capacity.

Rated Module Current (A) – The current output of a photovoltaic module measured at standard test conditions of 1000 w/m^2 and 25°C cell temperature.

Rated Power – Rated power of the inverter. However, some units can not produce rated power continuously. See duty rating.

Reactive Power – The sine of the phase angle between the current and voltage waveforms in an alternating current system. See power factor.

Recombination – The action of a free electron falling back into a hole. Recombination processes are either radiative, where the energy of recombination results in the emission of a photon, or nonradiative, where the energy of recombination is given to a second electron which then relaxes back to its original energy by emitting phonons. Recombination can take place in the bulk of the semiconductor, at the surfaces, in the junction region, at defects, or between interfaces.

Rectifier – A device that converts alternating current to direct current. See inverter.

Regulator – Prevents overcharging of batteries by controlling charge cycle-usually adjustable to conform to specific battery needs.

Remote Systems – See stand-alone systems.

Reserve Capacity – The amount of generating capacity a central power system must maintain to meet peak loads.

Resistance (R) – The property of a conductor, which opposes the flow of an electric current resulting in the generation of heat in the conducting material. The measure of the resistance of a given conductor is the electromotive force needed for a unit current flow. The unit of resistance is ohms.

Resistive Voltage Drop – The voltage developed across a cell by the current flow through the resistance of the cell.

Reverse Current Protection – Any method of preventing unwanted current flow from the battery to the photovoltaic array (usually at night). See blocking diode.

Ribbon (Photovoltaic) Cells – A type of photovoltaic device made in a continuous process of pulling material from a molten bath of photovoltaic material, such as silicon, to form a thin sheet of material.

Root Mean Square (RMS) – The square root of the average square of the instantaneous values of an ac output. For a sine wave the RMS value is 0.707 time the peak value.

The equivalent value of alternating current, I, that will produce the same heating in a conductor with resistance, R, as a dc current of value I.

Sacrificial Anode – A piece of metal buried near a structure that is to be protected from corrosion. The metal of the sacrificial anode is intended to corrode and reduce the corrosion of the protected structure.

Satellite Power System (SPS) – Concept for providing large amounts of electricity for use on the Earth from one or more satellites in geosynchronous Earth orbit. A very large array of solar cells on each satellite would provide electricity, which would be converted to microwave energy and beamed to a receiving antenna on the ground. There, it would be reconverted into electricity and distributed the same as any other centrally generated power, through a grid.

Schottky Barrier – A cell barrier established as the interface between a semiconductor, such as silicon, and a sheet of metal.

Scribing – The cutting of a grid pattern of grooves in a semiconductor material, generally for the purpose of making interconnections.

Sealed Battery – A battery with a captive electrolyte and a resealing vent cap, also called a valve-regulated battery. Electrolyte cannot be added.

Seasonal Depth of Discharge – An adjustment factor used in some system sizing procedures which "allows" the battery to be gradually discharged over a 30- to 90-day period of poor solar insolation. This factor results in a slightly smaller photovoltaic array.

Secondary Battery – A battery that can be recharged.

Self-Discharge – The rate at which a battery, without a load, will lose its charge.

Semiconductor – Any material that has a limited capacity for conducting an electric current. Certain semiconductors, including silicon, gallium arsenide, copper indium diselenide, and cadmium telluride, are uniquely suited to the photovoltaic conversion process.

Semicrystalline – See multicrystalline.

Series Connection – A way of joining photovoltaic cells by connecting positive leads to negative leads; such a configuration increases the voltage.

Series Controller – A charge controller that interrupts the charging current by open-circuiting the photovoltaic (PV) array. The control element is in series with the PV array and battery.

Series Regulator – Type of battery charge regulator where the charging current is controlled by a switch connected in series with the photovoltaic module or array.

Series Resistance – Parasitic resistance to current flow in a cell due to mechanisms such as resistance from the bulk of the semiconductor material, metallic contacts, and interconnections.

Shallow-Cycle Battery – A battery with small plates that cannot withstand many discharges to a low state of charge.

Shelf Life of Batteries – The length of time, under specified conditions, that a battery can be stored so that it keeps its guaranteed capacity.

Short-Circuit Current (Isc) – The current flowing freely through an external circuit that has no load or resistance; the maximum current possible.

Shunt Controller – A charge controller that redirects or shunts the charging current away from the battery. The controller requires a large heat sink to dissipate the current from the short-circuited photovoltaic array. Most shunt controllers are for smaller systems producing 30 amperes or less.

Shunt Regulator – Type of a battery charge regulator where the charging current is controlled by a switch connected in parallel with the photovoltaic (PV) generator. Shorting the PV generator prevents overcharging of the battery.

Siemens Process – A commercial method of making purified silicon.

Silicon (Si) – A semi-metallic chemical element that makes an excellent semiconductor material for photovoltaic devices. It crystallizes in face-centered cubic lattice like a diamond. It's commonly found in sand and quartz (as the oxide).

Sine Wave – A waveform corresponding to a single-frequency periodic oscillation that can be mathematically represented as a function of amplitude versus angle in which the value of the curve at any point is equal to the sine of that angle.

Sine Wave Inverter – An inverter that produces utility-quality, sine wave power forms.

Single-Crystal Material – A material that is composed of a single crystal or a few large crystals.

Single-Crystal Silicon – Material with a single crystalline formation. Many photovoltaic cells are made from single-crystal silicon.

Single-Stage Controller – A charge controller that redirects all charging current as the battery nears full state of charge.

Solar Cell – See photovoltaic (PV) cell.

Solar Constant – The average amount of solar radiation that reaches the earth's upper atmosphere on a surface perpendicular to the Sun's rays; equal to 1353 watts per square meter or 492 Btu per square foot.

Solar Cooling – The use of solar thermal energy or solar electricity to power a cooling appliance. Photovoltaic systems can power evaporative coolers ("swamp" coolers), heat pumps, and air conditioners.

Solar Energy – Electromagnetic energy transmitted from the Sun (solar radiation). The amount that reaches the Earth is equal to one billionth of total solar energy generated, or the equivalent of about 420 trillion kilowatt-hours.

Solar-Grade Silicon – Intermediate-grade silicon used in the manufacture of solar cells. Less expensive than electronic-grade silicon.

Solar Insolation – See insolation.

Solar Irradiance – See irradiance.

Solar Noon – The time of the day, at a specific location, when the Sun reaches its highest, apparent point in the sky; equal to true or due, geographic south.

Solar Panel – See photovoltaic (PV) panel.

Solar Resource – The amount of solar insolation a site receives, usually measured in kWh/m²/day, which is equivalent to the number of peak Sun hours.

Solar Spectrum – The total distribution of electromagnetic radiation emanating from the Sun. The different regions of the solar spectrum are described by their wavelength range. The visible region extends from about 390 to 780 nanometers (a nanometer is one billionth of 1 meter). About 99 percent of solar radiation is contained in a wavelength region from 300 nm (ultraviolet) to 3000 nm (near-infrared). The combined radiation in the wavelength region from 280 nm to 4,000 nm is called the broadband, or total, solar radiation.

Solar Thermal Electric Systems – Solar energy conversion technologies that convert solar energy to electricity, by heating a working fluid to power a turbine that drives a generator. Examples of these systems include central receiver systems, parabolic dish, and solar trough.

Space Charge – See cell barrier.

Specific Gravity – The ratio of the weight of the solution to the weight of an equal volume of water at a specified temperature. Used as an indicator of battery state of charge.

Spinning Reserve – Electric power plant or utility capacity online and running at low power in excess of actual load.

Split-Spectrum Cell – A compound photovoltaic device in which sunlight is first divided into spectral regions by optical means. Each region is then directed to a different photovoltaic cell optimized for converting that portion of the spectrum into electricity. Such a device achieves significantly greater overall conversion of incident sunlight into electricity. See multijunction device.

Sputtering – A process used to apply photovoltaic semiconductor material to a substrate by a physical vapor deposition process where high-energy ions are used to bombard elemental sources of semiconductor material, which eject vapors of atoms that are then deposited in thin layers on a substrate.

Square Wave – A waveform that has only two states, (i.e., positive or negative). A square wave contains a large number of harmonics.

Square Wave Inverter – A type of inverter that produces square wave output. It consists of a direct current source, four switches, and the load. The switches are power semiconductors that can carry a large current and withstand a high voltage rating. The switches are turned on and off at a correct sequence, at a certain frequency.

Staebler-Wronski Effect – The tendency of the sunlight to electricity conversion efficiency of amorphous silicon photovoltaic devices to degrade (drop) upon initial exposure to light.

Stand-Alone System – An autonomous or hybrid photovoltaic system not connected to a grid. May or may not have storage, but most stand-alone systems require batteries or some other form of storage.

Stand-Off Mounting – Technique for mounting a photovoltaic array on a sloped roof, which involves mounting the modules a short distance above the pitched roof and tilting them to the optimum angle.

Standard Reporting Conditions (SRC) – A fixed set of conditions (including meteorological) to which the electrical performance data of a photovoltaic module are translated from the set of actual test conditions.

Standard Test Conditions (STC) – Conditions under which a module is typically tested in a laboratory.

Standby Current – This is the amount of current (power) used by the inverter when no load is active (lost power). The efficiency of the inverter is lowest when the load demand is low.

Starved Electrolyte Cell – A battery containing little or no free fluid electrolyte.

State-of-Charge (SOC) – The available capacity remaining in the battery, expressed as a percentage of the rated capacity.

Storage Battery – A device capable of transforming energy from electric to chemical form and vice versa. The reactions are almost completely reversible. During discharge, chemical energy is converted to electric energy and is consumed in an external circuit or apparatus.

Stratification – A condition that occurs when the acid concentration varies from top to bottom in the battery electrolyte. Periodic, controlled charging at voltages that produce gassing will mix the electrolyte. See equalization.

String – A number of photovoltaic modules or panels interconnected electrically in series to produce the operating voltage required by the load.

Substrate – The physical material upon which a photovoltaic cell is applied.

Subsystem – Any one of several components in a photovoltaic system (i.e., array, controller, batteries, inverter, load).

Sulfation – A condition that afflicts unused and discharged batteries; large crystals of lead sulfate grow on the plate, instead of the usual tiny crystals, making the battery extremely difficult to recharge.

Superconducting Magnetic Energy Storage (SMES) – SMES technology uses the superconducting characteristics of low-temperature materials to produce intense magnetic fields to store energy. It has been proposed as a storage option to support large-scale use of photovoltaics as a means to smooth out fluctuations in power generation.

Superconductivity – The abrupt and large increase in electrical conductivity exhibited by some metals as the temperature approaches absolute zero.

Superstrate – The covering on the sunny side of a photovoltaic (PV) module, providing protection for the PV materials from impact and environmental degradation while allowing maximum transmission of the appropriate wavelengths of the solar spectrum.

Surge Capacity – The maximum power, usually 3–5 times the rated power, that can be provided over a short time.

System Availability – The percentage of time (usually expressed in hours per year) when a photovoltaic system will be able to fully meet the load demand.

System Operating Voltage – The photovoltaic array output voltage under load. The system operating voltage is dependent on the load or batteries connected to the output terminals.

System Storage – See battery capacity.

Tare Loss – Loss caused by a charge controller. One minus tare loss, expressed as a percentage, is equal to the controller efficiency.

Temperature Compensation – A circuit that adjusts the charge controller activation points depending on battery temperature. This feature is recommended if the battery temperature is expected to vary more than ±5°C from ambient temperature.

Temperature Factors – It is common for three elements in photovoltaic system sizing to have distinct temperature corrections: a factor used to decrease battery capacity at cold temperatures; a factor used to decrease PV module voltage at high temperatures; and a factor used to decrease the current carrying capability of wire at high temperatures.

Thermophotovoltaic Cell (TPV) – A device where sunlight concentrated onto a absorber heats it to a high temperature, and the thermal radiation emitted by the absorber is used as the energy source for a photovoltaic cell that is designed to maximize conversion efficiency at the wavelength of the thermal radiation.

Thick-Crystalline Materials – Semiconductor material, typically measuring from 200–400 microns thick, that is cut from ingots or ribbons.

Thin Film – A layer of semiconductor material, such as copper indium diselenide or gallium arsenide, a few microns or less in thickness, used to make photovoltaic cells.

Thin Film Photovoltaic Module – A photovoltaic module constructed with sequential layers of thin film semiconductor materials. See amorphous silicon.

Tilt Angle – The angle at which a photovoltaic array is set to face the sun relative to a horizontal position. The tilt angle can be set or adjusted to maximize seasonal or annual energy collection.

Tin Oxide – A wide band-gap semiconductor similar to indium oxide; used in heterojunction solar cells or to make a transparent conductive film, called NESA glass when deposited on glass.

Total AC Load Demand – The sum of the alternating current loads. This value is important when selecting an inverter.

Total Harmonic Distortion – The measure of closeness in shape between a waveform and it's fundamental component.

Total Internal Reflection – The trapping of light by refraction and reflection at critical angles inside a semiconductor device so that it cannot escape the device and must be eventually absorbed by the semiconductor.

Tracking Array – A photovoltaic (PV) array that follows the path of the sun to maximize the solar radiation incident on the PV surface. The two most common orientations are (1) one axis where the array tracks the sun east to west and (2) two-axis tracking where the array points directly at the sun at all times. Tracking arrays use both the direct and diffuse sunlight. Two-axis tracking arrays capture the maximum possible daily energy.

Transformer – An electromagnetic device that changes the voltage of alternating current electricity.

Tray Cable (TC) – may be used for interconnecting balance-of-systems.

Trickle Charge – A charge at a low rate, balancing through self-discharge losses, to maintain a cell or battery in a fully charged condition.

Two-Axis Tracking – A photovoltaic array tracking system capable of rotating independently about two axes (e.g., vertical and horizontal).

Tunneling – Quantum mechanical concept whereby an electron is found on the opposite side of an insulating barrier without having passed through or around the barrier.

Ultraviolet – Electromagnetic radiation in the wavelength range of 4 to 400 nanometers.

Underground Feeder (UF) – May be used for photovoltaic array wiring if sunlight resistant coating is specified; can be used for interconnecting balance-of-system components but not recommended for use within battery enclosures.

Underground Service Entrance (USE) – May be used within battery enclosures and for interconnecting balance-of-systems.

Uninterruptible Power Supply (UPS) – The designation of a power supply providing continuous uninterruptible service. The UPS will contain batteries.

Utility-Interactive Inverter – An inverter that can function only when tied to the utility grid, and uses the prevailing line-voltage frequency on the utility line as a control parameter to ensure that the photovoltaic system's output is fully synchronized with the utility power.

Vacuum Evaporation – The deposition of thin films of semiconductor material by the evaporation of elemental sources in a vacuum.

Vacuum Zero – The energy of an electron at rest in empty space; used as a reference level in energy band diagrams.

Valence Band – The highest energy band in a semiconductor that can be filled with electrons.

Valence Level Energy/Valence State – Energy content of an electron in orbit about an atomic nucleus. Also called bound state.

Varistor – A voltage-dependent variable resistor. Normally used to protect sensitive equipment from power spikes or lightning strikes by shunting the energy to ground.

Vented Cell – A battery designed with a vent mechanism to expel gases generated during charging.

Vertical Multijunction (VMJ) Cell – A compound cell made of different semiconductor materials in layers, one above the other. Sunlight entering the top passes through successive cell barriers, each of which converts a separate portion of the spectrum into electricity, thus achieving greater total conversion efficiency of the incident light. Also called a multiple junction cell. See multijunction device and split-spectrum cell.

Volt (V) – A unit of electrical force equal to that amount of electromotive force that will cause a steady current of one ampere to flow through a resistance of 1 ohm.

Voltage – The amount of electromotive force, measured in volts, that exists between two points.

Voltage at Maximum Power (Vmp) – The voltage at which maximum power is available from a photovoltaic module.

Voltage Protection – Many inverters have sensing circuits that will disconnect the unit from the battery if input voltage limits are exceeded.

Voltage Regulation – This indicates the variability in the output voltage. Some loads will not tolerate voltage variations greater than a few percent.

Wafer – A thin sheet of semiconductor (photovoltaic material) made by cutting it from a single crystal or ingot.

Watt – The rate of energy transfer equivalent to 1 ampere under an electrical pressure of 1 volt. One watt equals 1/746 horsepower, or 1 joule per second. It is the product of voltage and current (amperage).

Waveform – The shape of the phase power at a certain frequency and amplitude.

Wet Shelf Life – The period of time that a charged battery, when filled with electrolyte, can remain unused before dropping below a specified level of performance.

Window – A wide band gap material chosen for its transparency to light. Generally used as the top layer of a photovoltaic device, the window allows almost all of the light to reach the semiconductor layers beneath.

Wire Types – See Article 300 of National Electric Code for more information.

Work Function – The energy difference between the Fermi level and vacuum zero. The minimum amount of energy it takes to remove an electron from a substance into the vacuum.

Zenith Angle – the angle between the direction of interest (of the sun, for example) and the zenith (directly overhead)

California Solar Initiative – PV Incentives

Last DSIRE Review: 01/04/2010

Program Overview:

State:	California
Incentive Type:	State Rebate Program
Eligible Renewable/Other Technologies:	Solar Space Heat, Solar Thermal Electric, Solar Thermal Process Heat, Photovoltaics
Applicable Sectors:	Commercial, Industrial, Residential, Nonprofit, Schools, Local Government, State Government, Fed. Government, Multi-Family Residential, Low-Income Residential, Agricultural, Institutional, (All customers of PG&E, SDG&E, SCE; Bear Valley eligible only for NSHP)
Amount:	Varies by sector and system size (see below)
Equipment Requirements:	System components must be on the CEC's list of eligible equipment. Systems must be grid-connected. Inverters and modules must each carry a 10-year warranty. PV modules must be UL 1703-certified Inverters must be UL 1741-certified, and tested by the Energy Commission
Installation Requirements:	Systems must be installed by appropriately licensed California solar contractors or self-installed by the system owner. Installer certification by NABCEP is encouraged.
Ownership of Renewable Energy Credits:	Remains with customer-generator

Program Administrator: SCE, CCSE, PG&E

Program Budget: $3.2 billion over 10 years

Start Date: 7/1/2009

Web Site: http://www.cpuc.ca.gov/PUC/energy/solar

Authority 1:
SB 1
Date Enacted:
8/21/2006

Authority 2:
CSI Handbook (2009)
Date Effective:
7/1/2009

Authority 3:
CPUC decision 06–01–024

Authority 4:
CPUC Proceeding R0803008

Summary:

In January 2006, the California Public Utilities Commission (CPUC) adopted a program – the California Solar Initiative (CSI) – to provide more than $3 billion in incentives for solar-energy projects with the objective of providing 3,000 megawatts (MW) of solar capacity by 2016. The CPUC manages the solar program for non-residential projects and projects on existing homes ($2+ billion), while the CEC oversees the *New Solar Homes Partnership*, targeting the residential new construction market (~$400 million). Together, these two programs comprise the effort to expand the presence of photovoltaics (PV) throughout the state, Go Solar California.

Originally limited to customers of the state's investor-owned utilities, the CSI was expanded in August 2006, as a result of Senate Bill 1, to encompass municipal utility territories as well. Municipal utilities are required to offer incentives beginning in 2008 (nearly $800 million); many already offer PV rebates.

CSI Incentives for Non-residential Buildings and Existing Homes:

The CSI includes a transition to performance-based and expected performance-based incentives (as opposed to capacity-based buydowns), with the aim of promoting effective system design and installation. CSI incentive levels will automatically be reduced over the duration of the program in 10 steps based on the aggregate capacity of solar installed. In this way, incentive reductions are linked to levels of solar demand rather than an arbitrary timetable.

Expected Performance-Based Buydowns for systems under 30 kW began in 2007 at $2.50/W AC for residential and commercial systems (adjusted based on expected performance) and $3.25/W AC for government entities and nonprofits (adjusted based on expected performance). The incentive levels decline as the aggregate capacity of PV installations increases. Incentives will be awarded as a one-time, up-front payment based on expected performance, which is calculated using equipment ratings and installation factors such as geographic location, tilt, orientation and shading. Systems under 30 kW also have the option of opting for a performance-based incentive rather than the incentive based on expected performance.

Performance-Based Incentives (PBI) for systems 30 kW and larger began in 2007 at \$0.39/kWh for the first five years for taxable entities, and \$\$0.50/kWh for the first five years for government entities and nonprofits. The incentive levels decline as the aggregate capacity of PV installations increases. PBI will be paid monthly based on the actual amount of energy produced for a period of five years. Residential and small commercial projects under the 30 kW threshold can also choose to opt in to the PBI rather than the upfront Expected Performance-Based Buydown approach. However, all installations of 30 kW or larger must take the PBI.

The program is managed by the Pacific Gas and Electric Company (PG&E), Southern California Edison (SCE), and the California Center for Sustainable Energy.

Low-Income Programs

Ten percent of the CSI Program budget (\$216 million) has been allocated to two low-income solar incentive programs. As of March 2009, the single family low income program is still being developed; but SCE, PG&E and CCSE are accepting applications for Track 1 of the multi-family affordable solar housing (MASH) program. Rebates are available through Track 1 in the amount of \$3.30/W for PV systems offsetting common area loads, and \$4.00/W for systems offsetting tenant loads. As required by the CPUC, the utilities are developing *virtual* net energy metering (VNEM) tariffs which will allow MASH participants to allocate the kWh credits from a single solar system across several electric accounts at the same building complex.

Incentives for Other Solar Electric Generating Technologies

The CSI Handbook released in January 2008 clarified the eligibility of other solar electric generating technologies which either produce electricity or displace electricity. Incentives for other solar electric generating technologies are available for CSI incentives effective October 1, 2008. The CPUC specifically recognizes electric generating solar thermal as including dish stirling, solar trough, and concentrating solar technologies, while technologies that displace electricity include solar forced air heating, and solar cooling or air-conditioning. The budget for electric displacing technologies is capped at \$100.8 million. While solar water heaters can also displace electricity, the CPUC excludes them from the CSI because they plan to offer incentives for solar water heaters through a separate program based on the pilot program currently in operation within the service territory of San Diego Gas and Electric. Future CSI rulemaking activities will address energy-efficiency requirements, additional affordable housing incentives, and other program elements.

CPUC Program Administrators:

Pacific Gas & Electric (PG&E)

Web Site: www.pge.com/solar
E-mail Address: solar@pge.com

Contact Person: Program Manager, California Solar Initiative Program
Telephone: 877–743–4112
Fax: 415–973–2510
Mailing Address:
PG&E Integrated Processing Center
P.O. Box 7265
San Francisco, CA 94120–7265

California Center for Sustainable Energy (CCSE)

Web Site: www.energycenter.org
E-mail Address: csi@energycenter.org
Contact Person: John Supp, Program Manager
Telephone: 858–244–1177/(866)-sdenergy
Fax: 858–244–1178
Mailing Address:
California Center for Sustainable Energy
Attn: SELFGEN Program Manager
8690 Balboa Avenue Suite 100
San Diego, CA 92123

Southern California Edison (SCE)

Web Site: www.sce.com/rebatesandsavings/CaliforniaSolarInitiative
E-mail Address: greenh@sce.com
Contact Person: Program Manager, California Solar Initiative Program
Telephone: 1–800–799–4177
Fax: 626–302–6253
Mailing Address:
Southern California Edison
6042A Irwindale Avenue
Irwindale, CA 91702

California State Energy Code

Last DSIRE Review: 01/14/2010

Program Overview:

State:	California
Incentive Type:	Building Energy Code
Eligible Efficiency Technologies:	Comprehensive Measures/Whole Building
Eligible Renewable/Other Technologies:	Passive Solar Space Heat, Solar Water Heat, Photovoltaics
Applicable Sectors:	Commercial, Residential
Residential Code:	State developed code, Title 24, Part 6, exceeds 2006 IECC, and is mandatory statewide.
Commercial Code:	State developed code, Title 24, Part 6, meets or exceeds ASHRAE/IESNA 90.1–2004, and is mandatory statewide.

| Code Change Cycle: | Three-year code change cycle. The 2005 California Energy Efficiency Standards became effective October 1, 2005. The Commission adopted the new 2008 Standards on April 23, 2008. They will be effective January 1, 2010. |
| Web Site: | http://bcap-ocean.org/state-country/ california |

Summary:

Much of the information presented in this summary is drawn from the U.S. Department of Energy's (DOE) Building Energy Codes Program and the Building Codes Assistance Project (BCAP). For more detailed information about building energy codes, visit the DOE and BCAP websites.

The California Building Standards Commission (BSC) is responsible for administering California's building standards adoption, publication, and implementation. Since 1989, the BSC has published triennial editions of the code, commonly referred to as Title 24, in its entirety every three years. On July 17, 2008 the BSC unanimously approved the nation's first statewide voluntary green building code. In January 2010, the BSC adopted a final version of the new building code, CALGreen, which becomes mandatory on January 1, 2011. The new code includes provisions to ensure the reduction water use by 20%, require separate water meters for nonresidential building's indoor and outdoor water use, improve indoor air quality, require the diversion of 50% of construction waste from landfills, and a number of other green building principles.

Title 24 applies to all buildings that are heated and/or mechanically cooled and are defined under the Uniform Building Code as A, B, E, H, N, R, or S occupancies, except registered historical buildings. Additions and renovations are also covered by the code. Institutional building's which include hospitals and prisons are not covered.

For residential low-rise buildings the current code provision include compliance credits for high performance ducts and building envelope features. The size of credit depends on the action taken. For example simply designing ducts to Air Conditioner Contractor's Association guidelines or properly sealing duct joints provided lower levels of credit than having the HVAC system tested for duct leaks. To take credit for these measures the installer and inspector must be trained and certified.

Local governmental agencies can modify the state energy code to be more stringent when documentation is provided to the California Energy Commission.

AB 1103, passed in October 2007, requires annual energy-use reporting for all of California's nonresidential buildings effective January 2009. Beginning in 2010, owners of commercial buildings must disclose their energy usage and Energy Star rating to potential buyers, leasers, and financiers. The state Department of General Services has been working closely with utilities to streamline the reporting process.

Federal Incentives/Policies for Renewables & Efficiency

Business Energy Investment Tax Credit (ITC)

Last DSIRE Review: 06/10/2009

Program Overview:

State:	Federal
Incentive Type:	Corporate Tax Credit
Eligible Renewable/Other Technologies:	Solar Water Heat, Solar Space Heat, Solar Thermal Electric, Solar Thermal Process Heat, Photovoltaics, Wind, Biomass, Geothermal Electric, Fuel Cells, Geothermal Heat Pumps, CHP/Cogeneration, Solar Hybrid Lighting, Microturbines, Geothermal Direct-Use
Applicable Sectors:	Commercial, Industrial, Utility
Amount:	30% for solar, fuel cells and small wind;** 10%** for geothermal, microturbines and CHP
Maximum Incentive:	Fuel cells: $1,500 per 0.5 kW Microturbines: $200 per kW Small wind turbines placed in service 10/4/08–12/31/08: $4,000 Small wind turbines placed in service after 12/31/08: no limit All other eligible technologies: no limit
Eligible System Size:	Small wind turbines: 100 kW or less** Fuel cells: 0.5 kW or greater Microturbines: 2 MW or less CHP: 50 MW or less**
Equipment Requirements:	Fuel cells, microturbines and CHP systems must meet specific energy-efficiency criteria
Program Administrator:	U.S. Internal Revenue Service

Authority 1:
26 USC § 48

Authority 2:
Instructions for IRS Form 3468

Authority 3:
IRS Form 3468

Summary:

Note: The American Recovery and Reinvestment Act of 2009 (H1.R. 1) allows taxpayers eligible for the federal renewable electricity production tax credit (PTC)** to take the federal business energy investment tax credit (ITC) or to receive a grant from the U.S. Treasury Department instead of taking the PTC for new installations. The new law also allows taxpayers eligible for the business ITC to receive a grant from the U.S. Treasury Department instead of taking the business ITC for new installations. The Treasury Department issued Notice 2009–52 in June 2009, giving limited guidance on how to take the federal business energy investment tax credit instead of the federal renewable electricity production tax credit. The Treasury Department will issue more extensive guidance at a later time.

The federal business energy investment tax credit available under 26 USC § 48 was expanded significantly by the *Energy Improvement and Extension Act of 2008* (H.R. 1424), enacted in October 2008. This law extended the duration – by eight years – of the existing credits for solar energy, fuel cells and microturbines; increased the credit amount for fuel cells; established new credits for small wind-energy systems, geothermal heat pumps, and combined heat and power (CHP) systems; extended eligibility for the credits to utilities; and allowed taxpayers to take the credit against the alternative minimum tax (AMT), subject to certain limitations. The credit was further expanded by *The American Recovery and Reinvestment Act of 2009*, enacted in February 2009.

In general, credits are available for eligible systems placed in service on or before December 31, 2016:*

- *Solar.* The credit is equal to 30% of expenditures, with no maximum credit. Eligible solar energy property includes equipment that uses solar energy to generate electricity, to heat or cool (or provide hot water for use in) a structure, or to provide solar process heat. Hybrid solar-lighting systems, which use solar energy to illuminate the inside of a structure using fiber-optic distributed sunlight, are eligible. Passive solar systems and solar pool-heating systems are *not* eligible. (Note that the Solar Energy Industries Association has published a *three-page document* that provides answers to frequently asked questions regarding the federal tax credits for solar energy.)
- *Fuel Cells.* The credit is equal to 30% of expenditures, with no maximum credit. However, the credit for fuel cells is capped at $1,500 per 0.5 kilowatt (kW) of capacity. Eligible property includes fuel cells with a minimum capacity of 0.5 kW that have an electricity-only generation efficiency of 30% or higher. (Note that the credit for property placed in service before October 4, 2008, is capped at $500 per 0.5 kW.)
- *Small Wind Turbines.*** The credit is equal to 30% of expenditures, with no maximum credit for small wind turbines placed in service after December 31, 2008. Eligible small wind property includes wind turbines up to 100 kW in capacity. (In general, the maximum credit is $4,000 for eligible property placed in service after October 3, 2008, and before January 1, 2009. *The American Recovery and Reinvestment Act of 2009* removed the $4,000 maximum credit limit for small wind turbines.)
- *Geothermal Systems.*** The credit is equal to 10% of expenditures, with no maximum credit limit stated. Eligible geothermal energy property includes geothermal heat pumps and equipment used to produce, distribute or use energy derived from a geothermal deposit. For electricity produced by geothermal power, equipment qualifies only up to, but not including, the electric transmission

stage. For geothermal heat pumps, this credit applies to eligible property placed in service after October 3, 2008.

- *Microturbines.* The credit is equal to 10% of expenditures, with no maximum credit limit stated (explicitly). The credit for microturbines is capped at *$200* per kW of capacity. Eligible property includes microturbines up to two megawatts (MW) in capacity that have an electricity-only generation efficiency of 26% or higher.
- *Combined Heat and Power (CHP).* ** The credit is equal to 10% of expenditures, with no maximum limit stated. Eligible CHP property generally includes systems up to 50 MW in capacity that exceed 60% energy efficiency, subject to certain limitations and reductions for large systems. The efficiency requirement does not apply to CHP systems that use biomass for at least 90% of the system's energy source, but the credit may be reduced for less-efficient systems. This credit applies to eligible property placed in service after October 3, 2008.

In general, the original use of the equipment must begin with the taxpayer, or the system must be constructed by the taxpayer. The equipment must also meet any performance and quality standards in effect at the time the equipment is acquired. The energy property must be operational in the year in which the credit is first taken.

Significantly, *The American Recovery and Reinvestment Act of 2009* repealed a previous limitation on the use of the credit for eligible projects also supported by "subsidized energy financing." For projects placed in service after December 31, 2008, this limitation no longer applies. Businesses that receive other incentives are advised to consult with a tax professional regarding how to calculate this federal tax credit.

Federal Incentives/Policies for Renewables & Efficiency

U.S. Department of Treasury – Renewable Energy Grants

Last DSIRE Review: 03/31/2010

Program Overview:

State:	Federal
Incentive Type:	Federal Grant Program
Eligible Renewable/Other Technologies:	Solar Water Heat, Solar Space Heat, Solar Thermal Electric, Solar Thermal Process Heat, Photovoltaics, Landfill Gas, Wind, Biomass, Hydroelectric, Geothermal Electric, Fuel Cells, Geothermal Heat Pumps, Municipal Solid Waste, CHP/Cogeneration, Solar Hybrid Lighting, Hydrokinetic, Anaerobic Digestion, Tidal Energy, Wave Energy, Ocean Thermal, Microturbines
Applicable Sectors:	Commercial, Industrial, Agricultural

Amount: 30% of property that is part of a qualified facility,
                                 qualified fuel cell property, solar property, or
                                 qualified small wind property
                                 10% of all other property

Maximum Incentive: $1,500 per 0.5 kW for qualified fuel cell property
                                 $200 per kW for qualified microturbine property
                                 50 MW for CHP property, with limitations for large
                                 systems

Program Administrator: U.S. Department of Treasury

Funding Source: The American Recovery and Reinvestment Act
                                 (ARRA)

Start Date: 1/1/2009

Expiration Date: 12/31/2010 (construction must begin by this date)

Web Site: http://www.treas.gov/recovery/1603.shtml

Authority 1:
H.R. 1: Div. B, Sec. 1104 & 1603 (The American Recovery and Reinvestment Act of
 2009)
Date Enacted:
2/17/2009
Date Effective:
1/1/2009

Authority 2:
U.S. Department of Treasury: Grant Program Guidance
Date Enacted:
07/09/2009, subsequently amended

Summary:

Note: The American Recovery and Reinvestment Act of 2009 (H.R. 1) allows
taxpayers eligible for the federal business energy investment tax credit (ITC) to take
this credit or to receive a grant from the U.S. Treasury Department instead of taking
the business ITC for new installations. The new law also allows taxpayers eligible
for the renewable electricity production tax credit (PTC) to receive a grant from the
U.S. Treasury Department instead of taking the PTC for new installations. (It does
not allow taxpayers eligible for the residential renewable energy tax credit to receive
a grant instead of taking this credit.) Taxpayers may not use more than one of these
incentives. Tax credits allowed under the ITC with respect to progress expenditures
on eligible energy property will be recaptured if the project receives a grant. The
grant is not included in the gross income of the taxpayer.

The *American Recovery and Reinvestment Act of 2009* (H.R. 1), enacted in
February 2009, created a renewable energy grant program that will be administered
by the U.S. Department of Treasury. This cash grant may be taken in lieu of the
federal business energy investment tax credit (ITC). In July 2009 the Department of
Treasury issued documents detailing guidelines for the grants, terms and conditions

and a sample application. There is an online application process, and applications are currently being accepted. See the US Department of Treasury program web site for more information, including answers to frequently asked questions.

Grants are available to eligible property* placed in service in 2009 or 2010, or placed in service by the specified credit termination date,** if construction began in 2009 or 2010. The guidelines include a "safe harbor" provision that sets the beginning of construction at the point where the applicant has incurred or paid at least 5% of the total cost of the property, excluding land and certain preliminary planning activities. Generally, construction begins when "physical work of a significant nature" begins. Below is a list of important program details as they apply to each different eligible technology.

- *Solar.* The grant is equal to 30% of the basis of the property for solar energy. Eligible solar-energy property includes equipment that uses solar energy to generate electricity, to heat or cool (or provide hot water for use in) a structure, or to provide solar process heat. Passive solar systems and solar pool-heating systems are *not* eligible. Hybrid solar-lighting systems, which use solar energy to illuminate the inside of a structure using fiber-optic distributed sunlight, are eligible.
- *Fuel Cells.* The grant is equal to 30% of the basis of the property for fuel cells. The grant for fuel cells is capped at $1,500 per 0.5 kilowatt (kW) in capacity. Eligible property includes fuel cells with a minimum capacity of 0.5 kW that have an electricity-only generation efficiency of 30% or higher.
- *Small Wind Turbines.* The grant is equal to 30% of the basis of the property for small wind turbines. Eligible small wind property includes wind turbines up to 100 kW in capacity.
- *Qualified Facilities.* The grant is equal to 30% of the basis of the property for qualified facilities that produce electricity. Qualified facilities include wind energy facilities, closed-loop biomass facilities, open-loop biomass facilities, geothermal energy facilities, landfill gas facilities, trash facilities, qualified hydropower facilities, and marine and hydrokinetic renewable energy facilities.
- *Geothermal Heat Pumps.* The grant is equal to 10% of the basis of the property for geothermal heat pumps.
- *Microturbines.* The grant is equal to 10% of the basis of the property for microturbines. The grant for microturbines is capped at $200 per kW of capacity. Eligible property includes microturbines up to two megawatts (MW) in capacity that have an electricity-only generation efficiency of 26% or higher.
- *Combined Heat and Power (CHP).* The grant is equal to 10% of the basis of the property for CHP. Eligible CHP property generally includes systems up to 50 MW in capacity that exceed 60% energy efficiency, subject to certain limitations and reductions for large systems. The efficiency requirement does not apply to CHP systems that use biomass for at least 90% of the system's energy source, but the grant may be reduced for less-efficient systems.

It is important to note that only tax-paying entities are eligible for this grant. Federal, state and local government bodies, non-profits, qualified energy tax credit bond lenders, and cooperative electric companies are not eligible to receive this grant. Partnerships or pass-thru entities for the organizations described above are also not eligible to receive this grant, except in cases where the ineligible party only owns an indirect interest in the applicant through a taxable C corporation. Grant applications must be submitted by October 1, 2011. The U.S. Treasury Department will make payment of the grant within 60 days of the grant application date or the date the property is placed in service, whichever is later.

U.S. Department of Energy – Loan Guarantee Program

Last DSIRE Review: 12/14/2009

Program Overview:

State:	Federal
Incentive Type:	Federal Loan Program
Eligible Efficiency Technologies:	Yes; specific technologies not identified
Eligible Renewable/Other Technologies:	Solar Thermal Electric, Solar Thermal Process Heat, Photovoltaics, Wind, Hydroelectric, Geothermal Electric, Fuel Cells, Daylighting, Tidal Energy, Wave Energy, Ocean Thermal, Biodiesel
Applicable Sectors:	Commercial, Industrial, Nonprofit, Schools, Local Government, State Government, Agricultural, Institutional, Any non-federal entity, Manufacturing Facilities
Amount:	Varies. Program focuses on projects with total project costs over $25 million.
Maximum Incentive:	Not specified.
Terms:	Full repayment is required over a period not to exceed the lesser of 30 years or 90% of the projected useful life of the physical asset to be financed
Program Administrator:	U.S. Department of Energy
Web Site:	http://www.lgprogram.energy.gov

Authority 1:
42 USC § 16511 et seq.

Authority 2:
10 CFR 609

Summary:

Innovative Technology Loan Guarantee Program:

Title XVII of the federal *Energy Policy Act of 2005* (EPAct 2005) authorized the U.S. Department of Energy (DOE) to issue loan guarantees for projects that "avoid, reduce or sequester air pollutants or anthropogenic emissions of greenhouse gases;

and employ new or significantly improved technologies as compared to commercial technologies in service in the United States at the time the guarantee is issued." The loan guarantee program has been authorized to offer more than $10 billion in loan guarantees for energy efficiency, renewable energy and advanced transmission and distribution projects.

DOE actively promotes projects in three categories: (1) manufacturing projects, (2) stand-alone projects, and (3) large-scale integration projects that may combine multiple eligible renewable energy, energy efficiency and transmission technologies in accordance with a staged development scheme. Under the original authorization, loan guarantees were intended to encourage early commercial use of new or significantly improved technologies in energy projects. The loan guarantee program generally does not support research and development projects.

In July 2009, the U.S. DOE issued a new solicitation for projects that employ innovative energy efficiency, renewable energy, and advanced transmission and distribution technologies. Proposed projects must fit within the criteria for "New or Significantly Improved Technologies" as defined in 10 CFR 609. The solicitation provides for a total of $8.5 billion in funding and is to remain open until that amount is fully obligated. The initial due date for applicants was September 16, 2009.

Temporary Loan Guarantee Program:
The American Recovery and Reinvestment Act of 2009 (ARRA) (H.R. 1), enacted in February 2009, extended the authority of the DOE to issue loan guarantees and appropriated $6 billion for this program. Under this act, the DOE may enter into guarantees until September 30, 2011. The act amended EPAct 2005 by adding a new section defining eligible technologies for new loan guarantees. Eligible projects include renewable energy projects that generate electricity or thermal energy and facilities that manufacture related components, electric power transmission systems, and innovative biofuels projects. Funding for biofuels projects is limited to $500 million. Davis-Bacon wage requirements apply to any project receiving a loan guarantee.

In July 2009, the U.S. DOE issued a solicitation for innovative energy efficiency, renewable energy, transmission and distribution technologies. The solicitation is expected to support as much as $8.5 billion in lending to eligible projects.

Modified Accelerated Cost-Recovery System (MACRS) + Bonus Depreciation (2008–2009)

Last DSIRE Review: 01/20/2010

Program Overview:

State:	Federal
Incentive Type:	Corporate Depreciation
Eligible Renewable/Other Technologies:	Solar Water Heat, Solar Space Heat, Solar Thermal Electric, Solar Thermal Process Heat, Photovoltaics, Landfill Gas, Wind, Biomass,

	Geothermal Electric, Fuel Cells, Geothermal Heat Pumps, Municipal Solid Waste, CHP/ Cogeneration, Solar Hybrid Lighting, Anaerobic Digestion, Microturbines, Geothermal Direct-Use
Applicable Sectors:	Commercial, Industrial
Program Administrator:	U.S. Internal Revenue Service
Start Date:	1986

Authority 1:
26 USC § 168
Date Effective:
1986

Authority 2:
26 USC § 48

Summary:

Note: While the general Modified Accelerated Cost Recovery System (MACRS) remains in effect, the provision authorizing additional first-year bonus depreciation of 50% of eligible costs expired December 31, 2009. Although it is possible that bonus depreciation could be renewed for projects placed in service in 2010, as of this writing no such renewal had been enacted.

Under the federal Modified Accelerated Cost-Recovery System (MACRS), businesses may recover investments in certain property through depreciation deductions. The MACRS establishes a set of class lives for various types of property, ranging from three to 50 years, over which the property may be depreciated. A number of renewable energy technologies are classified as five-year property (26 USC § 168(e)(3)(B)(vi)) under the MACRS, which refers to 26 USC § 48(a)(3)(A), often known as the energy investment tax credit or ITC to define eligible property. Such property currently includes:

- a variety of solar electric and solar thermal technologies
- fuel cells and microturbines
- geothermal electric
- direct-use geothermal and geothermal heat pumps
- small wind (100 kW or less)
- combined heat and power (CHP).
- The provision which defines ITC technologies as eligible also adds the general term "wind" as an eligible technology, extending the five-year schedule to large wind facilities as well.

In addition, for certain other biomass property, the MACRS property class life is seven years. Eligible biomass property generally includes assets used in the conversion of biomass to heat or to a solid, liquid or gaseous fuel, and to equipment and structures used to receive, handle, collect and process biomass in a waterwall, combustion system, or refuse-derived fuel system to create hot water, gas, steam and electricity.

The 5-year schedule for most types of solar, geothermal, and wind property has been in place since 1986. The federal *Energy Policy Act of 2005* (EPAct 2005) classified fuel cells, microturbines and solar hybrid lighting technologies as five-year property as well by adding them to § 48(a)(3)(A). This section was further expanded in October 2008 by the addition of geothermal heat pumps, combined heat and power, and small wind under *The Energy Improvement and Extension Act of 2008.*

The federal *Economic Stimulus Act of 2008,* enacted in February 2008, included a 50% first-year bonus depreciation (26 USC § 168(k)) provision for eligible renewable-energy systems acquired and placed in service in 2008. This provision was extended (retroactively to the entire 2009 tax year) under the same terms by *The American Recovery and Reinvestment Act of 2009*, enacted in February 2009. To qualify for bonus depreciation, a project must satisfy these criteria:

- the property must have a recovery period of 20 years or less under normal federal tax depreciation rules;
- the original use of the property must commence with the taxpayer claiming the deduction;
- the property generally must have been acquired during 2008 or 2009; and
- the property must have been placed in service during 2008 or 2009

If property meets these requirements, the owner is entitled to deduct 50% of the adjusted basis of the property in 2008 and 2009. The remaining 50% of the adjusted basis of the property is depreciated over the ordinary depreciation schedule. The bonus depreciation rules do not override the depreciation limit applicable to projects qualifying for the federal business energy tax credit. Before calculating depreciation for such a project, including any bonus depreciation, the adjusted basis of the project must be reduced by one-half of the amount of the energy credit for which the project qualifies.

For more information on the federal MACRS, see *IRS Publication 946, IRS Form 4562: Depreciation and Amortization*, and *Instructions for Form 4562.* The IRS web site provides a search mechanism for forms and publications. Enter the relevant form, publication name or number, and click "GO" to receive the requested form or publication.

Florida Incentives/Policies for Renewables & Efficiency [Insert Unnumbered Figure 1 here]

Renewable Energy Production Tax Credit

Last DSIRE Review: 06/17/2009

Program Overview:

State:	Florida
Incentive Type:	Corporate Tax Credit
Eligible Renewable/Other Technologies:	Solar Thermal Electric, Photovoltaics, Wind, Biomass, Hydroelectric, Geothermal Electric, CHP/ Cogeneration, Hydrogen, Tidal Energy, Wave Energy, Ocean Thermal

Applicable Sectors:	Commercial
Amount:	$0.01/kWh for electricity produced from 1/1/2007 through 6/30/2010
Maximum Incentive:	No maximum specified for individual projects; Maximum of $5 million per state fiscal year for all credits under this program
Carryover Provisions:	Unused credit may be carried forward for up to 5 years
Program Administrator:	Florida Department of Revenue
Start Date:	7/1/2006
Expiration Date:	6/30/2010
Web Site:	http://www.myfloridaclimate.com/ climate_quick_links/florida_energ…

Authority 1:
Fla. Stat. § 220.193
Date Enacted:
6/19/2006
Date Effective:
7/1/2006
Expiration Date
6/30/2010

Summary:

In June 2006, SB 888 established a renewable energy production tax credit to encourage the development and expansion of renewable energy facilities in Florida. This annual corporate tax credit is equal to $0.01/kWh of electricity produced and sold by the taxpayer to an unrelated party during a given tax year. For new facilities (placed in service after May 1, 2006) the credit is based on the sale of the facility's entire electrical production. For an expanded* facility, the credit is based on the increases in the facility's electrical production that are achieved after May 1, 2006.

For the purposes of this credit, renewable energy is defined as "electrical, mechanical, or thermal energy produced from a method that uses one or more of the following fuels or energy sources: hydrogen, biomass, solar energy, geothermal energy, wind energy, ocean energy, waste heat, or hydroelectric power."

The credit may be claimed for electricity produced and sold on or after January 1, 2007 through June 30, 2010. Beginning in 2008 and continuing until 2011, each taxpayer claiming a credit under this section must first apply to the Department of Revenue (DOR) by February 1 of each year for an allocation of available credit. If the credit granted is not fully used in one year because of insufficient tax liability, the unused amount may be carried forward for up to 5 years.

The combined total amount of tax credits which may be granted for all taxpayers under this program is limited to $5 million per state fiscal year. If the amount of credits applied for each year exceeds $5 million, the DOR will award a prorated amount based on each applicant's increased production and sales.

A taxpayer cannot claim both this production tax credit and Florida's Renewable Energy Technologies Investment Tax Credit. In June 2008, Florida enacted HB 7135 which specified that a taxpayer's use of the credit does not reduce the amount of the Florida alternative minimum tax available to the taxpayer.

Florida Incentives/Policies for Renewables & Efficiency

[Insert Unnumbered figure 1 here]

Renewable Energy Technologies Investment Tax Credit

Last DSIRE Review: 09/15/2009

Program Overview:

State:	Florida
Incentive Type:	Corporate Tax Credit
Eligible Renewable/Other Technologies:	Fuel Cells, Hydrogen, Ethanol, Biodiesel
Applicable Sectors:	Commercial
Amount:	75% of all capital costs, operation and maintenance costs, and research and development costs
Maximum Incentive:	Varies by application
Carryover Provisions:	Unused amount may be carried forward and used in tax years beginning 1/1/2007 and ending 12/31/2012
Program Administrator:	Executive Office of the Governor
Start Date:	7/1/2006
Expiration Date:	6/30/2010 (tax credit provision)
Web Site:	http://myfloridaclimate. com/climate_quick_links/ florida_energy_cl...

Authority 1:
Fla. Stat. § 220.192
Date Enacted:
6/19/2006
Date Effective:
7/1/2006
Expiration Date
6/30/2010 (tax credit provision)

Michigan Incentives/Policies for Renewables & Efficiency

Consumers Energy – Photovoltaic Purchase Tariff

Last DSIRE Review: 10/26/2009

Program Overview:

State: Michigan

Incentive Type: Production Incentive

Eligible Renewable/Other Technologies: Photovoltaics

Applicable Sectors: Commercial, Industrial, Residential, Nonprofit, Schools, Local Government, State Government, Fed. Government, Multi-Family Residential, Institutional

Amount: Residential: $0.65/kWh or $0.525/kWh (see summary for details)
Non-Residential: $0.45/kWh or $0.375/kWh (see summary for details)

Maximum Incentive: None specified

Terms: Fixed rate contract for up to 12 years; participants pay system access charge ($6 – $50 per month) to cover additional metering costs.

Program Administrator: Consumers Energy

Start Date: 08/27/2009

Expiration Date: 12/31/2010

Web Site: http://www.consumersenergy.com/welcome.htm?/products/index.asp?AS...

Authority 1:
Experimental Advanced Renewables Program Tariff (multiple sheets)
Date Effective:
08/27/2009
Expiration Date
12/31/2010

Authority 2:
EARP Application Instructions

Authority 3:
EARP Program Application

Summary:
Note: While the overall program limit is set at 2 megawatts (MW), Consumer's Energy has already received applications for more than 6.3 MW of capacity. New applications continue to be accepted, but they will be placed in a queue for funding should it become available due to project failures or application withdrawals.

In addition, in response to concerns from installers and developers, the in-service deadline for residential and non-residential projects seeking the higher incentive amount has been extended to May 1, 2010. The higher incentive level is intended apply to the first 250 kW (residential) or 750 kW (non-residential) of capacity that comes on-line by the May deadline. Projects that miss this deadline will be eligible for the lower incentive. If a portion of a project's capacity will exceed the 250 kW or 750 kW limit, that project will be offered the rate that corresponds to the largest proportion of the project capacity above or below the limit.

Beginning in August 2009, Consumers Energy of Michigan is offering its residential and non-residential customers an experimental buy-back tariff – termed the Experimental Advanced Renewables Program (EARP) – for electricity produced by solar photovoltaic (PV) systems. Owners of residential systems from 1–20 kilowatts (kW) and non-residential systems from 20–150 kW are eligible to participate in the program. The minimum system size is 1 kW. Residential customers must receive electric service on tariff rate RS or RT in order to be eligible for the program. Non-residential customers on tariff rates RS, RT, GS, GSD, GP, and GPD are eligible for the program. The overall program is limited to 2 megawatts (MW) with 500 kW reserved for residential sites.

It is important to note that this is not a net metering program and program participants are not eligible for net metering. Under the program, Consumers Energy will purchase all of the electricity produced by the system through a fixed-rate contract of 1 to 12 years. Electricity production is metered separately from the customer's existing electricity source (i.e., the grid). Participants are assessed a monthly System Access Charge equivalent to the existing distribution account used to qualify for the program to cover metering costs.* Systems with battery backup or any other type of energy storage capability are not eligible to participate in this program. Purchase rates are as follows:

- Residential: $0.65/kilowatt-hour (kWh) for systems available by May 1, 2010 (up to roughly 250 kW of capacity); $0.525/kWh for systems that do not qualify for the higher level.
- Non-residential: $0.45 kWh for systems available by May 1, 2010 (up to roughly 750 kW of aggregate capacity); $0.375/kWh for systems available in 2010 that do not qualify for the higher incentive level.

Solar systems that receive the residential tariff rate may not be located on property that is used for commercial purposes, such as rental properties, warehouses, workshops, office buildings, etc.. Tax exempt entities are not eligible to participate in the program under the residential rates, although they are eligible under the non-residential rates. Third-party ownership structures are not eligible for the program. The applicant must be a Consumers Energy customer on one of the qualifying electricity rates and must own the generating system. If the generation system is located on property that is not owned by the applicant, the applicant must have a lease or other instrument that permits them to construct, own, and operate the system throughout the term of the contract. Systems installed on newly constructed buildings are eligible for this program as long as the applicant will receive electric service from the utility at that site, or an adjacent site.

In order to be eligible for the program, solar equipment must be manufactured in the state of Michigan or constructed by a Michigan workforce. The manufacturing requirement can be met if 50% or more of the equipment and material costs associated with the system are attributable to components manufactured or assembled in Michigan. In order to qualify as a system constructed by a Michigan workforce, at least 60% of the total labor hours associated with installing the system must be performed by Michigan residents. All systems must meet the requirements of UL 1741 and IEEE 1547.1 and be installed in compliance with all current local and state electric and construction code requirements. The utility owns all renewable energy credits (RECs) associated with electricity purchased under this program, including all Michigan RECs, Michigan Incentive RECs, and Federal RECs.

The program began accepting formal applications on August 3, 2009 and is scheduled to run through December 31, 2010, subject to the limitations on overall enrollment described above. For further information on this program, please see the EARP tariff and application documents above, and contact program personnel.

Nevada Incentives/Policies for Renewables & Efficiency

NV Energy – RenewableGenerations Rebate Program

Last DSIRE Review: 04/27/2010

Program Overview:

State: Nevada

Incentive Type: State Rebate Program

Eligible Renewable/ Photovoltaics, Wind, Small Hydroelectric
Other
Technologies:

Applicable Sectors: Commercial, Residential, Nonprofit, Schools, Local Government, State Government, Agricultural, Other Public Buildings

Amount: Solar (Step 1, 2010–2011 program year):
Schools and public and other property, including non-profits and churches: $5.00 per watt AC
Residential and small business property: $2.30 per watt AC
Wind (Step 1, 2010–2011 program year):
Residential, small business, agriculture: $3.00 per watt
Schools and Public Buildings: $4.00 per watt
Small Hydro (Step 1, 2010–2011 program year):
Non-net metered systems: $2.80/W
Net metered systems: $2.50/W

Maximum Solar (Step 1, 2010–2011 program year):
Incentive: Public and other property, including non-profits and churches: $500,000
Schools: $250,000, or up to $500,000 if given permission by the utilities commission

Residential: $23,000
Small business property: $115,000
Wind (Step 1, 2010–2011 program year):
Residential: $180,000
Small Business: $750,000
Schools: $1.0 million
Agriculture: $1.5 million
Public Buildings: $2.0 million
Small Hydro (Step 1, 2010–2011 program year):
Non-net metered systems: $560,000
Net metered systems: $500,000

Eligible System Size:	Maximum of 1 MW
Equipment Requirements:	Solar: Systems must be in compliance with all applicable standards; Must carry a minimum 7-year warranty on inverters, 20-year warranty on panels, and 2-year warranty on labor; Modules and inverters must be on the California Energy Commission (CEC) approved equipment list. Wind: Systems must be in compliance with all applicable standards; Generator must be listed or certified by at least one of the following organizations: American Wind Energy Association (AWEA), British Wind Energy Association (BWEA), California Energy Commission (CEC), New York State Energy and Research Development Authority (NYSERDA), Small Wind Certification Council (SWCC). Hydro: Systems must be in compliance with all applicable standards;
Installation Requirements:	Installations must comply with all federal, state, and local codes and meet detailed siting criteria specified in program guidelines. Systems must be grid-connected and net metered. Solar systems must be installed by a Nevada-licensed electrical C-2 or C-2g electrical contractor. Wind and microhydro systems must be installed by a Nevada-licensed C-2 electrical contractor.
Ownership of Renewable Energy Credits:	NV Energy
Program Administrator:	NV Energy
Web Site:	http://www.Nvenergy.com/renewablegenerations

Authority 1:
NRS § 701B.010 et. seq.

Authority 2:
Senate Bill 358
Date Enacted:
5/28/2009

Authority 3:
LCB File R175–07

Summary:
Note: In January 2010, the Public Utilities Commission of Nevada (PUCN) approved new regulations for the Renewable Generations programs that included changes to the application process and a new step system for reservations. The solar incentive program began accepting applications on April 21, 2010 for program year 2010–2011. Solar Generations had 13.4 megawatts (MW) of solar capacity available for this program year, and had received 34.8 MW worth of applications within the first six hours. The Solar Generations program is now closed for this program year and NV Energy is no longer accepting applications. Applications are currently being accepted for the wind and hydro incentive programs. Only customers of NV Energy are eligible to participate in the program.

NV Energy (formerly Sierra Pacific Power and Nevada Power) administers the Renewable Generations Rebate Program for photovoltaic (PV) systems and small wind and hydroelectric systems on behalf of the Nevada Task Force on Energy Conservation and Renewable Energy. With rebates originally available only for PV, the SolarGenerations Rebate Program was established in 2003 as a result of AB 431 ("the Solar Energy Systems Demonstration Program") and began in August 2004. Rebates are now available for grid-connected PV installations on residences, small businesses,* public buildings, non-profits and schools; small wind systems on residences, small businesses, agricultural sites, schools and public buildings; and small hydroelectric systems installed at grid-connected agricultural sites. Participants must be current Nevada customers of NV Energy to participate.

SB 358 of 2009 made adjustments to the administration of the RenewableGenerations program. After the utility approves the applicant, the utility will have 30 days to notify them in writing. Further, applicants will have 12 months to complete a project following their initial approval. If projects that have been approved miss the 12-month target date, they can become eligible again after the project is complete, but will receive an incentive at the current rate, rather than the rate when they received initial authorization.

Including three years as a demonstration program, SolarGenerations is now in its sixth program year. In June 2007 the program was made permanent (the planned end date had been June 2010 for a total of six years of demonstration program funding). As demonstrated above, incentive levels vary by technology type, customer class and program year, with incentive levels stepping down with each program year. Each program year has a designated amount of installed capacity set aside for each customer class. Applications received after one step is fully subscribed for that customer class may be reserved for the next incentive step. However, applications reserved for a future program year will have 12 months to be installed following the date of approval, but will not receive the rebate until that program year commences.

NV Energy's website will be frequently updated to indicate current subscription levels.

There are no size restrictions for participating systems, aside from the net metering limit of 1 megawatt, but rebates will be limited to certain system sizes corresponding to the customer class and the technology.

NV Energy takes ownership of the renewable energy credits (RECs) associated with the electricity produced by a customer's PV, wind or small hydro system. The RECs count towards the utility's' goals under Nevada's renewable portfolio standards (RPS).

Renewable Energy Sales and Use Tax Abatement

Last DSIRE Review: 07/07/2009

Program Overview:

State:	Nevada
Incentive Type:	Sales Tax Incentive
Eligible Renewable/ Other Technologies:	Solar Thermal Electric, Solar Thermal Process Heat, Photovoltaics, Landfill Gas, Wind, Biomass, Hydroelectric, Geothermal Electric, Fuel Cells, Municipal Solid Waste, Facilities for the transmission of electricity produced from renewable energy or geothermal resources located in Nevada, Anaerobic Digestion, Fuel Cells using Renewable Fuels
Applicable Sectors:	Commercial, Industrial, Utility, Agricultural, (Renewable Energy Power Producers)
Amount:	Purchaser is only required to pay sales and use taxes imposed in Nevada at the rate of 2.6 % (effective through June 30, 2011) and at the rate of 2.25 % (effective July 01, 2011 – June 30, 2049)
Equipment Requirements:	Systems must have a generating capacity of at least 10 megawatts.
Program Administrator:	Nevada State Office of Energy
Start Date:	7/1/2009
Expiration Date:	6/30/2049
Web Site:	http://renewableenergy.state.nv.us/TaxAbatement.htm

Authority 1:
AB 522
Date Enacted:
5/30/2009
Date Effective:
7/1/2009
Expiration Date
6/30/2049

Summary:

New or expanded businesses in Nevada may apply to the Director of the State Office of Energy for a sales and use tax abatement for qualifying renewable energy technologies. Purchaser is only required to pay sales and use taxes imposed in Nevada at the rate of 2.6 % (effective through June 30, 2011) and at the rate of 2.25 % (effective July 01, 2011 – June 30, 2049). The start date begins when the first piece of equipment is delivered to the designated facility or taxes are paid on the equipment.

The abatement applies to property used to generate electricity from renewable energy resources including solar, wind, biomass*, fuel cells, geothermal or hydro. Generation facilities must have a capacity of at least 10 megawatts (MW). Facilities that use solar energy to generate at least 25,840,000 British thermal units of process heat per hour can also qualify for an abatement.

There are several job creation and job quality requirements that must be met in order for a project to receive an abatement. Depending on the population of the county or city where the project will be located, the project owners must:

- Employ a certain number of full-time employees during construction, a percentage of whom must be Nevada residents
- Ensure that the hourly wage paid to the facility's employees and construction workers is a certain percentage higher than the average statewide hourly wage
- Make a capital investment of a specified amount in the state of Nevada
- Provide the construction workers with health insurance, which includes coverage for the worker's dependents

Note that this exemption does not apply to residential property. A facility that is owned, operated, leased or controlled by a governmental entity is also ineligible for this abatement. Note that this exemption does not apply to residential property, or property that is owned, operated, leased or controlled by a governmental entity.

History

This abatement went through significant revisions with AB 522, signed in May 2009. Notably, AB 522 raised the capacity minimum for eligible projects from 10 kilowatts (kW) to 10 MW. It also changed the abatement such that the purchaser is only required to pay sales and use taxes imposed in Nevada at the rate of 2.6 % (effective through June 30, 2011) and at the rate of 2.25 % (effective July 01, 2011 – June 30, 2049), extended it to additional technologies, and increased the qualification requirements to ensure that incentivized projects result in more high quality jobs. These changes took effect on July 1, 2009. AB 522 also created a property tax abatement for renewable energy producers.

New Mexico Incentives/Policies for Renewables & Efficiency [Insert Unnumbered figure 1 here]

Renewable Energy Production Tax Credit (Corporate)

Last DSIRE Review: 05/12/2010

Program Overview:

State: New Mexico

Incentive Type: Corporate Tax Credit

Eligible Renewable/Other   Solar Thermal Electric, Photovoltaics, Landfill
  Technologies: Gas, Wind, Biomass, Municipal Solid Waste,
                                      Anaerobic Digestion

Applicable Sectors: Commercial, Industrial

Amount: $0.01/kWh for wind and biomass
                                    $0.027/kWh (average) for solar (see below)

Maximum Incentive: Wind and biomass: First 400,000 MWh annually for
                                      10 years (i.e. $4,000,000/year)
                                    Solar electric: First 200,000 MWh annually for 10
                                      years (annual amount varies)
                                    Statewide cap: 2,000,000 MWh plus an additional
                                      500,000 MWh for solar electric

Eligible System Size: Minimum of 1 MW capacity per facility

Equipment Requirements: System must be in compliance with all applicable
                                      performance and safety standards; generators
                                      must be certified by the New Mexico Energy,
                                      Minerals, and Natural Resources Department
                                      (EMNRD).

Carryover Provisions: Prior to 10/1/2007: Excess credit may be carried
                                      forward five years
                                    After 10/1/2007: Excess credit is refunded to the
                                      taxpayer

Program Administrator: Taxation and Revenue Department

Start Date: 7/1/2002

Expiration Date: 1/1/2018

Web Site: http://www.cleanenergynm.org

Authority 1:
N.M. Stat. § 7–2A-19
Date Enacted:
3/4/2002, amended 2003, 2007
Date Effective:
7/1/2002
Expiration Date
1/1/2018

Summary:

Enacted in 2002, the New Mexico Renewable Energy Production Tax Credit provides a tax credit against the corporate income tax of one cent per kilowatt-hour for companies that generate electricity from wind or biomass. Companies that generate electricity from solar energy receive a tax incentive that varies annually according to the following scale:

- Year 1: 1.5¢/kWh
- Year 2: 2¢/kWh
- Year 3: 2.5¢/kWh
- Year 4: 3¢/kWh
- Year 5: 3.5¢/kWh
- Year 6: 4¢/kWh
- Year 7: 3.5¢/kWh
- Year 8: 3¢/kWh
- Year 9: 2.5¢/kWh
- Year 10: 2¢/kWh

According to the EMNRD, this incentive averages 2.7¢/kWh annually.

For wind and biomass generators, the credit is applicable only to the first 400,000 megawatt-hours (MWh) of electricity in each of 10 consecutive taxable years. For solar, the credit is applicable only to the first 200,000 MWh of electricity in each taxable year. To qualify, an energy generator must have a capacity of at least 1 megawatt and be installed before January 2018.

Total generation from both the corporate and personal tax credit programs combined must not exceed two million megawatt-hours of production annually, plus an additional 500,000 MWh produced by solar energy. Taxpayers cannot claim both the corporate and the personal tax credit for the same renewable energy system.

For electricity generated prior to October 1, 2007, excess credit may be carried forward for up to five consecutive taxable years. For electricity generated on or after October 1, 2007, excess credit shall be refunded to the taxpayer in order to allow project owners with limited tax liability to fully utilize the credit.

PNM – Performance-Based Customer Solar PV Program

Last DSIRE Review: 04/05/2010

Program Overview:

State:	New Mexico
Incentive Type:	Production Incentive
Eligible Renewable/Other Technologies:	Photovoltaics
Applicable Sectors:	Commercial, Residential
Amount:	Systems up to 10 kW: $0.13/kWh for RECs
	Systems greater than 10 kW up to 1 MW: $0.15/kWh for RECs
Maximum Incentive:	None specified
Terms:	Systems up to 10 kW: 12-year contract
	Systems greater than 10 kW up to 1 MW: 20-year contract

	(System must be net-metered to be eligible.)
Program Administrator:	PNM
Start Date:	3/1/2006
Web Site:	http://www.pnm.com/customers/pv/program.htm

Date Effective:
3/1/2006

Summary:

In March 2006, PNM initiated a renewable energy credit (REC) purchase program as part of its plan to comply with New Mexico's renewable portfolio standard (RPS). PNM will purchase RECs from customers who install photovoltaic (PV) systems up to one megawatt (MW). PNM will then be able to apply these RECs towards their obligations under the state's RPS, which requires 4% of the total generation capacity to come from solar electricity by 2020, and 0.6% from distributed generation in 2020.

REC payments are based on the system's total output. PNM will purchase RECs from each participant as part of the regular monthly billing process. Participants will receive a monthly bill documenting the number of kilowatt-hours (kWh) produced by the PV system, the number of RECs purchased by PNM, the purchase price per REC and the total price of RECs purchased that billing period. REC purchase payments will be applied as a credit to the participant's electric bill on a monthly basis.

Systems up to 10 kW

PNM will purchase RECs generated by small PV systems at a rate of $0.13/kWh for 12 years of the system's operation. If the amount paid for the RECs is greater than the total of the customer's monthly electric service plus kWh charges, the balance of the REC payment will be carried forward as a credit for the following month's bill if $20 or less. If the REC payment balance is greater than $20 after credits to the customer's electric bill have been made, the entire REC payment balance will be paid directly to the customer. Program participants must pay an application fee of $100 for residential customers, which includes the cost of installing a second meter to monitor system output. Customers also must pay a net-metering application fee of $50 to establish an approved interconnection with PNM.

Systems greater than 10 kW up to 1 MW

PNM will purchase RECs associated with the electricity generated by large PV systems and used on-site at a rate of $0.15/kWh for 20 years of the system's operation. If the amount paid for the RECs is greater than the total of the customer's monthly electric service plus kWh charges, the balance of the REC payment will be carried forward as a credit for the following month's bill if $200 or less. If the REC payment balance is greater than $200 after credits to the customer's electric bill have been made, the entire REC payment balance will be paid directly to the customer. PNM does not pay for RECs associated with net excess generation. Program participants must pay an application fee of $350 for commercial customers, which includes the cost of installing a second meter to monitor system output. Customers also must pay an interconnection application fee of $100.00 up to 100 kW for interconnection + $1.00 for every kW above 100 kW up to 1 MW to establish an approved interconnection with PNM.

Oregon Incentives/Policies for Renewables & Efficiency

EWEB – Solar Electric Program (Production Incentive)

Last DSIRE Review: 09/30/2009

Program Overview:

State:	Oregon
Incentive Type:	Production Incentive
Eligible Renewable/Other Technologies:	Photovoltaics
Applicable Sectors:	Commercial, Industrial, Residential, Nonprofit, Schools, Local Government, State Government, Agricultural, Institutional
Amount:	$0.076-$0.12/kWh for 10 years (subject to annual review), actual rate depends on level of monthly generation and season
Maximum Incentive:	Available to systems sized 10 kW to 1 MW
Terms:	System must be greater than 10 kW in capacity. System owners must execute an EWEB interconnection agreement and program agreement. A building permit is required. Systems must be inspected by city or county building officials, and by EWEB. All system equipment must be UL-listed. All PV modules and inverters must be listed and rated by the CEC.
Program Administrator:	Eugene Water & Electric Board
Start Date:	1/25/2008
Web Site:	http://www.eweb.org/content.aspx/ee5003fe-cb03–484c-86e0–2bb16f5d…

Authority 1:
EWEB Solar Electric Program Information and Requirements
Date Effective:
1/25/2008

Summary:

The Eugene Water & Electric Board's (EWEB) Solar Electric Program offers financial incentives for residential and commercial customers who generate electricity using solar photovoltaic (PV) systems. Rebates are available to customers who choose to net meter, and a production incentive is available to customers with systems greater than 10 kilowatts (kW) in capacity who choose *not* to net meter. Under the latter arrangement, all electricity generated is fed into the grid.

The rebate for residential customers who choose to net meter is $2.00 per watt-AC, with a maximum incentive of $10,000. The rebate for commercial customers who choose to net meter is $1.00 per watt-AC, with a maximum incentive of $25,000. Rebate amounts are based on the electrical output of the system after equipment and site losses are calculated. Under the rebate program, customers retain ownership of all renewable-energy credits (RECs) associated with customer generation.

PV systems sized 10 kW to 1 megawatt (MW) in capacity that are designed to generate and feed electricity directly into the grid – an arrangement under which the customer uses none of the electricity generated by the PV system – are eligible for a production payment of $0.076-$0.12 per kilowatt-hour (kWh) generated, payable for 10 years (but subject to annual review). The level of the incentive varies, depending on the season and level of monthly kWh generation. These "direct generation" systems require a separate EWEB service and electric meter to measure the amount of kWh generated. Under this program, EWEB assumes ownership of all RECs associated with customer generation.

All system owners must execute an EWEB interconnection agreement and program agreement. A building permit is required, and all systems must be inspected first by city or county building officials and then by EWEB. All system equipment must be UL-listed. All PV modules and inverters must be listed and rated in the California Energy Commission's Emerging Renewables Program. This list is available on the California Solar Initiative's Eligible Solar Equipment website.

Energy Trust – Industrial Production Efficiency Program

Last DSIRE Review: 06/12/2009

Program Overview:

State:	Oregon
Incentive Type:	State Rebate Program
Eligible Efficiency Technologies:	Lighting, Lighting Controls/Sensors, Heat pumps, Compressed air, Motors, Motor-ASDs/VSDs, Agricultural Equipment, Custom/Others pending approval
Eligible Renewable/Other Technologies:	Geothermal Heat Pumps
Applicable Sectors:	Industrial, Agricultural, Manufacturing, Water/Wastewater Treatment

Amount:	Varies depending on technology; incentives awarded per kilowatt-hour saved by project
Maximum Incentive:	Non-lighting projects: $0.25/kWh, up to 60% of cost until end of 2009 and up to 50% of cost after 2009
	Lighting projects: $0.17/kWh, up to 50% of project cost. Custom lighting incentives are 35% of project cost
	NEMA Premium efficiency motors: $10 per horsepower, up to 200 horsepower
	Municipal/service district project: $0.32/kWh, up to 50% of cost
Equipment Requirements:	Minimum efficiency levels for all equipment is available on program web site
Program Administrator:	Energy Trust of Oregon
Web Site:	http://energytrust.org/Business/incentives/industrial/production-...

Summary:

Energy Trust of Oregon offers the Industrial Production Efficiency Program to industrial customers of Portland General Electric, Pacific Power, NW Natural or Cascade Natural Gas. In order to qualify for these rebates, customers must be contributing to the Public Purpose Charge. Energy Trust offers technical assistance and cash incentives for industrial processes of all kinds – including large industrial, manufacturing, agriculture, and water/wastewater treatment. Standard prescriptive incentives include lighting, premium motors, heat pumps, variable speed drives, and premium HVAC equipment. Other rebates that are designed to fit the needs of specific industrial processes also exist. For example, there are irrigation system rebates for agricultural customers and compressed air rebates for small manufacturing customers. Customers interested in participating in the program should contact a Production Service Representative to find out which rebates best fit their particular facility.

Steps for reserving incentive funds for projects vary depending on the type and magnitude of the project. Customers can

Tax Credit for Renewable Energy Equipment Manufacturers

Last DSIRE Review: 03/23/2010

Program Overview:

State:	Oregon
Incentive Type:	Industry Recruitment/Support
Eligible Renewable/Other Technologies:	Solar Water Heat, Solar Space Heat, Photovoltaics, Wind, Biomass, Geothermal Heat Pumps, Solar Pool Heating, Small Hydroelectric, Tidal Energy, Wave Energy

Applicable Sectors:	Commercial, Industrial
Amount:	50% of eligible costs (10% per year for 5 years)
Maximum Incentive:	$20 million
Program Administrator:	Oregon Department of Energy
Start Date:	6/20/2008
Web Site:	http://egov.oregon.gov/ENERGY/CONS/BUS/ BETC.shtml

Authority 1:
OAR 330–090–0105 to 330–090–0150
Date Effective:
6/20/2008

Summary:

Oregon's Business Energy Tax Credit (BETC) is for investments in energy conservation, recycling, renewable energy resources, sustainable buildings, and less-polluting transportation fuels. The Tax Credit for Renewable Energy Resource Equipment Manufacturing Facilities was enacted as a part of BETC in July 2007, with the passage of HB 3201. The tax credit equals 50% of the construction costs of a facility which will manufacture renewable energy systems, and includes the costs of the building, excavation, machinery and equipment which is used primarily to manufacture renewable energy systems. The credit may also be applied to the costs of improving an existing facility which will be used to manufacture renewable energy systems. The 50% credit is taken over the course of five years, at 10% each year. The original maximum credit of $10 million was expanded to $20 million (50% of a $40 million facility) upon the enactment of HB 3619 in March 2008. This legislation clarified the manufacturing credit and separated the revenue stream from the rest of BETC.

The credit applies to companies that manufacture systems that harness energy from wood waste or other wastes from farm and forest lands, non-petroleum plant or animal based biomass, the sun, wind, water, or geothermal resources. Prior to construction, a business must apply to the Oregon Department of Energy for preliminary certification. In addition to this preliminary certification, the manufacturing facility must apply for final certification. Another review required for manufacturing facilities is a financial feasibility review. The Oregon Department of Energy may establish other rules to govern the type of equipment, machinery or other manufactured products eligible for this credit, as well as minimum performance and efficiency standards for those manufactured products. The passage of HB 3680 in March 2010 set a sunset date for the tax credit. Renewable energy equipment manufacturing facilities must receive preliminary certification before January 1, 2014 in order to use the tax credit.

Incentives/Policies for Renewables & Efficiency

Solar and Wind Equipment Sales Tax Exemption

Last DSIRE Review: 05/13/2010

Program Overview:

State:	Arizona
Incentive Type:	Sales Tax Incentive
Eligible Renewable/Other Technologies:	Passive Solar Space Heat, Solar Water Heat, Solar Space Heat, Solar Thermal Electric, Photovoltaics, Wind, Solar Pool Heating, Daylighting
Applicable Sectors:	Commercial, Residential, General Public/Consumer
Amount:	100% of sales tax on eligible equipment
Maximum Incentive:	No maximum
Start Date:	1/1/1997
Expiration Date:	12/31/2016
Web Site:	http://www.azsolarcenter.com/economics/taxbreaks.html

Authority 1:
A.R.S. § 42–5061 (N)
Date Effective:
1/1/1997
Expiration Date
12/31/2016

Authority 2:
A.R.S. § 42–5075 (14)
Date Effective:
1/1/1997
Expiration Date
12/31/2016

Authority 3:
HB 2700
Date Enacted:
5/10/2010
Date Effective:
5/10/2010
Expiration Date
12/31/2016

Summary:
Arizona provides a sales tax exemption* for the retail sale of solar energy devices and for installation of solar energy devices by contractors. The statutory definition of "solar energy device" includes wind electric generators and wind-powered water pumps in addition to daylighting, passive solar heating, active solar space heating,

solar water heating, and photovoltaics. The sales tax exemption does not apply to batteries, controls, etc., that are not part of the system. (Note that HB 2429, enacted in June 2006, eliminated the $5,000 limit per device.)

To take advantage of these exemptions from tax, a solar energy retailer or a solar energy contractor must register with the Arizona Department of Revenue prior to selling or installing solar energy devices. (Arizona Form 6015, Solar Energy Devices – Application for Registration)

The Arizona Department of Commerce Energy Office has compiled a guide to the solar energy devices that qualify for exemption under the statutory definition. It is possible to petition the Arizona Department of Commerce to add additional items if they qualify per the statutory definition.

According to the Arizona Solar Center's website, another provision of Arizona sales tax exemption may apply without value limit to the basic power generating part of the system (consisting of at least PV modules, structure, array wiring and controls; the limits have not been clearly defined). This further exemption requires the filling out of form ADOR 5000 titled "Transaction Privilege Tax Exemption Certificate" and checking reason #15, "Machinery, equipment or transmission lines used directly in producing or transmitting electrical power, but not including distribution."

Most cities have a 0.5 to 2% city privilege ("sales") tax that is applicable to sales or installations of solar energy devices, unless a city specifically exempts such sales under its city tax code. Solar energy retailers should check with the city in which the retail business is located to find out whether city privilege tax is applicable. Solar energy contractors should check with the city in which the installation will be performed to find out whether city privilege tax is applicable.

Technically, the law allows retailers to deduct the amount received from the sale of solar energy devices from their transaction privilege tax base, and similarly, it allows prime contractors to deduct proceeds from a contract to provide and install a device from their transaction privilege tax base.

Non-Residential Solar & Wind Tax Credit (Personal)

Last DSIRE Review: 05/13/2010

Program Overview:

State:	Arizona
Incentive Type:	Personal Tax Credit
Eligible Renewable/Other Technologies:	Passive Solar Space Heat, Solar Water Heat, Solar Space Heat, Solar Thermal Electric, Solar Thermal Process Heat, Photovoltaics, Wind, Solar Cooling, Solar Pool Heating, Daylighting
Applicable Sectors:	Commercial, Industrial, Nonprofit, Schools, Local Government, State Government, Tribal Government, Fed. Government, Agricultural, Institutional
Amount:	10% of installed cost

Maximum Incentive:	$25,000 for any one building in the same year and $50,000 per business in total credits in any year
Eligible System Size:	No size restrictions specified
Carryover Provisions:	Unused credits may be carried forward for not more than five consecutive taxable years
Program Administrator:	Arizona Department of Revenue
Program Budget:	$1 million annually
Start Date:	1/1/2006
Expiration Date:	12/31/2018
Web Site:	http://www.azcommerce.com/BusAsst/Incentives/Solar+Energy+Tax+Inc…

Authority 1:
A.R.S. §43–1085
Date Effective:
1/1/2006
Expiration Date
12/31/2018

Authority 2:
A.R.S. §43–1164
Date Effective:
1/1/2006
Expiration Date
12/31/2018

Authority 3:
A.R.S. §41–1510.01

Authority 4:
HB 2700
Date Enacted:
5/10/2010
Date Effective:
5/10/2010
Expiration Date
12/31/2018

Authority 5:
Tax Credit Guidelines

Summary:

Arizona's tax credit for solar and wind installations in commercial and industrial applications was established in June 2006 (HB 2429). In May 2007, the credit was revised by HB 2491 to extend the credit to all non-residential entities, including those that are tax-exempt. Third parties who install or manufacture systems for non-residential applications are now eligible to claim the credit – not only those that finance a system as allowed in the original legislation. These provisions are retroactive to January 1, 2006.

The tax credit, which may be applied against corporate or personal taxes, is equal to 10% of the installed cost of qualified "solar energy devices" and applies to taxable years beginning January 1, 2006 and extending through December 31, 2018.

A solar energy device is defined as "a system or series of mechanisms designed primarily to provide heating, to provide cooling, to produce electrical power, to produce mechanical power, to provide solar daylighting or to provide any combination of the foregoing by means of collecting and transferring solar-generated energy into such uses either by active or passive means, including wind generator systems that produce electricity. Solar energy systems may also have the capability of storing solar energy for future use. Passive systems shall clearly be designed as a solar energy device, such as a trombe wall, and not merely as a part of a normal structure, such as a window."

The maximum credit per taxpayer is $25,000 for any one building in the same year and $50,000 in total credits in any year. If the allowable credit exceeds the taxpayer's income tax liability, the amount of the claim not used to offset taxes may be carried forward for not more than five consecutive taxable years as a credit against subsequent years' income tax liability.

To qualify for the tax credits, a business must submit an application to the Arizona Department of Commerce (DOC). The DOC will review applications, provide an initial certification to qualifying installations, and issue a credit certificate to the business once the installation is complete and approved. The Arizona Department of Revenue will also receive a copy of the credit certificate. The DOC may certify tax credits up to a total of $1 million each calendar year. The DOC and the Department of Revenue will collaborate in adopting rules to implement the tax credit program.

TEP – Renewable Energy Credit Purchase Program

Last DSIRE Review: 05/07/2010

Program Overview:

State:	Arizona
Incentive Type:	Utility Rebate Program
Eligible Renewable/ Other Technologies:	Solar Water Heat, Solar Space Heat, Photovoltaics, Landfill Gas, Wind, Biomass, Geothermal Electric, Solar Pool Heating, Daylighting, Anaerobic Digestion, Small Hydroelectric
Applicable Sectors:	Commercial, Residential
Amount:	Up-front incentives for PV may be de-rated based on expected performance
	Residential grid-tied PV: $3.00/W-DC
	Residential off-grid PV: $2.00/W-DC
	Non-Residential grid-tied PV (100 kW or less): $2.50/W-DC or production-based incentive

Non-Residential grid-tied PV (more than 100 kW):
production-based incentive

Non-Residential off-grid PV: $2.00/W-DC

Wind (grid-tied): $2.25/W-AC

Wind (off-grid): $1.80/W-AC or a performance-based
incentive

Daylighting (non-residential only): $0.18 per kWh savings
for 5 years;

Residential Solar Water Heater: $0.25/kWh-equivalent,
plus $750 up to a maximum amount of $1,750

Non-Residential Solar Water Heater: $0.50/kWh-
equivalent, plus $750

Non-Residential Solar Water Heater: performance base
incentive;

All other eligible technologies can receive performance
based incentives which vary depending on the
technology and the length of the contract. Contract
options for performance-based incentives are 10, 15 and
20 years.

Maximum Incentive:	Residential PV systems greater than 20 kW AC will receive rebates on the first 20 kW AC only
	TEP incentive for PV can not exceed 60% of the total project cost.
	TEP incentive for PV may be combined with other state and federal incentives, but combined they can not pay for more than 85% of the total project cost
Eligible System Size:	PV: Minimum size is 1.2 kW.
Equipment Requirements:	Photovoltaic modules must be covered by a manufacturer's warranty of at least 20 years to qualify for the up-front incentive.
	SWH systems must be certified to SRCC OG-300 standards
	Eligible small wind systems must be certified and nameplate rated by the CEC.
	Wind systems must have at least a 10-year manufacturer's warranty to qualify for the up-front incentive; wind systems must have at least a 5-year manufacturer's warranty to qualify for the performance-based incentive
Ownership of Renewable Energy Credits:	TEP
Program Administrator:	Tucson Electric Power
Web Site:	http://www.tep.com/Green/

Summary:

Tucson Electric Power (TEP) created the SunShare Program in 2001 to encourage residential and business customers to install new photovoltaic (PV) equipment. TEP transitioned away from their original incentive structure for PV and added incentives for a variety of other technologies in May 2008 under the new Renewable Energy Credit Purchase Program (RECPP). The technologies now eligible for funding through the RECPP all qualify under Arizona's renewable energy standard (RES). TEP offers these incentives in exchange for the renewable energy certificates they generate. Incentives for the 2010 program year are as follows:

- *Residential PV (grid-tied): $3.00/W up front for qualified systems. Systems with less than a 20 year warranty on the module or a 10 year warranty on the inverter, or with a building-integrated PV (BIPV) system over 5 kW must receive the 10, 15 or 20 year performance-based incentive (PBI) which is based on the actual metered electricity output.*
- *Residential PV (off-grid): $2.00/W*
- *Non-Residential PV (grid-tied): $2.50/W for systems 100 kW or less. Systems greater than 100kW must take the PBI*
- *Non-Residential PV (off-grid): $2.00/W*
- *Solar Domestic Water Heating and Solar Space Heating: $0.25/kWh equivalent, plus $750 up to a maximum incentive of $1,750.*
- *Non-Residential Solar Water Heating and Solar Space Heating: $0.50/kWh-equivalent, plus $750*
- *Daylighting (Non-residential only):$0.18/kWh equivalent for 5 years.*
- *Wind (grid-tied, up to 1 MW):$2.25/W-AC*
- *Wind (off-grid, up to 1 MW): $1.80/W-AC*
- *On-grid small hydro, biomass-biogas systems, pool heating (non-residential only), space cooling, and geothermal (electric, cooling and heating systems) are all eligible to receive PBIs. See program website for full details including contract options, equipment requirements and PBI amounts.*

Solar and Wind Energy Business Franchise Tax Exemption

Last DSIRE Review: 11/13/2009

Program Overview:

State:	Texas
Incentive Type:	Industry Recruitment/Support
Eligible Renewable/Other Technologies:	Solar Water Heat, Solar Space Heat, Solar Thermal Electric, Solar Thermal Process Heat, Photovoltaics, Wind
Applicable Sectors:	Commercial, Industrial
Amount:	All
Maximum Incentive:	None
Terms:	N/A

Program Administrator:	Comptroller of Public Accounts
Start Date:	1982
Web Site:	http://www.seco.cpa.state.tx.us/re_incentives-taxcode-statutes.ht...

Authority 1:
Texas Tax Code § 171.056
Date Enacted:
1981
Date Effective:
1982

Summary:

Companies in Texas engaged solely in the business of manufacturing, selling, or installing solar energy devices are exempted from the franchise tax. The franchise tax is Texas's equivalent to a corporate tax; their primary elements are the same. There is no ceiling on this exemption, so it is a substantial incentive for solar manufacturers.

For the purposes of this exemption, a solar energy device means "a system or series of mechanisms designed primarily to provide heating or cooling or to produce electrical or mechanical power by collecting and transferring solar-generated energy. The term includes a mechanical or chemical device that has the ability to store solar-generated energy for use in heating or cooling or in the production of power." Under this definition wind energy is also listed as an eligible technology.

Texas also offers a franchise tax deduction for solar energy devices which also includes wind energy as an eligible technology.

Texas Incentives/Policies for Renewables & Efficiency

Solar and Wind Energy Device Franchise Tax Deduction

Last DSIRE Review: 11/13/2009

Program Overview:

State:	Texas
Incentive Type:	Corporate Deduction
Eligible Renewable/Other Technologies:	Solar Water Heat, Solar Space Heat, Solar Thermal Electric, Solar Thermal Process Heat, Photovoltaics, Wind
Applicable Sectors:	Commercial, Industrial
Amount:	10% of amortized cost
Maximum Incentive:	None
Program Administrator:	Comptroller of Public Accounts

Start Date: 1982

Web Site: http://www.seco.cpa.state.tx.us/re_
                                         incentives-taxcode-statutes.ht…

Authority 1:
Texas Tax Code § 171.107
Date Enacted:
1981 (subsequently amended)
Date Effective:
1982

Summary:

Texas allows a corporation or other entity subject the state franchise tax to deduct the cost of a solar energy device from the franchise tax. Entities are permitted to deduct 10% of the amortized cost of the system from their apportioned margin. This treatment is effective January 1, 2008 and replaces prior tax law that allowed a company to deduct (1) the total cost of the system from the company's taxable capital; or, (2) 10% of the system's cost from the company's earned surplus (i.e., income). The franchise tax is Texas's equivalent to a corporate tax.

For the purposes of this deduction, a solar energy device means "a system or series of mechanisms designed primarily to provide heating or cooling or to produce electrical or mechanical power by collecting and transferring solar-generated energy. The term includes a mechanical or chemical device that has the ability to store solar-generated energy for use in heating or cooling or in the production of power." Under this definition wind energy is also included as an eligible technology.

Texas also offers a franchise tax exemption for manufacturers, seller, or installers of solar energy systems which also includes wind energy as an eligible technology.

LoanSTAR Revolving Loan Program

Last DSIRE Review: 04/21/2010

Program Overview:

State: Texas

Incentive Type: State Loan Program

Eligible Efficiency   Lighting, Lighting Controls/Sensors, Chillers, Furnaces,
  Technologies: Boilers, Heat pumps, Central Air conditioners, Heat
                                           recovery, Programmable Thermostats, Energy Mgmt.
                                           Systems/Building Controls, Building Insulation,
                                           Motors, Motor-ASDs/VSDs, Custom/Others pending
                                           approval, Led Exit Signs

Eligible Renewable/   Passive Solar Space Heat, Solar Water Heat, Solar Space
  Other Technologies: Heat, Photovoltaics, Wind, Geothermal Heat Pumps

Applicable Sectors: Schools, Local Government, State Government,
                                           Hospitals

Amount:	Varies
Maximum Incentive:	$5 million
Terms:	Current interest rates are 3% APR. Loans are repaid through energy cost savings. Projects must have an average payback of 10 years or less.
Program Administrator:	Comptroller of Public Accounts State Energy Conservation Office (SECO)
Funding Source:	Petroleum Violation Escrow Funds
Program Budget:	$98.6 million (revolving loan)
Start Date:	1989
Web Site:	http://seco.cpa.state.tx.us/ls/

Summary:

Through the State Energy Conservation Office, the LoanSTAR Program offers low-interest loans to all public entities, including state, public school, colleges, university, and non-profit hospital facilities for Energy Cost Reduction Measures (ECRMs). Such measures include, but are not limited to: HVAC, lighting, and insulation. Funds can be used for retrofitting existing equipment or, in the case of new construction, to finance the difference between standard and high efficiency equipment. The evaluation of on-site renewable energy options (e.g., solar water heating, photovoltaic panels, small wind turbines) is encouraged in the analysis of potential projects.

The LoanSTAR Program funds "Design, Bid, Built" or "Design, Built" projects. All projects are approved based on the Detailed Energy Assessment Report, which must be prepared according to LoanSTAR Technical Guidelines or the Performance Contracting Guidelines. SECO performs design specification review and on-site construction monitoring at the very minimum when the project is 100% complete. Repayment of the loans does not begin until after construction is 100% completed.

As of November 2007, LoanSTAR had funded a total of 191 loans totaling over $240 million dollars and resulting in approximately $212 million in energy savings. The National Association of State Energy Officials (NASEO) reports that the LoanSTAR program helped state agencies save more than $20 million in energy costs during 2008 and that the program had a waiting list of $28 million in proposed projects as of Winter 2009. Applications are available on the program website. The technical guidelines for the LoanSTAR program can be found on the program web site.

Index